Genetic Optimization Techniques for Sizing and Management of Modern Power Systems

Genetic Optimization Techniques for Sizing and Management of Modern Power Systems

Juan Lujano-Rojas, Rodolfo Dufo-López, and José A. Domínguez-Navarro
Department of Electrical Engineering, Universidad de Zaragoza, Zaragoza, Spain

ELSEVIER

Elsevier
Radarweg 29, PO Box 211, 1000 AE Amsterdam, Netherlands
The Boulevard, Langford Lane, Kidlington, Oxford OX5 1GB, United Kingdom
50 Hampshire Street, 5th Floor, Cambridge, MA 02139, United States

ISBN: 978-0-12-823889-9

For information on all Elsevier publications
visit our website at https://www.elsevier.com/books-and-journals

Publisher: Charlotte Cockle
Acquisitions Editor: Graham Nisbet
Editorial Project Manager: Aleksandra Packowska
Production Project Manager: Kamesh Ramajogi
Cover Designer: Miles Hitchen

Typeset by STRAIVE, India

Contents

Acknowledgment

The authors thank the editorial team for their kind support and patience during the making of this work.

Introduction to optimization techniques for sizing and management of integrated power systems

1

1.1 Heuristic optimization techniques

With the progress of science and technology, investigators from different fields have created a vast family of heuristics based on natural processes. These optimization techniques have been used in the solution of many problems in electrical engineering. In this section, we briefly describe the working principles of some of the most important algorithms. Most of the mathematical formulations presented in this section are used for illustrative purposes without following any strict format.

1.1.1 Ant Colony optimization algorithm

A popular method known as the ant system (Dorigo et al., 1996) is inspired by the behavior and aptitudes of ant colonies. Ant colony optimization (ACO) is a population-based algorithm with versatility and robustness. It can be applied to many combinatory problems with only a limited number of changes in its structure.

In the ACO, each ant is considered as an agent participating in the search process. Pheromone trails inspire information exchange among the agents. The ants use these trails in the colony to share information about the shortest route to the food source. Thus, an ant on the optimal path follows a strong pheromone trail left by the rest of the colony, while an ant following a weak trail moves randomly. It is essential to highlight that an ant on a determined route also leaves its own pheromone trail on the way. The accumulative quantity of pheromones on a specific route determines how attractive this route is for the colony.

The intensity of the trail (ζ_{ij}) laid by the ants when they travel from a point i to j at a determined time t after n iterations of the algorithm (every n iterations the ants complete a tour) is shown in Eq. (1.1) (Dorigo et al., 1996):

$$\zeta_{ij}(t+n) = \rho\zeta_{ij}(t) + \Delta\zeta_{ij} \tag{1.1}$$

where ρ is a parameter used to model the evaporation of the trail once the ants have completed a tour. This means between the intervals t and $t+n$. Furthermore, $\Delta\varsigma_{ij}$ is the cumulative amount of pheromones laid by the group between the steps t and $t+n$.

Regarding the agent model, this is very basic. Each ant or agent makes its next movement according to a probability determined by the distance to the next stage

Genetic Optimization Techniques for Sizing and Management of Modern Power Systems
https://doi.org/10.1016/B978-0-12-823889-9.00002-3

of the route and the trail left by other ants that previously stopped at that stage. Even when the mathematical model is simple, the work of a massive number of agents in a collective and consolidated manner makes ACO a powerful optimization tool. A positive feedback loop mathematically represents the cumulative effect of a pheromone trail over a specific route. The probability of following this route increases with the number of ants that have chosen it.

Although natural ant colonies inspire the algorithm, it employs artificial ones. Artificial ant colonies live in an environment with discrete time, have limited memory, and are not blind. Additionally, the feasibility of each route or solution to the optimization problem is forced using a tabu list.

As previously stated, the probability of an ant transitioning from one stage to another one depends essentially on the pheromone trail and the distance. To this end, the decision maker (DM) adjusts several parameters related to the trail, the visibility of the next stage, the persistence of the trail, and the amount of pheromone left by the other ants.

Eq. (1.2) (Dorigo et al., 1996) shows the probability ($p_{ijk}(t)$) of moving from a point i to j at time t for the ant k. This is the expression for valid transitions, which are included using a tabu list. The parameters α and β are used to represent the importance of the pheromone trail and route visibility (η_{ij}).

$$p_{ijk}(t) = \frac{\left[\zeta_{ij}(t)\right]^{\alpha}\left[\eta_{ij}\right]^{\beta}}{\sum_{k}\left[\zeta_{ik}(t)\right]^{\alpha}\left[\eta_{ik}\right]^{\beta}} \tag{1.2}$$

Depending on the values of these parameters, the DM could observe the stagnation of the optimization process and good and bad solutions. Specifically, bad solutions with stagnation appear when the trial's relative importance is adjusted to a high value. Under these circumstances, the algorithm stagnates rapidly without finding a good solution. Bad solutions without stagnation occur in the contrary case when the relative importance of the trial is adjusted to a low value. With these settings, the algorithm cannot find good solutions. Finally, the DM can find reasonable solutions when the importance of the trial and the visibility are set to fair values.

Researchers have compared ACO with other essential techniques such as simulated annealing, neural networks (NNs), self-organizing maps, and genetic algorithms (GAs) (Dorigo & Gambardella, 1997). According to the reported results in the literature, ACO offers solutions as good as those found using the methodologies above. Furthermore, scientists determined essential characteristics that could be improved to face larger optimization problems. First, ACO could be initialized with other local optimization heuristic results to increase algorithm efficiency. Second, ACO could be parallelized to distribute the computational burden over several processors, reducing the time and the dependence on the number of ants. Third, families of ants and specialized heuristics operators could be incorporated to increase the search capabilities.

Following these research directions, the original algorithm has been enriched by adding the subroutine daemon (Dorigo et al., 1999). Daemon collects relevant information for global optimization and modifies the pheromone trails to guide the colony in the global-optimum direction. Another improving characteristic is the pseudo-random-proportional rule and the implementation of a candidate list of preferred locations to visit from a specific point.

Stützle and Hoos (2000) introduced the max-min ant system (MMAS) to take advantage of the best solutions found during each iteration and avoid stagnation during the first iterations of the optimization process. Thus, after each iteration, in MMAS, only one ant adds a pheromone to the trail. This ant could be that with the best performance during the current iteration or with the best solution found until the present iteration. MMAS avoids stagnation at the beginning of the optimization process by limiting the pheromone trail to a determined interval. Finally, MMAS initializes the pheromone trail to the maximum allowed value to increase exploration during the early stages of the optimization process.

ACO and MMAS have been applied to solve many optimization problems, such as the traveling salesman problem (TSP), quadratic assignment problem, job-shop scheduling problem, and vehicle routing problem. In TSP, the optimal global solution can be found when considering up to 64 locations. However, the incorporation of parallel computing increases the capabilities of these techniques to solve problems of a higher dimension.

Recently, Skinderowicz (2020) presented a discussion related to the implementation of MMAS using parallel computing in a graphics processing unit. The author analyzed TSP considering dimensions between 198 and 3795 cities, observing a reduction in the computational time by a factor between 7.18 and 21.79.

1.1.2 Artificial bee colony algorithm

Another nature-inspired technique is the artificial bee colony (ABC) (Karaboga & Basturk, 2007) algorithm. It is based on the behavior of natural honey bees. This algorithm identifies three types of bees: the employed, onlookers, and scouts bees. The algorithm recognizes the dancing area in the hive as a place for information exchange and the location of the food source as the crucial aspect during the optimization process.

The onlooker bees remain in the dancing area, thinking and selecting the appropriate food source. The employed bees go to a food source chosen previously, and the scout bees are in charge of finding new food sources using a random search process. Initially, half of the bees in the colony are considered employed, while the remainder are considered onlookers. Each employed bee works on exploiting one food source; therefore, the number of employed bees is equal to the number of food sources. If a food source has been exhausted, the employed and onlooker bees become scouts to find a new food source.

According to the information stored in the bee's memory, ABC sequentially locates the employed and onlooker bees in a food source. Then, the algorithm places scout bees randomly to find new food sources. This process is repeated until a specific condition previously established by the DM is reached.

During the early stage of the algorithm, ABC requires an initialization process. This process consists of setting the position of the food sources randomly. Then, the employed bees explore these locations and determine the nectar quantity. After that, the employed bees go to the hive and inform the onlooker bees about the nectar quantities. Once an employed bee explores a route to the food source, the insect memorizes its position. Then, it goes back to that area and selects another food source based on the analysis of visual information.

The onlooker bees give priority to those food sources with high amounts of nectar. Thus, those food sources with high nectar concentrations will be more likely to be exploited during the subsequent iterations. The probability (p_i) of choosing a food source i is given by Eq. (1.3) (Karaboga & Basturk, 2007):

$$p_i = \frac{\text{fit}_i}{\sum_n \text{fit}_n} \quad (1.3)$$

where fit_i is the fitness of individual i in the population. When a specific location is exhausted, the scout bees randomly search to find new food sources.

Concerning a specific optimization problem, the location of a determined food source represents a solution to the optimization problem. The nectar quantity represents the objective function to be minimized or maximized. The number of employed and onlooker bees is equal to the number of solutions considered by DM.

ABC starts by randomly setting the location of the food source. This is carried out by adjusting the initial population with random numbers. The position of the food source is a vector whose dimension is equal to the number of the variables of the optimization problem. Once ABC has been initialized, the position of the food sources will continuously change while the optimization process evolves. Specifically, the interaction between employed and onlooker bees creates a displacement in the location of the food source. ABC uses Eq. (1.4) (Karaboga & Basturk, 2007) to determine the position of the food source (v_{ij}) at a determined iteration from the previous one.

$$v_{ij} = x_{ij} + \Phi_{ij}\left(x_{ij} - x_{kj}\right) \quad (1.4)$$

The indices k and j are chosen randomly. The term Φ_{ij} is a random number used to specify a new position close to x_{ij}.

ABC offers high-quality solutions similar to those obtained from the implementation of other well-known approaches such as differential evolution (DE), particle swarm optimization (PSO), and GA (Karaboga & Akay, 2009; Karaboga & Basturk, 2008). Relevant problems of science and engineering related to data clustering have been effectively solved using ABC (Karaboga & Ozturk, 2011).

1.1.3 Bacterial foraging algorithm

The bacterial foraging algorithm (BFA) is inspired by the behavior of a specific bacteria known as *Escherichia coli*. *E. coli* is provided with pili and flagella. The pili allow the transfer of genes from one bacterium to another, while the flagella are used for bacterium mobility (Passino, 2002).

From an engineer's perspective, *E. coli* has a control system with the flagellum working as an actuator, the receptor proteins working as sensors, and a closed-loop to determine how the bacterium moves in different environments. Thus, the bacterium carries out a search process looking for food and avoiding areas with harmful substances. Under realistic conditions, if *E. coli* is located in an alkaline or acid environment, it swims toward a neutral environment using its abilities (Passino, 2002).

Each flagellum is a helix connected to the cell. When the base of the flagellum (the connection point with the cell) rotates counterclockwise, the flagella form a bundle that works as a propeller, resulting in a force that pushes the cell. Then, the bacterium swims in a specific direction. If the flagella rotate clockwise, they work independently of each other, pulling at the cell, the net result being that the bacterium tumbles. In general, *E. coli* can swim for a specific period, and it can tumble for another period (Passino, 2002).

When bacteria are in an environment with chemical attractants and repellants, they perform motion patterns known as chemotaxes. In *E. coli*, serine or aspartate produces attractant attitudes, while the metal ions Ni and Co, modifications in pH, and amino acids have repellant attitudes (Passino, 2002).

If *E. coli* is located in a neutral environment for a long time, the flagella will rotate clockwise and counterclockwise. In other words, the bacterium will tumble and swim, resulting in a random movement of the bacterium. Suppose *E. coli* is located in an environment with a homogeneous concentration of serine, the bacterium tends to increase the swimming length and speed while the tumble period decreases. If *E. coli* is located in an environment where nutrient concentrations increase, the bacterium will swim for more time, and the tumbling period will be short. However, if the bacterium moves down a gradient, it will perform periodical transitions between swimming and tumbling. This describes how the bacterium continuously searches for areas where nutrient concentrations are abundant (Passino, 2002).

Now we have described the foraging process of *E. coli*, the next step consists of discussing the relationship between bacterial swarm foraging and the optimization process. We assume a minimization problem of the objective function $J(\theta)$, where θ is the vector of the p variables under optimization. Furthermore, we do not have any analytical expression for the gradient of $J(\theta)$, so we employ the reasoning behind the bacterial swarm foraging to solve the optimization problem (Passino, 2002).

The effect of attractants and repellants is represented by $J(\theta)$ and the position of the bacteria is represented by θ. Thus, $J(\theta) < 0$, $J(\theta) = 0$, and $J(\theta) > 0$ represent the areas with a high concentration of nutrients, and neutral and harmful substances, respectively (Passino, 2002).

BFA interprets chemotaxis as a natural process that uses optimization principles so that the bacteria find those points of high nutrient concentration, avoiding harmful substances and neutral areas. Other functions related to reproduction, elimination, and dispersal of natural bacterial systems are considered in the optimization model (Passino, 2002).

Assuming a population of S individuals, the position ($P(j,k,l)$) at a determined instant (j, k, l) is given by Eq. (1.5) (Passino, 2002):

$$P(j,k,l) = \{\theta_i(j,k,l) \mid i = 1, \ldots, S\} \tag{1.5}$$

The chemotactic step is defined as the period of tumble followed by a tumble or a tumble followed by a swim. In Eq. (1.5), j is the index for the chemotactic step,

k is the index for the reproduction step, and l is the index for the elimination-dispersal process. Under this notation, the term $J(i,j,k,l)$ represents the cost (value of the nutrient surface) of locating the bacterium i in the position $\theta_i(i,j,k)$ (Passino, 2002).

Tumbling is modeled using Eq. (1.6), where $C(i)$ is the size of the chemotactic step of a bacterium i and $\phi(j)$ is used to represent the random variation in the bacterium's direction (Passino, 2002):

$$\theta_i(j+1,k,l)=\theta_i(j,k,l)+C(i)\phi(j) \tag{1.6}$$

In Eq. (1.7), $J_{CC}(\theta,P(j,k,l))$ represents the effects of attraction and repulsion in a combined manner. The term $\theta_{i,\ m}$ is the element m of the vector position of a bacterium i, and the term θ_m refers to a point in the optimization domain. The repellant and attractant phenomena are modeled using the constant terms $d_{attract}$, $w_{attract}$, $h_{repellant}$, and $w_{repellant}$ (Passino, 2002).

$$\begin{aligned}J_{CC}(\theta P(j,k,l))=&\sum_{i=1}^{S}\left[-d_{\text{attract}}\times\exp\left(-w_{\text{attract}}\sum_{m=1}^{p}(\theta_m-\theta_{i,m})^2\right)\right]\\&+\sum_{i=1}^{S}\left[h_{\text{repellant}}\times\exp\left(-w_{\text{repellant}}\sum_{m=1}^{p}(\theta_m-\theta_{i,m})^2\right)\right]\end{aligned} \tag{1.7}$$

It is important to highlight that a determined bacterium i will escalate over the surface $J(i,j,k,l)+J_{CC}(\theta,P)$ instead $J(i,j,k,l)$. The term $J_{CC}(\theta,P)$ changes the objective function surface as a result of swarming. Thus, the development of swarming is the minimization process (Passino, 2002).

Once the chemotactic steps have been completed, the reproduction process starts. Eq. (1.8) presents the number of bacteria (S_r) that have a high level of nutrients and consequently will be reproduced (Passino, 2002):

$$S_r=\frac{S}{2} \tag{1.8}$$

BFA has been successfully applied in harmonic estimation (Mishra, 2005), parameter setting of an automatic voltage regulator (Kim et al., 2007), and many other problems (Mavrovouniotis et al., 2017).

1.1.4 Cat swarm optimization

Cat swarm optimization (CSO) was created from cats' attitudes and group behavior, specifically, from their curiosity about moving objects and their attention to the surrounding area even when they are resting. CSO is implemented using two subroutines, one to model the seeking attitudes of cats and the other one to represent their tracking behavior. During the optimization process, the DM combines these characteristics, mixing them in various proportions (Chu et al., 2006; Chu & Tsai, 2007).

In CSO, any possible solution to the optimization problem is represented by a cat swarm. Thus, we first need to decide how many cats to use in CSO implementation. The behavior of a particular cat is defined by its position, speed in each dimension, and fitness. A multidimensional variable represents the position, while the fitness models how well a determined cat adjusts to the objective under optimization. Cats' attitudes alternate between seeking and tracing, which are modeled using a flag (Chu et al., 2006; Chu & Tsai, 2007).

During the seeking mode, the cat is supposed to be resting or analyzing its situation to decide its next movement. Significant definitions are introduced in this mode: the seeking memory pool (SMP), seeking range of the selected dimension (SRD), count of dimension to change (CDC), and self-position considering (SPC). SMP specifies the seeking memory available for each cat, SRD establishes the mutation ratio per dimension, while CDC represents the number of dimensions to be modified during the seeking process. Finally, SPC defines if a determined cat is willing to move. This is a Boolean variable. Seeking mode is implemented so that the cats decide where to move to, based on a list of preferred locations and their corresponding objective function values (FS_i). The place (i) selection depends on a probability value (P_i) defined according to Eq. (1.9) (Chu et al., 2006; Chu & Tsai, 2007):

$$P_i = \frac{|FS_i - FS_b|}{FS_{\max} - FS_{\min}} \tag{1.9}$$

where $FS_{\max}$ and $FS_{\min}$ are the maximum and minimum fitness values. FS_b is set to $FS_{\max}$ for minimization and $FS_{\min}$ for maximization (Chu et al., 2006; Chu & Tsai, 2007).

During the tracing mode, the cat is supposed to be tracing a determined target. Each cat displaces at a concrete speed in each dimension. The velocity and position of a specific cat are calculated during each iteration, following Eqs. (1.10) and (1.11), respectively (Chu et al., 2006; Chu & Tsai, 2007).

$$v_{k,d} = v_{k,d} + r_1 c_1 \left(x_{\text{best},d} - x_{k,d}\right) \tag{1.10}$$

$$x_{k,d} = x_{k,d} + v_{k,d} \tag{1.11}$$

where k is the index for each cat, and d is the index for each dimension. $v_{k,d}$ is the speed of cat k at dimension d, while $x_{k,d}$ is its corresponding position. Finally, the best solution is stored in the variable $x_{\text{best},d}$ (Chu et al., 2006; Chu & Tsai, 2007).

Drawbacks related to computational speed affect many swarm-based algorithms. This problem has been overcome by using parallel computing. In this regard, Tsai et al. (2008, 2012) have proposed the parallel implementation of the CSO algorithm. In addition, Pradhan and Panda (2012) have improved CSO to solve multiobjective problems.

CSO has been applied to solve many technical problems. Santosa and Ningrum (2009) proposed a clustering algorithm. Guo et al. (2016) performed the solar cell parameter identification, and Lin et al. (2016) developed a methodology to perform feature selection.

1.1.5 Chaotic swarming of particles

Chaos theory has been used to enhance the capabilities of PSO. In such an approach, the sensitive dependence on the initial conditions of the logistic equation is combined with the traditional PSO formulation to improve its search characteristics. The chaotic local search is defined in Eq. (1.12) through the logistic map (Liu et al., 2005):

$$cx_{i,k+1} = 4cx_{i,k}(1 - cx_{i,k}) \tag{1.12}$$

where $cx_{i,\,k}$ is the chaotic variable i, during iteration k. The term $cx_{i,k}$ is defined in the interval (0, 1). However, some restrictions on the initial values ($k=0$) apply. Using the value of the decision variable $x_{i,k}$, we need to calculate the value of the chaotic variable ($cx_{i,k}$) using Eq. (1.13) (Liu et al., 2005):

$$cx_{i,k} = \frac{x_{i,k} - x_{\min,i}}{x_{\max,i} - x_{\min,i}} \tag{1.13}$$

where $x_{\max,\,i}$ and $x_{\min,\,i}$ are the maximum and minimum values of the decision variable, respectively. Once the chaotic variable for iteration k has been calculated, the corresponding value for the next iteration ($cx_{i,k+1}$) is estimated using Eq. (1.12). Then, the decision variable for the next iteration ($x_{i,k+1}$) is obtained using Eq. (1.14) (Liu et al., 2005):

$$x_{i,k+1} = x_{\min,i} + cx_{i,k+1}(x_{\max,i} - x_{\min,i}) \tag{1.14}$$

After that, the objective function for the decision variable at iteration $k+1$ is calculated, and the appropriate solution is selected (Liu et al., 2005). Important improvements have been made to the original chaotic PSO previously described. Alatas et al. (2009) proposed 12 variations of chaotic PSO and several chaotic maps (Eq. 1.12), such as the tent, sinusoidal, Gauss, circle, Arnold's cat, Sinai, and Zaslavskii maps. Gandomi, Yun, et al. (2013) proposed the chaos-enhanced accelerated PSO, in which the chaotic and the accelerated PSO are combined. In addition, different chaotic maps, including the Chebyshev and Liebovitch, among many others, were analyzed. Recently, Bingol and Alatas (2020) combined the optics-inspired optimization (OIO) algorithm with chaos theory. OIO is based on the particular features of convex and concave mirrors. The authors proposed three different versions of the chaotic OIO.

Chaotic swarming of particles (CSP) has been successfully used in the optimal design of truss structures (Kaveh et al., 2014). Specifically, optimization problems between 25 and 942 bars were analyzed. The problem of feature selection has also been solved using CSP. Some representative examples have been proposed by Chuang et al. (2011) and Sayed et al. (2019). CSP has been applied to problems related to power system engineering. Some works related to the solution of economic dispatch (ED) have been published by Cai et al. (2007, 2009), Adarsh et al. (2016), and dos Santos Coelho and Lee (2008). Hong (2009) developed a technique based on a support vector machine combined with CSP regarding load forecasting.

1.1.6 Cuckoo search algorithm

The cuckoo search algorithm (CSA) was developed from a preliminary analysis of cuckoo breeding behavior. Specifically, CSA results from the combination of the brood parasitic conduct of cuckoos with Lévy flight. Some species, such as the *ani* and *Guira*, use communal nests to incubate their eggs, and in certain circumstances, cuckoos remove the eggs of other birds to increase their survival probability. Other species lay their eggs in other birds' nests. In this sense, the host bird could recognize the foreign eggs and expulse them or abandon the nest and build a new one in another location. For example, some cuckoo species (New World *Tapera*) have evolved to deal with this problem, imitating the color of some specific bird species used as hosts. Thus, the survival probabilities are increased (Yang & Deb, 2009).

CSA also bases its search capabilities on the flight behavior of fruit flies. *Drosophila melanogaster* analyses the surfaces following a combination of straight routes and right-angle turns, which could be studied using the Lévy flight theory (Yang & Deb, 2009).

Cuckoos have a particular manner of breeding, accurate modeling of which is challenging. Thus, CSA is based on some specific assumptions. First, CSA assumes that each cuckoo lays only one egg at a time in a nest randomly selected. Second, those nests whose eggs have a high performance will survive to the next iteration. Third, the number of nests remains constant during the entire optimization process. Once the cuckoos have laid their eggs in the nests, the nest owners discover them following a determined probability. Then, the corresponding proportion of nests are randomly replaced by new ones according to the probability mentioned above (Yang & Deb, 2009).

In CSA, each egg within a nest represents a possible solution to the optimization problem, while the eggs added by the cuckoos are supposed to be better solutions, which would replace those solutions with a lower performance (Yang & Deb, 2009).

Concerning the searching method, CSA performs a search process similar to the first-order Markov process. However, the new solutions are generated using Lévy flight theory (Yang & Deb, 2009).

Regarding its implementation, CSA is initialized by creating a population with several host nests. Then, a determined cuckoo is randomly selected using Lévy flights, and the corresponding objective function value is determined. After that, a specific nest is randomly chosen, the fitness value is calculated, and the result is compared with the value previously obtained. From such a comparison, the solution with the best fitness value is selected (Yang & Deb, 2009).

Finally, a proportion of the nests are chosen and abandoned in order to build new nests in other locations. This procedure is repeated, fulfilling a stopping condition, storing the best solution at each iteration (Yang & Deb, 2009).

This CSA version has been improved over time by many authors. Walton et al. (2011) proposed incorporating two significant changes in order to increase computational performance. In the original CSA implementation contributed by Yang and Deb (2009), the parameter of the Lévy flight was assumed to be constant. Walton et al. introduced a dynamic parameter in the modeling of the Lévy flight, which reduces with the CSA iterations. This variation on the original CSA formulation improves the local searching when the algorithm is closer to the global solution. In addition to this change, Walton et al. proposed incorporating communication between those eggs with high fitness. This change increases the computational efficiency of CSA (Walton et al., 2011). Yang and Deb (2013) introduced a CSA for the solution to multiobjective problems. The effectiveness of the algorithm was tested by designing a welded beam and a disc brake, among other general-purpose problems.

Other important problems have been solved using CSA. A representative example is the work of Gandomi, Yang, et al. (2013), where several structural problems, including the design of a pressure vessel and of car side-impact, among many other technical problems, were successfully solved. Indeed, the authors reported better results than those obtained from the implementation of GA and PSO. Similarly, Civioglu and Besdok (2013) discussed CSA, PSO, DE, and ABC, analyzing a large number of general-purpose test problems, concluding that solutions provided by CSA and DE are very similar and robust. Additionally, Yang and Deb (2014) presented a comparative discussion. They revealed the necessity of including a more significant number of optimization variables in the benchmark problems and developing a theoretical background to improve our understanding of the working principles of heuristic methods.

1.1.7 Cultural algorithm

The cultural algorithm (CA) is based on the premise that societal evolution allows us to adapt to the surrounding environment at higher rates than other natural processes based on the transmission of genetic information from one generation to another. Based on this reasoning, Reynolds (1994) created a conceptual framework for an optimization technique for analyzing cultural and societal interactions. Specifically, the author aimed to solve large-scale optimization problems, using the reasons behind the high rates of cultural evolution.

To this end, Reynolds starts with the definition of culture, emphasizing the capabilities developed by human society to create, represent, and transmit information through time on social and historical scales (Reynolds, 1994).

Human society owes its development and cultural progress to its capacity to create categories. In other words, this reasoning suggests that humans perform a categorization process over their experiences, and consequently, they build symbols. The result of this process can be summarized in a sort of world map or *mappa*, which can be specialized and joined with another *mappa* to obtain a group *mappa*. The concept of group *mappa* is helpful because it can guide the attitudes of a specific society during the analysis of its cultural evolution (Reynolds, 1994).

In CA, cultural evolution is modeled considering the individual and general perspectives of the whole population. The information exchange among the individuals and the general viewpoint of the population are incorporated into the optimization process (Reynolds, 1994).

On the one hand, information exchange among individuals is understood as a method for behavior transmission. Indeed, it is a micro-evolutionary process. On the other hand, creating a shared perspective for the entire population is represented by a macro-evolutionary procedure. Thus, the macro-evolutionary process establishes concrete limitations to the operations of the micro-evolutionary method. It models the fact that the accepted behavior of the social group constrains the way information is exchanged among its members (Reynolds, 1994).

Regarding CA implementation, the micro-evolutionary process can be modeled using GA. In contrast, the macro-evolutionary subroutine can be implemented using a general representation of the chromosomes according to their attitudes (Reynolds, 1994).

CA has been improved by combining it with other techniques. Becerra and Coello (2006) combined CA with DE, reducing the number of objective function evaluations required, and increasing the computational efficiency. Similarly, Sun et al. (2012) combined CA and PSO, obtaining an efficient method to avoid local optima with fast convergence. In addition, the authors incorporated the factorial design technique to determine the appropriated values for the optimization parameters. Awad et al. (2017) combined CA and DE, evaluating both methods using parallel computing. In addition, the influence of CA and DE on the whole performance is controlled through a participation ratio. Qin et al. (2010) combined CA with DE for the solution to multiobjective problems. They apply their method to reservoir management, obtaining a satisfactory description of the Pareto optimal front in a reasonable computational time. Regarding power system engineering, CA and DE have been combined for solving ED (economic dispatch) problems, and hydrothermal coordination in mono-objective and multiobjective optimization frameworks (Coelho et al., 2009; Lu et al., 2011; Zhang et al., 2013).

1.1.8 Differential evolution algorithm

DE is a widely used optimization technique. DE is a population-based algorithm in which the solutions are encoded as vectors. The population is initialized using random numbers uniformly distributed. As in many evolutionary algorithms, DE employs the mutation, crossover, and selection operators (Storn & Price, 1997).

During the mutation process, three different individuals, with the indices r_1, r_2, and r_3 of the current generation (g) are required to estimate the mutant of the next generation ($g+1$). This reasoning is expressed in Eq. (1.15) (Storn & Price, 1997):

$$v_{i,g+1} = x_{r1,g} + F\left(x_{r2,g} - x_{r3,g}\right) \tag{1.15}$$

where $x_{r1,\ g}$, $x_{r2,\ g}$, and $x_{r3,\ g}$ are the individuals of the current generation involved in the mutation process. The indices r_1, r_2, and r_3 are randomly selected from the population in order to estimate the characteristics of individual i for the next generation ($g+1$). The parameter F is a positive factor between 0 and 2, which adjusts the influence of the difference $x_{r2,\ g} - x_{r3,\ g}$ on the resulting individual. The crossover operation was introduced to enhance the diversity of individuals according to a specific rate. To this end, it combines the results of the mutation operation with a random procedure. Finally, the selection operator is in charge of taking the ultimate decisions about the individuals of the next generation using the objective function value as the principal criterion. Other versions of DE have also been analyzed, showing promising results. An example consists of changing the formulation of Eq. (1.15) to the one shown in Eq. (1.16) (Storn & Price, 1997):

$$v_{i,g+1} = x_{best,g} + F\left(x_{r1,g} + x_{r2,g} - x_{r3,g} - x_{r4,g}\right) \tag{1.16}$$

where $x_{best,g}$ represents the individual with the best performance.

DE has been widely studied, and it has undergone numerous improvements since its early days. Vesterstrøm and Thomsen (2004) compared DE with PSO and other algorithms using many test functions, finding that DE outperforms PSO and other techniques in most cases. However, this trend is reverted when noisy functions need to be optimized.

Qin and Suganthan (2005) developed a self-adaptive differential evolution algorithm to overcome the problems related to the selection of factor F of Eqs. (1.15) and (1.16), the choice of the crossover rate, and the population size. The authors start their analysis by creating a methodology to select the appropriate learning strategy, similar to those previously mentioned in Eqs. (1.15) and (1.16). Then, they estimate the probability of employing a concrete learning strategy using the results obtained at the end of a specific number of generations (learning period). In fact, this is an adaptive approach in which the decisions related to the learning strategy are taken based on the probabilistic analysis of the optimization results after the learning period (Qin & Suganthan, 2005).

Concerning the selection of the factor F and the crossover rate, they were represented using random numbers. The factor F directly impacts the convergence speed. According to the authors' experience, this parameter is frequently adjusted in the interval (0, 1). However, it was set using a random number in the interval (0, 2) using a normal distribution with a mean equal to 0.5 and a standard deviation equal to 0.5. The performance of DE is susceptible to the crossover rate. Using their expertise, the authors suggested adjusting the crossover rate according to a normal distribution with a mean equal to 0.5 and a standard deviation equal to 0.1 (Qin & Suganthan, 2005).

Liu and Lampinen (2005) developed a fuzzy differential evolution (FDE) algorithm as a self-adaptive methodology for DE implementation. FDE employs a fuzzy logic controller to determine the appropriate values of the factor F and the crossover rate. To this end, FDE estimates the mean square root of the variations between individuals and their corresponding objective function value, obtaining the parameter vector change and the function value change, respectively. These parameters are subsequently updated and used as input information to the fuzzy controller, obtaining the improved values of F and the crossover rate. The fuzzy rules are adjusted so that F and the crossover rate become large or small depending on the algorithm's performance through the generations.

Brest et al. (2006) proposed an improved self-adaptive algorithm in which F and the crossover rate are included in the structure of each individual. Each individual is formulated to incorporate the structure of the optimization problem studied, the parameter F, and the crossover rate. Thus, the optimization problem can be solved by including the values of the optimal parameters. The authors propose updating the F factor ($F_{i,g}$) and the crossover rate ($CR_{i,g}$) using Eqs. (1.17) and (1.18) (Brest et al., 2006):

$$F_{i,g+1} = \begin{cases} F_l + r_1 F_u; r_2 < \zeta_1 \\ F_{i,g}; r_2 \geq \zeta_1 \end{cases} \tag{1.17}$$

$$CR_{i,g+1} = \begin{cases} r_3; r_4 < \zeta_2 \\ CR_{i,g}; r_4 \geq \zeta_2 \end{cases} \tag{1.18}$$

where F_l and F_u are equal to 0.1 and 0.9, respectively. The parameters r_1, r_2, r_3, and r_4 are random numbers uniformly distributed over the interval [0, 1]. The variables ζ_1 and ζ_2 represent the probability of changing the values of the parameters. They were adjusted to 0.1 following the authors' experience (Brest et al., 2006).

Many improvements can be found in the literature. Rahnamayan et al. (2008) combined DE with opposition-based learning, Zhang and Sanderson (2009) introduced an optional external archive on DE implementation. Das et al. (2009) incorporated a neighborhood-based mutation operator. Vrugt et al. (2009) improved the computational efficiency of Markov chain Monte Carlo simulation using a self-adaptive DE algorithm. Mallipeddi et al. (2011) proposed working with a set of mutation strategies and promising values for each parameter, obtaining new individuals from competitive activity.

1.1.9 Electromagnetic field optimization

The electromagnetic field optimization (EFO) (Abedinpourshotorban et al., 2016) algorithm is motivated by the concepts behind electromagnetic theory. Its formulation is based on the attraction and repulsion forces of electromagnets and the ratio between them, which is modeled using the golden ratio. EFO implementation requires building a population of individuals, where a group of electromagnets represents each individual. Specifically, each individual is an electromagnetic particle composed of several electromagnets. Thus, a possible solution or individual is an electromagnetic particle, while the optimization variables are the group of electromagnets. In addition, the electromagnets belonging to the same particle have the same polarity.

As stated before, EFO is a population-based algorithm, so that the first step in its computational implementation consists of creating a population of electromagnetic particles using random numbers. Then, the objective function value is calculated for each individual (electromagnetic particle), and using these results, the entire population is sorted according to their fitness. The sorted population is then divided into three groups according to polarity (positive, negative, and neutral). The group of positive polarity is composed of those particles with the highest fitness, the group with negative polarity is those particles with the lowest performance, and the group of neutral particles is composed of the remainder (Abedinpourshotorban et al., 2016).

During each iteration, a new particle is created and evaluated in the objective function. If its fitness is better than the worst, the new particle is included in the population, and the particle with the worst performance is discarded. Besides these steps, EFO is also provided with a random procedure to give diversity to the individuals in the population, avoiding local optima (Abedinpourshotorban et al., 2016).

The EFO search strategy is a crucial structure in its working principle. In this sense, the neutral electromagnets are affected by the positive and negative ones in a virtual manner. The position of a new particle ($EMP_{\mathrm{NEW},\ j}$) is shown in Eq. (1.19) (Abedinpourshotorban et al., 2016):

$$\begin{aligned} EMP_{\mathrm{NEW},j} = EMP_{Kj,j} + \{(\Phi \times r)(EMP_{Pj,j} - EMP_{Kj,j})\} \\ -\{r \times (EMP_{Nj,j} - EMP_{Kj,j})\} \end{aligned} \tag{1.19}$$

where r is a random number in the interval [0, 1], j is the index for each variable, K_j is the index for those neutral electromagnets, P_j is the index for positive electromagnets, and N_j is the index for negative electromagnets. Finally, ϕ is the golden ratio (Abedinpourshotorban et al., 2016).

Among the most critical optimization parameters of EFO are the number of electromagnetic particles and the portion of the particles included in the group with positive and negative polarities. In this regard, the authors suggest using a portion between 5% and 10% for the group of positive polarity particles (Abedinpourshotorban et al., 2016). Concerning the groups of particles with negative polarity, a portion between 40% and 50% was suggested.

In addition, other parameters related to the probability of randomly changing several electromagnets are also important. Thus, the probability of selecting some electromagnets with positive polarity without performing any modification is recommended as 10%. On the other hand, the probability of changing one electromagnet should be set to a value of around 40% (Abedinpourshotorban et al., 2016).

Regarding its applications in power system engineering, EFO has been successfully employed on the optimal coordination of protective equipment (Bouchekara et al., 2017).

1.1.10 Genetic algorithm

The genetic algorithm (GA) is based on the fundamentals of heredity. Heredity is a concept related to the natural procedures and mechanisms to transfer biochemical information from one generation to the next. These mechanisms are implemented using genes, and they play a crucial role in species evolution. Specifically, the physical and behavioral features of the members of a determined species are transmitted from parents to offspring using heredity.

Genetics is the knowledge field dedicated to heredity. Concretely, it recognizes the relationship between the genes and their corresponding traits. It is vital because humanity interprets the changes in a particular habitat or ecosystem through generations as a process of continuous evolution and progress. In other words, the generational modifications of biodiversity are motivated by evolutionary forces, which are constrained by natural laws and random events.

Supported by the reasoning previously explained, Holland (1975) proposed GA as an optimization tool, understanding adaptation as an optimization process. In this sense, adaptation is related to the capacity of a living being to improve its structure and behavior to increase its survival probability. Holland analyzed essential features of ecosystems from a behavioral perspective to define the crucial parameters of GA. As a result, the author introduced the crossover and mutation operators.

Goldberg (1989) contributed outstandingly to GA theory and applications. As a consequence, GA was adopted by scientists and practitioners of many fields. In his book, this author presented the mathematical and computational instruments required to implement GA efficiently.

Deb follows the efforts of Holland and Goldberg, contributing outstandingly with GA theory to solve multiobjective optimization problems, enabling GA as an instrument to analyze more realistic situations. In his book (Deb, 2001), Deb explained the concepts and algorithms to solve optimization problems with multiple variables transparently. Thus, it is a reference for researchers in many fields interested in the GA application. Additionally, Deb et al. (2002) and Deb and Jain (2014) developed essential studies to solve problems with multiple and many objectives using a theory based on nondominated sorting.

GA has also been implemented using parallel-computing structures to deal with the problems related to the computational burden, among other issues. In this sense, Cantú-Paz (1999) studied the complex relationship between the solution, computational efficiency, and parameter settings. As a result, the author contributed crucial information about carrying out the correct design of a parallel GA.

GA has been successfully implemented to design energy systems. It is highly recommended to obtain a system design minimizing the total costs (mono-objective optimization). In addition, it has been implemented for multiobjective optimization, where minimizing the total costs and the environmental impact simultaneously is required. A detailed description of GA is presented in Chapter 2.

1.1.11 Gravitational search algorithm

The law of gravity inspires the gravitational search algorithm (GSA) (Rashedi et al., 2009). Specifically, it is based on the physical behavior of a system of masses, where each mass is supposed to be a possible solution or an individual, and its weight is directly related to its objective-function value or fitness.

The masses belonging to the system are subject to the laws of motion. Thus, the gravitational force serves as a communication channel between the system's masses. Under these circumstances, the lightest mass feels attraction toward the heaviest mass. In other words, the solution with high fitness attracts those objects with lower performance. This process guarantees the algorithm's evolution toward a promising solution (Rashedi et al., 2009).

Concretely, the relevant features for each mass are its position, inertial mass, active gravitational mass, and passive gravitational mass. The position refers to the possible solution, and the gravitational and inertial masses are intrinsically related to fitness (Rashedi et al., 2009).

As previously mentioned, the system is subject to the laws of gravity and motion. According to the law of gravity, the gravitational force between two masses is proportional to the product of both masses and inversely proportional to the squared value of the distance between them. However, to guarantee the precision of the solution, GSA is implemented using the distance without considering its squared value. Under this assumption, the force acting between masses i and j at time t is described in Eqs. (1.20) and (1.21) (Rashedi et al., 2009):

$$F_{ij}(t) = G(t)\frac{M_{pi}(t) \times M_{aj}(t)}{R_{ij}(t) + \varepsilon}\left(x_j(t) - x_i(t)\right) \tag{1.20}$$

$$G(t) = G(G_0, t) \tag{1.21}$$

where $x_i(t)$ and $x_j(t)$ are the positions of both masses, $R_{ij}(t)$ is the distance between them, $M_{aj}(t)$ is the active gravitational mass of the object j, $M_{pi}(t)$ is the passive gravitational mass of object i, $G(t)$ is the gravitational constant, and G_0 is the gravitational constant at the first cosmic quantum-interval. The parameter ε is a small value and the value $R_{ij}(t)$ is determined as the Euclidean distance between the objects i and j. Eq. (1.21) represents the changes of the gravitational constant with the age of the universe (Rashedi et al., 2009).

The position ($x_i(t)$) and velocity ($v_i(t)$) of each mass are updated according to Eqs. (1.22) and (1.23) (Rashedi et al., 2009):

$$v_i(t+1) = r_i \times v_i(t) + a_i(t) \tag{1.22}$$

$$x_i(t+1) = x_i(t) + v_i(t+1) \tag{1.23}$$

where r_i is a random number in the interval [0, 1] and $a_i(t)$ is the acceleration of the mass i (Rashedi et al., 2009).

Concerning GSA implementation, it starts by identifying the search space and initialization. Then, the fitness is calculated for all masses, and the gravitational constant is also calculated. After that, we calculate the total force and acceleration. Using this information, we can update the position of each mass for the next iteration (Rashedi et al., 2009).

GSA has been widely applied for the solution to technical problems. Mirjalili et al. (2012) developed a hybrid algorithm combining PSO and GSA to train an NN. Another fundamental application in power system engineering is the optimal power flow problem, which Duman et al. (2012) solved using GSA.

1.1.12 Grey wolf optimizer

The grey wolf optimizer (GWO) is an algorithm based on the behavior of the *Canis lupus*. In GWO, the social relationship of wolves is analyzed, deriving a family of mathematical properties for the exploration and exploitation of the objective function of interest. Grey wolves are recognized as predators at the top of the food chain. Usually, they live in a pack, with between 5 and 12 members, approximately. In addition, grey wolves interact with each other following a specific set of rules and social hierarchy. The social hierarchy includes alpha, beta, delta, and omega wolves (Mirjalili et al., 2014).

The alpha wolf, which can be a male or a female, is in charge of making crucial decisions related to hunting, selection of the sleeping location, and the time to wake up, among other activities. Generally, decisions dictated by the alpha have to be followed by the pack. However, democratic behaviors where the alpha follows the other members have also been observed. Something important is that the alpha is not always the strongest member, but the wolf with solid leadership skills, or in other words, it is the member with the most suitable qualities to manage the pack (Mirjalili et al., 2014).

The beta wolves support the alpha during the daily activities of the pack. The beta can be male or female, and it would be the best substitute for the alpha member. The beta wolf is one step down from alpha, so that the beta must follow the alpha's directions. However, the beta wolf can guide those members lower in the hierarchy. The beta wolf serves as an advisor for the alpha, strengthening its authority in the remainder of the pack (Mirjalili et al., 2014).

The omega wolf lives under the authority of alpha and beta wolves. Even when omega wolves have a low position in the hierarchy, they are essential to keeping the social balance between the pack members. If the omega wolves are missing, violence between the pack members can be observed due to frustration. Thus, omega wolves are important to support the hierarchy of the entire pack (Mirjalili et al., 2014).

The delta wolves have a lower position in the hierarchy than alpha and beta wolves but a higher one than omega wolves. Thus, the wolves working as scouts, sentinels, elders, and hunters are included on this hierarchy level (Mirjalili et al., 2014).

The working principle of GWO is based on modeling the social hierarchy described above and the group attitudes such as hunting, including encircling, attacking, and searching for prey. Social hierarchy is defined according to fitness. The solution with the highest fitness is the alpha wolf. Then, those individuals with the second and third best performances are classified as beta and delta wolves, respectively. Finally, the remainder of the individuals in the population are omega wolves. In GWO, the optimization process is understood as hunting, and the omega wolves must follow the alpha, beta, and delta wolves. This mechanism guides the algorithm during the searching process (Mirjalili et al., 2014).

Encircling prey procedure mathematically represents the pack behavior using vector equations for the position of the prey and grey wolves. A specific wolf updates its status following the prey location. Thus, many places close to the alpha wolf can be analyzed from the current situation by adjusting the encircling prey model properly (Mirjalili et al., 2014).

During the hunting process, we do not have the specific location of the prey, which represents the global optimum. However, the alpha, beta, and delta wolves know the closest location. For this reason, we force the remainder of wolves to update their positions to those of the best solutions (Mirjalili et al., 2014).

The attacking prey process occurs when the hunting is finished and the prey is not in motion. From an algorithmic perspective, it represents the exploitation process. However, GWO still requires a concrete operator to perform exploration (Mirjalili et al., 2014).

Exploration is inspired by the idea that alpha, beta, and delta wolves rule the optimization process. However, the difference between them relies on the way they search and attack the prey. This divergence motivates the exploration operator of GWO (Mirjalili et al., 2014).

GWO has been improved to solve multiobjective problems (Mirjalili et al., 2016). In addition, it has been used to train an NN (Mirjalili, 2015), to enhance the operation of a maximum power tracking algorithm (Yang et al., 2017), and to determine optimal active and reactive power dispatch (Jayabarathi et al., 2016; Sulaiman et al., 2015).

1.1.13 Imperialist competitive algorithm

The imperialist competitive algorithm (ICA) (Atashpaz-Gargari & Lucas, 2007) is inspired by the imperialist competition observed throughout human history. In ICA, the concept of the population frequently used in most of the swarm-based heuristics is understood as a set of empires. Each empire is composed of a country and its associated colonies. As previously mentioned, the reasoning behind ICA is the competitive activities among kingdoms, where weaker empires are discarded, losing their territories to stronger ones. After some time, the whole process converges to a state in which only one empire exists. Such an empire would represent the near-optimal solution found by ICA.

The instrument of ICA is the desire or need of an empire to spread its power beyond its limits: an empire tries to conquer other countries to sustain its existence and to guarantee its long-term survival. Those countries under the government of a determined empire are known as colonies. As a result of cultural spread, the colonies increase their development (Atashpaz-Gargari & Lucas, 2007).

Concerning ICA implementation, an initialization process is required. The initialization consists of randomly creating several solutions, classified as empires and colonies, depending on their fitness. ICA understands the performance of a solution as a measure of power (Atashpaz-Gargari & Lucas, 2007).

Once the imperialists and colonies have been recognized, the colonies move toward the imperialist countries. It is essential to highlight that the power of an empire also depends on the power of its colonies. This fact is considered by ICA using Eq. 1.24 (Atashpaz-Gargari & Lucas, 2007):

$$TC_n = \text{Cost}(\text{imperialist}_n) + \xi \times \text{mean}\{\text{Cost}(\text{colonies of empire}_n)\} \tag{1.24}$$

where TC_n is the total cost of an empire n and ξ is a position parameter lower than one.

During imperialistic competition, each empire works to control the territories of another kingdom. As a consequence, the powers of the conqueror empires increase, while the influence of the powerless kingdoms is reduced. ICA represents this situation by taking the weakest colonies of the weakest empires and allowing all empires to conquer these territories. The probability of victory of a determined empire (n) will be determined by its total power (TC_n). In other words, it is not guaranteed that the most powerful empire conquers the weakest colonies, but they have the highest probability. The probability of possession (P_n) of an empire n is estimated using Eqs. (1.25) and (1.26) (Atashpaz-Gargari & Lucas, 2007):

$$NTC_n = TC_n - \max\{TC_i\} \tag{1.25}$$

$$P_n = \left| \frac{NTC_n}{\sum_i NTC_i} \right| \tag{1.26}$$

where NTC_n is the normalized value of the total cost (TC_n). Finally, the powerless empires disappear, and their colonies are divided between the most powerful ones. ICA represents this situation by discarding the corresponding solutions (Atashpaz-Gargari & Lucas, 2007).

ICA has received a positive reception from researchers worldwide. In this regard, Kaveh and Talatahari (2010) successfully applied ICA in the optimal design of steel structures. Shabani et al. (2013) employed ICA to design a PID controller operating on a load-frequency control system. Ahmadi et al. (2013) combined fuzzy logic, NN, and ICA to develop a technique for oil rate prediction of wells. Regarding power system engineering, Ghasemi et al. (2014) solved the optimal power flow problem using a multiobjective version of ICA.

1.1.14 Particle swarm optimization

Particle swarm optimization (PSO) (Kennedy & Eberhart, 1995) was designed by studying bird flocking and fish schooling. It is analytically supported by the swarming theory. The initial idea was to simulate the particular behavior of a bird flock, considering spatial and geographical factors. PSO takes advantage of conclusions drawn from sociobiology and other knowledge fields. Specifically, a central idea behind PSO is that individuals within a swarm are beneficiaries of the previous experience of the remaining members.

In its early version, PSO employed two primary variations: the nearest-neighbor velocity matching and craziness, and the cornfield vector. In the first variation, PSO is first randomly initialized. Then the nearest neighbor of each agent was determined. After that, the corresponding velocity of each agent was estimated, resulting in the synchronized movement of the group. However, such a movement is made in a fixed direction. It requires the incorporation of a stochastic parameter to represent the craziness degree of the agents. In the second variation, each agent determines its current position and remembers its corresponding fitness. It also calculates its velocity in a simplified manner for each dimension. Additionally, each agent has information about the best position of the entire swarm (Kennedy & Eberhart, 1995).

Based on their experience, Kennedy and Eberhart refined the original concept, removing the craziness and the nearest neighbor velocity matching, giving more importance to the agents' particular and best positions. Remembering the best position of each agent is recognized as using autobiographical memory. When an agent adjusts its velocity utilizing this information, this agent feels simple nostalgia. Concerning the group's best position, it is recognized as publicly available knowledge or standard information. The influence of these two features is controlled through two parameters, which are set to equal values to improve the algorithm's efficiency. In addition, acceleration and velocity updates are performed considering the distance from the best position (Kennedy & Eberhart, 1995).

Development of PSO produced an algorithm able to analyze functions in a multidimensional space in a specific number of iterations. In addition, the swarm members can remember and take advantage of personal and global experiences. They can be allocated in a diverse manner and change their behavior according to the changes in the publicly available information (Kennedy & Eberhart, 1995).

Later, Eberhart and Shi (2000) assessed PSO with an inertia weight and a constriction factor using several benchmark functions. According to the observed results, the authors suggested implementing PSO with the constriction factor and recommended establishing certain limits for the maximum velocity and position.

Clerc and Kennedy (2002) carried out a notable effort to improve the understanding of PSO. The authors analyzed the agents' trajectories under discrete and continuous time and developed a complete mathematical model in a limited number of dimensions. The model allowed them to understand the convergence tendencies and suggest relevant changes in the original model to increase the algorithm capabilities to find the global optimum.

Coello and Lechuga (2002) and Coello et al. (2004) proposed a methodology for solving multiobjective problems using PSO. The authors employ the concept of Pareto dominance to estimate in which direction an agent should move. In addition, the nondominated solutions are stored in a repository accessible by all the members in the swarm to guide their own movements. Similarly, Trelea (2003) assessed PSO performance using dynamic system theory, obtaining a graphical methodology for parameter selection. In addition, the balance between exploitation and exploration was discussed using the analysis of several benchmark optimization problems.

Zhan et al. (2009) developed an adaptive PSO. The method is implemented in two main stages. During the first stage, the fitness value of each agent is calculated, and a state estimation process is performed. The state estimation considers several conditions such as exploration, exploitation, convergence, and jumping out. Thus, inertia weight, and acceleration coefficients, among other parameters, are adjusted by automatic control to increase computational efficiency. Finally, when the algorithm is in the convergence state, the algorithm tries to find the global optimum.

PSO has been successfully used to solve technical problems such as ED (Gaing, 2003), control systems design (Gaing, 2004), and many others.

1.1.15 Quantum-inspired optimization algorithms

The quantum-inspired genetic algorithm (QIGA) (Han et al., 2001; Narayanan & Moore, 1996) enhances the capabilities of classical GA using knowledge of quantum mechanics. Specifically, QIGA employs the quantum bits and the superposition of states for its working principle. A quantum bit (Q-bit) may be in a state equal to one, in a condition equivalent to zero, or the superposition of these states. Representing the state of a concrete quantum bit requires two complex numbers to specify the probability of being in a determined state. These characteristics make using the quantum representation an appropriate framework, because the superposition of states can be considered.

On the one hand, a single quantum bit chromosome can represent eight states. On the other hand, the classical representation would require a chromosome with eight states. These features provide essential advantages related to diversity when using QIGA. This fact also influences convergence because the diversity property disappears during the algorithm's evolution, and the quantum bits converge to a single state. Thus, the quantum bit representation allows us to take advantage of exploration and exploitation simultaneously (Han et al., 2001; Narayanan & Moore, 1996).

Quantum mechanics have also influenced evolutionary algorithms, resulting in the quantum-inspired evolutionary algorithm (QIEA). In this sense, Han and Kim (2002) proposed the quantum gate (Q-gate) operator to increase the QIEA search power. To improve the capabilities of QIEA, the authors (Han & Kim, 2004) developed a termination criterion to enhance understanding of the convergence process of the Q-bit individuals. Furthermore, they contributed with a modified version of the rotation gate operator and a two-phase structure as a variation operator.

Following these trends, Coelho (2008, 2010) proposed enriching PSO with the theory behind quantum mechanics and chaos, obtaining quantum-inspired particle swarm optimization (QIPSO). This contribution has produced essential changes in how PSO is understood because, in its classical version, it assumes that the velocity and position of each particle can be estimated however, this is not the case for quantum mechanics, in which the speed and location cannot be calculated simultaneously. Thus, the state of a determined particle is described using a wave function. Additionally, the author adopted a chaotic mutation operator to avoid convergence to a local minimum. Similarly, Soleimanpour-Moghadam et al. (2014) combined quantum mechanics with GSA, resulting in a quantum-inspired gravitational search algorithm (QIGSA).

QIGA, QIEA, QIPSO, and QIGSA have been used to solve relevant problems of power system engineering, including ED (Meng et al., 2010) and unit commitment (UC) (Jeong et al., 2010; Lau et al., 2009), among many others.

1.1.16 Simulated annealing and other optimization techniques

Simulated annealing (SA) (Kirkpatrick et al., 1983) is a legendary optimization technique widely applied in many fields. SA is based on statistical mechanics theory, which analyzes a system with many degrees of freedom in thermal equilibrium at a specific temperature. Thus, a computational method designed to analyze a group of atoms in equilibrium at a determined temperature is used to create a set of possible configurations for the solution of an optimization problem, considering the energy as the corresponding cost function.

Since its early days, SA has been improved (Corana et al., 1987) and its effectiveness has also been verified (Goffe et al., 1994). It has been successfully applied in essential fields such as genetics (Dupanloup et al., 2002) and data science (Maulik & Bandyopadhyay, 2002).

It is crucial to highlight that many other optimization techniques have not been commented on here because of the lack of space. An example is tabu search (Glover, 1990), which has a vast history. In addition to the existing methods, new approaches are continuously under development. An example is the case of the remora optimization algorithm (Jia et al., 2021), which has recently been published.

PSO is a widely used algorithm in science and engineering. Thus, we will briefly describe some key concepts related to its implementation to solve an optimization problem with continuous and binary variables.

1.2 Basic implementation of particle swarm optimization

As stated before, PSO is one of the most widely applied algorithms. For this reason, we are going to give a brief explanation of its computational implementation. The optimization process minimizes a specific function ($f(\boldsymbol{x})$) that depends on several variables represented by a vector ($\boldsymbol{x}$). The vector $\boldsymbol{x}$ has I elements, as shown in Eq. (1.27):

$$\boldsymbol{x} = \left[x_{(1)} \quad \cdots \quad x_{(i)} \quad \cdots \quad x_{(I)}\right] \tag{1.27}$$

In each dimension, the value $\boldsymbol{x}$ has to be within a specific lower and upper bound. These bounds are represented by vectors as shown in Eqs. (1.28) and (1.29), and the associated constraint is described in Eq. (1.30):

$$\boldsymbol{x}_{\mathbf{min}} = \left[x_{(1)}^{\min} \quad \cdots \quad x_{(i)}^{\min} \quad \cdots \quad x_{(I)}^{\min}\right] \tag{1.28}$$

$$\boldsymbol{x}_{\mathbf{max}} = \left[x_{(1)}^{\max} \quad \cdots \quad x_{(i)}^{\max} \quad \cdots \quad x_{(I)}^{\max}\right] \tag{1.29}$$

$$x_{(i)}^{\min} \le x_{(i)} \le x_{(I)}^{\max} \forall i = 1, \ldots, I \tag{1.30}$$

This general structure is the basis for the benchmark problems we are going to discuss next. In Section 1.2.1, we briefly comment on the fundamentals of PSO implementation when the vector $\boldsymbol{x}$ is composed of real numbers within limits imposed by the vector $\boldsymbol{x}_{\mathbf{min}}$ and $\boldsymbol{x}_{\mathbf{max}}$. Later, in Section 1.2.2, we discuss the solution to the same optimization problem but considering the values of $\boldsymbol{x}$ as binary numbers. Concretely, $x_{(i)} \in \{0, 1\} \forall 1, \ldots, I$.

1.2.1 Optimization of continuous variables

Basic PSO implementation keeps tracking the particle's location with the best performance in the swarm and the position of each particle where it has the best result. In other words, it stores performance information of the swarm as a whole and the performance of each particle, and uses this knowledge to guide the swarm movement in the right direction.

The position of each particle is stored in the matrix $\boldsymbol{s}$, which has I rows and N columns. The matrix $\boldsymbol{s}$ can be understood as N vectors, similar to that shown in Eq. (1.27), transposed and concatenated. In addition, the vectors of minimum ($\boldsymbol{x}_{\mathbf{min}}$) and maximum ($\boldsymbol{x}_{\mathbf{max}}$) limits specified in Eqs. (1.28) and (1.29) are also considered as $s_{(i)}^{\min}$ and $s_{(i)}^{\max}$.

The elements of this matrix are referred to by the index $i = 1, \ldots, I$ for the rows and $n = 1, \ldots, N$ for the columns. The index i represents the dimension of the problem, and the index n represents each particle. The index $j = 1, \ldots, J$ is used to represent each iteration of the optimization process.

On the one hand, the particle with the best performance until the current iteration (j) is stored in the variable $\boldsymbol{s}_{\mathbf{best}}^{\mathbf{global}}$, which has I rows and one column. On the other hand, the position in which a specific particle reached its best performance is stored in a column of the matrix $\boldsymbol{s}_{\mathbf{best}}^{\mathbf{indv}}$. The matrix $\boldsymbol{s}_{\mathbf{best}}^{\mathbf{indv}}$ has I rows and N columns.

Besides the position, fitness is also stored. The fitness of the best-positioned particle is stored in the variable $f_{\text{best}}^{\text{global}}$. Regarding the history of each particle, the best position reached until the current iteration is stored in the vector $\boldsymbol{f}_{\mathbf{best}}^{\mathbf{indv}}$; this vector has N elements. The detailed implementation is explained as follows (Lee & El-Sharkawi, 2008):

Step 1: Set $j \leftarrow 1$.

Step 2: If $j \leq J$, go to Step 3, else end.

Step 3: Calculate the parameter W using Eq. (1.31) with $W_{\min} = 0.4$ and $W_{\max} = 0.9$.

$$W = \left(\frac{W_{\min} - W_{\max}}{J - 1}\right) \times j + \left(\frac{JW_{\max} - W_{\min}}{J - 1}\right) \tag{1.31}$$

Step 4: If $j = 1$, initialize the algorithm randomly. Create the matrix $\boldsymbol{s}(I \times N)$ using random numbers in the interval [0, 1], else go to Step 11.

Step 5: Calculate the objective function value ($f(\boldsymbol{s})$) for each of the particles.

Step 6: Using the results obtained in Step 5, determine the best particle and store it in the variable $\boldsymbol{s}_{\mathbf{best}}^{\mathbf{global}}$. $\boldsymbol{s}_{\mathbf{best}}^{\mathbf{global}}$ is a vector with I elements. Additionally, store the corresponding objective function value (fitness) in the variable f_{best}^{global}.

Step 7: Set the best position of each individual ($\boldsymbol{s}_{\mathbf{best}}^{\mathbf{indv}}$) to the population created in Step 4 ($\boldsymbol{s}$). This means $\boldsymbol{s}_{\mathbf{best}}^{\mathbf{indv}} \leftarrow \boldsymbol{s}$. $\boldsymbol{s}_{\mathbf{best}}^{\mathbf{indv}}$ is a matrix with I rows and N columns. Additionally, store the corresponding objective function value (fitness) in the vector $\boldsymbol{f}_{\mathbf{best}}^{\mathbf{indv}}$, which means $\boldsymbol{f}_{\mathbf{best}}^{\mathbf{indv}} \leftarrow f(\boldsymbol{s})$.

Step 8: Calculate the velocity of each particle, following these steps:

Step 8.1: Set $n \leftarrow 1$.

Step 8.2: If $n \leq N$, go to Step 8.3, else go to Step 9.
Step 8.3: Update the velocity ($\boldsymbol{v}$) of each particle using Eq. (1.32) where $\boldsymbol{c_2}$ is a vector with I elements all equal to two. Similarly, $\boldsymbol{r_2}$ is a random vector with I elements in the interval [0, 1] (Lee & El-Sharkawi, 2008).

$$\boldsymbol{v}_{(:,n)} \leftarrow \boldsymbol{v}_{(:,\mathrm{n})} \times W + \boldsymbol{c_2} \times \boldsymbol{r_2} \times \left(\boldsymbol{s}_{\mathbf{best}}^{\mathbf{global}} - \boldsymbol{s}_{(:,n)} \right) \tag{1.32}$$

Step 8.4: Set $n \leftarrow n+1$ and go to Step 8.2.

Step 9: Update the position of each particle using Eq. (1.33).

$$\boldsymbol{s} \leftarrow \boldsymbol{s} + \boldsymbol{v} \tag{1.33}$$

Step 10: Limit the value of the particle in the matrix $\boldsymbol{s}$ to be within the required interval $[s_{(i)}^{\min}, s_{(i)}^{\max}] \forall i=1, \ldots, I$. $s_{(i)}^{\min}$ and $s_{(i)}^{\max}$ are the minimum and maximum values for each dimension ($i=1, \ldots, I$) (Eq. 1.30).

Step 11: Calculate the objective function value ($f(\boldsymbol{s})$) for each of the particles.

Step 12: Using the results obtained in Step 11, determine the best particle of the entire population. If the fitness of the best particle is better than f_{best}^{global}, update the value of f_{best}^{global} to this fitness. Additionally, update the vector $\boldsymbol{s}_{\mathbf{best}}^{\mathbf{global}}$ to this particle.

Step 13: Using the results from Step 11 determines the best particle, which means considering each particle's history from an individual perspective. To this end, we need to follow the next steps:

Step 13.1: Set $n \leftarrow 1$.
Step 13.2: If $n \leq N$, go to Step 13.3, else go to Step 14.
Step 13.3: If the fitness of the particle n is better than $f_{best(n)}^{indv}$, substitute the particle $\boldsymbol{s}_{\boldsymbol{best}(n)}^{\boldsymbol{indv}}$ for the current one. In other words, assign $\boldsymbol{s}_{\boldsymbol{best}(n)}^{\boldsymbol{indv}} \leftarrow \boldsymbol{s}_{(:,n)}$.
Step 13.4: Set $n \leftarrow n+1$ and go to Step 13.2.

Step 14: Update the velocity ($\boldsymbol{v}$) of each particle ($n=1, \ldots, N$) using Eq. (1.34), where $\boldsymbol{c_1}$, $\boldsymbol{c_2}$, and $\boldsymbol{r_1}$, $\boldsymbol{r_2}$ are vectors with I elements each. All elements of $\boldsymbol{c_1}$ and $\boldsymbol{c_2}$ are equal to two, and the elements of $\boldsymbol{r_1}$ and $\boldsymbol{r_2}$ are random numbers in the interval [0, 1]. Vectors $\boldsymbol{c_1}$, $\boldsymbol{r_1}$ and $\boldsymbol{c_2}$, $\boldsymbol{r_2}$, among others, are multiplied element by element (Lee & El-Sharkawi, 2008).

$$\boldsymbol{v}_{(:,n)} \leftarrow \boldsymbol{v}_{(:,n)} \times W + \boldsymbol{c_1} \times \boldsymbol{r_1} \times \left(\boldsymbol{s}_{\boldsymbol{best}(n)}^{\boldsymbol{indv}} - \boldsymbol{s}_{(:,n)} \right) + \boldsymbol{c_2} \times \boldsymbol{r_2} \times \left(\boldsymbol{s}_{\mathbf{best}}^{\mathbf{global}} - \boldsymbol{s}_{(:,n)} \right) \tag{1.34}$$

Step 15: Update the position of each particle using Eq. (1.35).

$$\boldsymbol{s} \leftarrow \boldsymbol{s} + \boldsymbol{v} \tag{1.35}$$

Step 16: Limit the value of the particle in the matrix $\boldsymbol{s}$ to be within the required interval $[s_{(i)}^{\min}, s_{(i)}^{\max}] \forall i=1, \ldots, I$. $s_{(i)}^{\min}$ and $s_{(i)}^{\max}$ are the minimum and maximum values for each dimension ($i=1, \ldots, I$) (Eq. 1.30).

Step 17: Set $j \leftarrow j+1$ and go to Step 2.

1.2.2 Optimization of binary variables

Another relevant application of PSO is in the solution of optimization problems with binary variables. To this end, the theory described in Section 1.2.1 needs to be extended. In this section, the method recently proposed by Beheshti (2020) will be described. The technique transforms the continuous variables to binaries using a time-varying mirrored S-shaped (TVMS) transfer function.

For the current iteration (j), the parameter σ is calculated using Eq. (1.36):

$$\sigma = \left(\frac{\sigma_{\max} - \sigma_{\min}}{J-1}\right) \times j + \frac{J\sigma_{\max} - \sigma_{\min}}{J-1} \tag{1.36}$$

The parameters $\sigma_{\min}$ and $\sigma_{\max}$ are equal to 0.1 and 1, respectively, and J is the number of iterations. Once σ has been calculated, the matrices $\chi_{\mathbf{a}}$ and $\chi_{\mathbf{b}}$ are calculated using Eqs. (1.37) and (1.38). Both matrices have I rows and N columns.

$$\begin{aligned} \chi_{a(v\sigma)} = \chi_{b(in)} &= \frac{1}{1+\exp\left(\sigma\left(-v_{(in)}\right)\right)} \\ \forall i &= 1,\ldots,I; \\ n &= 1,..,N \end{aligned} \tag{1.37}$$

$$\begin{aligned} \chi_{b(v\sigma)} = \chi_{b(in)} &= \frac{1}{1+\exp\left(\sigma\left(+v_{(in)}\right)\right)} \\ \forall i &= 1,\ldots,I; \\ n &= 1,\ldots,N \end{aligned} \tag{1.38}$$

Using the matrices $\chi_{\mathbf{a}}$ and $\chi_{\mathbf{b}}$, we calculate the binary matrices $\boldsymbol{\Gamma}_{\mathbf{a}}$ and $\boldsymbol{\Gamma}_{\mathbf{b}}$ using Eqs. (1.39) and (1.40). The variables r_3 and r_4 are random numbers in the interval [0, 1].

$$\Gamma_{a(i,n)} = \begin{cases} 1; r_3 < \chi_{a(v,\sigma)} \\ 0; r_3 \geq \chi_{a(v,\sigma)} \end{cases} \forall i = 1,\ldots,I;\ n = 1,\ldots,N \tag{1.39}$$

$$\Gamma_{b(i,n)} = \begin{cases} 1; r_4 > \chi_{b(v,\sigma)} \\ 0; r_4 \leq \chi_{b(v,\sigma)} \end{cases} \forall i = 1,\ldots,I;\ n = 1,\ldots,N \tag{1.40}$$

Using the information form matrices $\boldsymbol{\Gamma}_{\mathbf{a}}$ and $\boldsymbol{\Gamma}_{\mathbf{b}}$, we fulfill the binary matrix s considering the fitness values. For $n=1, \ldots, N$, if $f(\boldsymbol{\Gamma}_{a(:,n)})$ is better than $f(\boldsymbol{\Gamma}_{b(:,n)})$, the vector $\boldsymbol{\Gamma}_{a(:,n)}$ is selected. This means $\mathbf{s}_{(:,n)} \leftarrow \boldsymbol{\Gamma}_{a(:,n)}$. On the contrary, if $f(\boldsymbol{\Gamma}_{b(:,n)})$ is better than $f(\boldsymbol{\Gamma}_{a(:,n)})$, the vector $\boldsymbol{\Gamma}_{b(:,n)}$ is selected. This means $\mathbf{s}_{(:,n)} \leftarrow \boldsymbol{\Gamma}_{b(:,n)}$.

Once we have explained the basic implementation of the PSO algorithm, we will analyze its performance by studying some general examples and ED, which is a crucial problem of electrical engineering.

1.3 Case studies

In this section, we present some illustrative examples of PSO implementation. In Section 1.3.1, we present some benchmark optimization problems. In Section 1.3.2, we discuss the ED problem. Finally, in Section 1.3.3, we discuss the day-ahead power systems management.

We have implemented the optimization algorithm on a computer provided with i7-3630QM CPU at 2.40GHz, 8GB of RAM, and a 64-bit operating system. The computing language used was MATLAB®.

1.3.1 General-purpose optimization problems

In this section, we discuss several testing functions taken from Karaboga and Basturk (2007). The Griewank function is shown in Eq. (1.41). In this function $x_{(i)}^{\min} = -600$ and $x_{(i)}^{\max} = +600$ for $I = 30$ ($i = 1, \ldots, 30$). The global optimum is 0, and it is reached when $x_{(i)} = 0$.

$$f(\mathbf{x}) = \frac{1}{4000}\left(\sum_{i=1}^{I}\left\{x_{(i)}^2\right\}\right) - \left(\prod_{i=1}^{I}\left\{\cos\left(\frac{x_{(i)}}{\sqrt{i}}\right)\right\}\right) + 1 \tag{1.41}$$

The Rastrigin function is shown in Eq. (1.42). In this function $x_{(i)}^{\min} = -15$ and $x_{(i)}^{\max} = +15$ for $I = 30$ ($i = 1, \ldots, 30$). The global optimum is 0, and it is reached when $x_{(i)} = 0$.

$$f(\mathbf{x}) = \sum_{i=1}^{I}\left\{x_{(i)}^2 - 10\cos\left(2\pi x_{(i)}\right) + 10\right\} \tag{1.42}$$

The Rosenbrock function is shown in Eq. (1.43). In this function $x_{(i)}^{\min} = -15$ and $x_{(i)}^{\max} = +15$ for $I = 30$ ($i = 1, \ldots, 30$). The global optimum is 0, and it is reached when x $x_{(i)} = 1$.

$$f(\mathbf{x}) = \sum_{i=1}^{I}\left\{100\left(x_{(i)}^2 - x_{(i+1)}\right)^2 + \left(1 - x_{(i)}\right)^2\right\} \tag{1.43}$$

The Ackley function is shown in Eq. (1.44). In this function $x_{(i)}^{\min} = -32.768$ and $x_{(i)}^{\max} = +32.768$ for $I = 30$ ($i = 1, \ldots, 30$). The global optimum is 0, and it is reached when $x_{(i)} = 0$.

$$f(\boldsymbol{x}) = 20 + e - 20 \times \exp\left(-0.2\sqrt{\frac{1}{I}\sum_{i=1}^{I} x_{(i)}^2}\right) - \exp\left(\frac{1}{I}\sum_{i=1}^{I} \cos\left(2\pi x_{(i)}\right)\right) \tag{1.44}$$

The Schwefel function is shown in Eq. (1.45). In this function $x_{(i)}^{\min} = -500$ and $x_{(i)}^{\max} = +500$ for $I = 30$ ($i = 1, \ldots, 30$). The global optimum is 0, and it is reached when $x_{(i)} = 420.9867$.

$$f(\boldsymbol{x}) = I \times 418.9829 + \sum_{i=1}^{I} -x_{(i)} \sin\left(\sqrt{\left|x_{(i)}\right|}\right) \tag{1.45}$$

The algorithm described in Section 1.2.1 was implemented considering 1500 particles ($N = 1500$), 30 dimensions ($I = 30$), and 10,000 iterations ($J = 10{,}000$). The evolution of PSO for each function (Eqs. 1.41–1.45) is shown in Fig. 1.1. The objective function value for the near-optimal solutions found is reported in Table 1.1, including reference values taken from Srinivasan and Seow (2003). According to these results, the basic PSO implemented provides good results. Regarding the computational time, it was around 9 s.

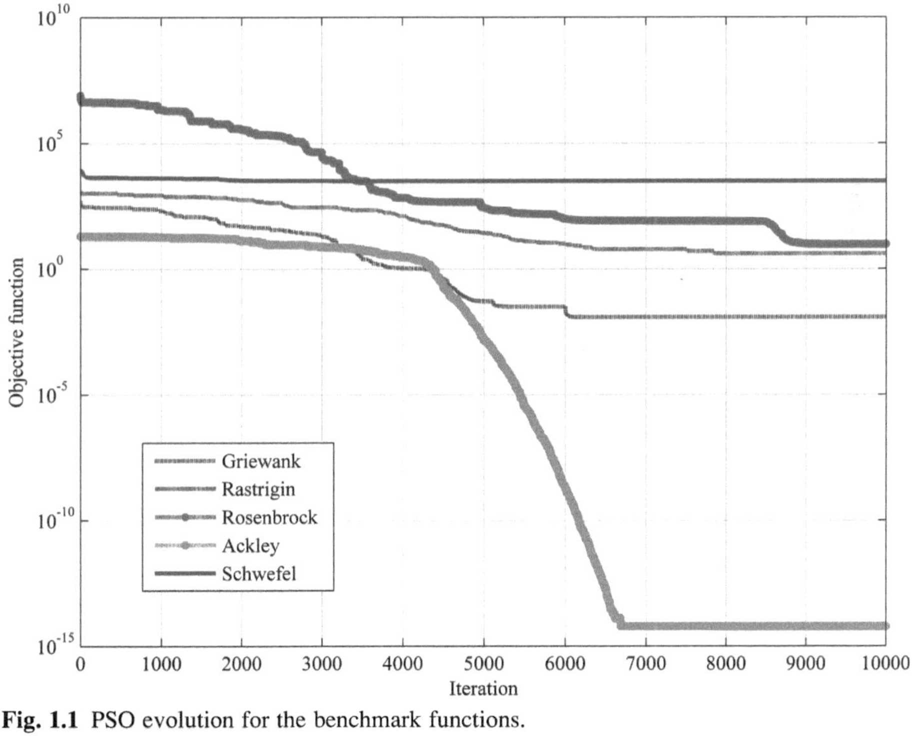

Fig. 1.1 PSO evolution for the benchmark functions.

Table 1.1 Objective function for each benchmark function.

Function	Objective function	Reference
Griewank	0.0123	0.011151
Rastrigin	3.9798	32.476
Rosenbrock	9.3946	402.54
Ackley	6.21725E−15	1.4917E−6
Schwefel	3099.1188	990.77

Comparative analysis with a reference method for each benchmark function.

1.3.2 Economic dispatch of power demand

Power dispatch of a group of generating units is a fundamental problem of electrical engineering. In its most straightforward formulation, the problem consists of finding out the power production of each unit in order to minimize the total operating cost. The power production has to be within a specific interval set by the manufacturer. In this chapter, we solve this problem using PSO, and in Chapter 4, we solve it using a real-coded GA.

Considering a system with I units, the power generation is represented as a vector ($\boldsymbol{p}$) according to Eq. (1.46). The upper and lower limits on power production are presented in Eqs. (1.47)–(1.49).

$$\boldsymbol{p} = \left[p_{(1)} \quad \cdots \quad p_{(i)} \quad \cdots \quad p_{(I)}\right] \tag{1.46}$$

$$\boldsymbol{p}_{\mathbf{min}} = \left[p_{(1)}^{\min} \quad \cdots \quad p_{(i)}^{\min} \quad \cdots \quad p_{(I)}^{\min}\right] \tag{1.47}$$

$$\boldsymbol{p}_{\mathbf{max}} = \left[p_{(1)}^{\max} \quad \cdots \quad p_{(i)}^{\max} \quad \cdots \quad p_{(I)}^{\max}\right] \tag{1.48}$$

$$p_{(i)}^{\min} \le p_{(i)} \le p_{(i)}^{\max} \forall i = 1,\ldots,I \tag{1.49}$$

Some generating technologies impose some prohibited operating zones (POZs) in which the unit cannot be operated. In such a case, the constraints on power generation are mathematically represented using the group of restrictions shown in Eq. (1.50):

$$\begin{aligned} & p_{(i)}^{\min} \le p_{(i)} \le p_{(i1)}^{\text{lower}} \\ & \vdots \\ & p_{(ik-1)}^{\text{upper}} \le p_{(i)} \le p_{(ik)}^{\text{lower}} \ \forall k = 2,\ldots,K \\ & \vdots \\ & p_{(iK+1)}^{\text{upper}} \le p_{(i)} \le p_{(i)}^{\max} \end{aligned} \tag{1.50}$$

where K is the number of POZs. The constraint of Eq. (1.51) represents the power balance, where p_L is the power demand.

$$\sum_{i=1}^{I} p_{(i)} = p_L \tag{1.51}$$

Eq. (1.52) is the cost related to fuel consumption, considering the effects of valve-point loading. Note that these effects can be avoided by setting the coefficients $e_{(i)}$ and $f_{(i)}$ to zero.

$$f(\mathbf{p}) = \sum_{i=1}^{I} \left\{ a_{(i)} + b_{(i)}\left(p_{(i)}\right) + c_{(i)}\left(p_{(i)}\right)^2 + \left| e_{(i)} \times \sin\left(f_{(i)}\left(p_{(i)}^{\min} - p_{(i)}\right)\right)\right| \right\} \tag{1.52}$$

The procedure explained in Section 1.2.1 was implemented to analyze several examples published in the literature. During the implementation, 100 particles ($N=100$) and 300 iterations ($J=300$) were considered.

1.3.2.1 Case study 1

In this section, we analyze the power system of 15 units described in Tables 1.2 and 1.3 (Orero & Irving, 1996), which has a power demand of 2650 MW ($p_L=2650$ MW) with a reserve of 265 MW ($SR=265$ MW). This example does not consider the effects of valve-point loading. However, it takes into account the prohibited operating zones.

Table 1.2 Power system of case study 1.

i	$p_{(i)}^{\min}$(MW)	$p_{(i)}^{\max}$(MW)	$a_{(i)}$(\$/h)	$b_{(i)}$(\$/MWh)	$c_{(i)}$(\$/MW^{2}h)
1	150	455	671.03	10.07	0.000299
2	150	455	574.54	10.22	0.000183
3	20	130	374.59	8.8	0.001126
4	20	130	374.59	8.8	0.001126
5	105	470	461.37	10.4	0.000205
6	135	460	630.14	10.1	0.000301
7	135	465	548.2	9.87	0.000364
8	60	300	227.09	11.21	0.000338
9	25	162	173.72	11.21	0.000807
10	20	160	175.95	10.72	0.001203
11	20	80	186.86	10.21	0.003586
12	20	80	230.27	9.9	0.005513
13	25	85	225.28	13.12	0.000371
14	15	55	309.03	12.12	0.001929
15	15	55	323.79	12.41	0.004447

Cost curve for the units of case study 1.

Table 1.3 Prohibited operating zones for case study 1.

i	Zone 1		Zone 2		Zone 3	
	$p_{(i)}^{min}$ (MW)	$p_{(i)}^{max}$ (MW)	$p_{(i)}^{min}$ (MW)	$p_{(i)}^{max}$ (MW)	$p_{(i)}^{min}$ (MW)	$p_{(i)}^{max}$ (MW)
2	185	225	305	335	440	450
5	180	200	260	335	390	420
6	230	255	365	395	430	455
12	30	55	65	75	–	–

Forbidden operating conditions for the units of case study 1.

1.3.2.2 Case study 2

This section analyzes the power system of 3 units described in Table 1.4 (Sinha et al., 2003). This example does not consider the effect of POZs. However, it takes into account the influence of valve-point loading. The power demand is 850 MW ($p_L = 850$ MW) with a reserve of 85 MW ($SR = 85$ MW).

1.3.2.3 Case study 3

The system in this case has 13 units. The case does not consider the effect of POZs. However, it takes into account the influence of valve-point loading. The power demand is 1800 MW ($p_L = 1800$ MW) with a reserve of 180 MW ($SR = 180$ MW). The technical data is reported in Table 1.5 (Sinha et al., 2003).

1.3.2.4 Case study 4

In this section, a system with 40 units (Sinha et al., 2003) is analyzed. As in the previous examples, only the influence of valve-point loading is considered. The power demand is 10,500 MW ($p_L = 10{,}500$ MW) with a reserve of 1050 MW ($SR = 1050$ MW). The complete data is shown in Table 1.6.

Table 1.4 Power system of case study 2.

i	$p_{(i)}^{min}$(MW)	$p_{(i)}^{max}$(MW)	$a_{(i)}$($/h)	$b_{(i)}$($/MWh)	$c_{(i)}$($/MW^{2}h)	$e_{(i)}$($/h)	$f_{(i)}$(1/MW)
1	100	600	561	7.92	0.001562	300	0.0315
2	50	200	78	7.97	0.00482	150	0.063
3	100	400	200	7.85	0.00194	200	0.042

Cost curve for the units of case study 2.

Table 1.5 Power system of case study 3.

i	$p_{(i)}^{\min}$(MW)	$p_{(i)}^{\max}$(MW)	$a_{(i)}$ (\$/h)	$b_{(i)}$ (\$/MWh)	$c_{(i)}$ (\$/MW²h)	$e_{(i)}$ (\$/h)	$f_{(i)}$ (1/MW)
1	0	680	550	8.1	0.00028	300	0.035
2	0	360	309	8.1	0.00056	200	0.042
3	0	360	307	8.1	0.00056	200	0.042
4	60	180	240	7.74	0.00324	150	0.063
5	60	180	240	7.74	0.00324	150	0.063
6	60	180	240	7.74	0.00324	150	0.063
7	60	180	240	7.74	0.00324	150	0.063
8	60	180	240	7.74	0.00324	150	0.063
9	60	180	240	7.74	0.00324	150	0.063
10	40	120	126	8.6	0.00284	100	0.084
11	40	120	126	8.6	0.00284	100	0.084
12	55	120	126	8.6	0.00284	100	0.084
13	55	120	126	8.6	0.00284	100	0.084

Cost curve for the units of case study 3.

1.3.2.5 Results

The optimization results for all cases are reported in Table 1.7. For comparative purposes, some reference results taken from Orero and Irving (1996) and Sinha et al. (2003) have been added. As can be observed, reasonable operating points were found. We can observe a difference of between 0.9% and 10.96% with respect to the reference solution; the highest difference corresponds to the case of highest size.

Regarding the computational time, it was around 2.4 s. The evolution of the optimization process is presented in Fig. 1.2.

1.3.3 Day-ahead management of power systems

Day-ahead scheduling of power systems is another crucial problem in electrical engineering. For instance, the unit commitment (UC) problem has been widely studied, and many techniques have been proposed for its solution. Binary PSO can be used to solve this type of problem, in which the goal is determining whether a specific unit or device should be connected.

The authors (Lujano-Rojas et al., 2019) recently solved this problem to determine the optimal operation of a battery energy storage system (BESS) connected to an isolated power system. The technique explained in Section 1.2.2 for optimizing with binary variables using PSO was implemented to manage a BESS daily. Specifically, the TVMS transfer function was used as aforementioned.

The employed algorithm takes advantage of essential features of the TVMS transfer function, such as its capabilities for global exploration and local exploitation. This was observed from the comparative analysis with a GA, which also offered good performance. The reader can find all the details in Lujano-Rojas et al. (2019), which is publicly available on the internet.

Table 1.6 Power system of case study 4.

i	$p_{(i)}^{\min}$(MW)	$p_{(i)}^{\max}$(MW)	$a_{(i)}$ (\$/h)	$b_{(i)}$ (\$/MWh)	$c_{(i)}$ (\$/MW^{2}h)	$e_{(i)}$ (\$/h)	$f_{(i)}$ (1/MW)
1	36	114	94.705	6.73	0.0069	100	0.084
2	36	114	94.705	6.73	0.0069	100	0.084
3	60	120	309.54	7.07	0.02028	100	0.084
4	80	190	369.03	8.18	0.00942	150	0.063
5	47	97	148.89	5.35	0.0114	120	0.077
6	68	140	222.33	8.05	0.01142	100	0.084
7	110	300	287.71	8.03	0.00357	200	0.042
8	135	300	391.98	6.99	0.00492	200	0.042
9	135	300	455.76	6.6	0.00573	200	0.042
10	130	300	722.82	12.9	0.00605	200	0.042
11	94	375	635.2	12.9	0.00515	200	0.042
12	94	375	654.69	12.8	0.00569	200	0.042
13	125	500	913.4	12.5	0.00421	300	0.035
14	125	500	1760.4	8.84	0.00752	300	0.035
15	125	500	1728.3	9.15	0.00708	300	0.035
16	125	500	1728.3	9.15	0.00708	300	0.035
17	220	500	647.85	7.97	0.00313	300	0.035
18	220	500	649.69	7.95	0.00313	300	0.035
19	242	550	647.83	7.97	0.00313	300	0.035
20	242	500	647.81	7.97	0.00313	300	0.035
21	254	550	785.96	6.63	0.00298	300	0.035
22	254	550	785.96	6.63	0.00298	300	0.035
23	254	550	794.53	6.66	0.00284	300	0.035
24	254	550	794.53	6.66	0.00284	300	0.035
25	254	550	801.32	7.1	0.00277	300	0.035
26	254	550	801.32	7.1	0.00277	300	0.035
27	10	150	1055.1	3.33	0.52124	120	0.077
28	10	150	1055.1	3.33	0.52124	120	0.077
29	10	150	1055.1	3.33	0.52124	120	0.077
30	47	97	148.89	5.35	0.0114	120	0.077
31	60	190	222.92	6.43	0.0016	150	0.063
32	60	190	222.92	6.43	0.0016	150	0.063
33	60	190	222.92	6.43	0.0016	150	0.063
34	90	200	107.87	8.95	0.0001	200	0.042
35	90	200	116.58	8.62	0.0001	200	0.042
36	90	200	116.58	8.62	0.0001	200	0.042
37	25	110	307.45	5.88	0.0161	80	0.098
38	25	110	307.45	5.88	0.0161	80	0.098
39	25	110	307.45	5.88	0.0161	80	0.098
40	242	550	647.83	7.97	0.00313	300	0.035

Cost curve for the units of case study 4.

Table 1.7 Operating cost for each case study.

Case	Objective function	Reference
1	32,808.82	32,514.00
2	8411.01	8234.07
3	18,879.62	18,127.06
4	137,036.34	123,489.74

Comparative analysis of case studies.

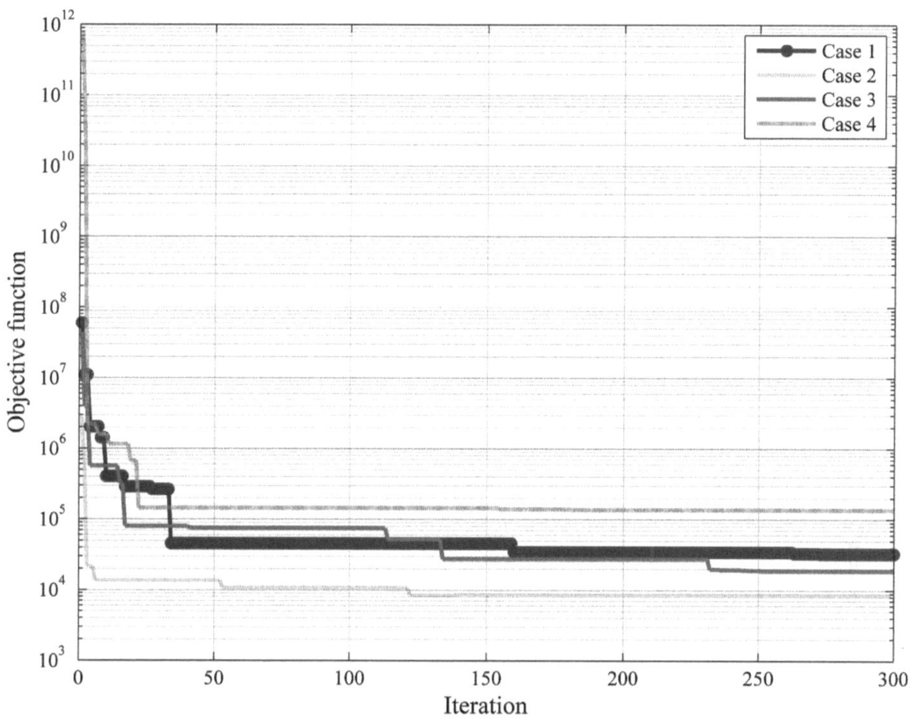

Fig. 1.2 PSO evolution for the economic dispatch problems.

1.4 Conclusions

In this chapter, the authors have provided a general overview of some of the most popular heuristic optimization techniques. Their working principle has been explained in brief, so that the reader can understand how they were created.

Some examples have also been described using PSO in its continuous and binary forms. They included the optimization of some well-known benchmark functions and the ED problem, which is a crucial problem of electrical engineering. In addition, the daily management of BESS was commented on as a relevant problem involving binary variables.

The test functions considered to illustrate the PSO capabilities were Griewank, Rastrigin, Rosenbrock, Ackley, and Schwefel. We have found a good agreement between the solutions offered by PSO and the reference solutions reported in the literature. Concerning the ED problem, we observed a difference of between 0.9% and 10.96% with respect to the reference solutions reported in other studies. It is important to highlight that the highest error corresponds to the power system with the highest number of generators. Thus, the relationship between the problem complexity and system size is revealed. This chapter offers the reader a broad perspective on swarm-based optimization techniques and their computational implementation.

References

Abedinpourshotorban, H., Shamsuddin, S. M., Beheshti, Z., & Jawawi, D. N. A. (2016). Electromagnetic field optimization: A physics-inspired metaheuristic optimization algorithm. *Swarm and Evolutionary Computation*, *26*, 8–22. https://doi.org/10.1016/j.swevo.2015.07.002.

Adarsh, B. R., Raghunathan, T., Jayabarathi, T., & Yang, X. S. (2016). Economic dispatch using chaotic bat algorithm. *Energy*, *96*, 666–675. https://doi.org/10.1016/j.energy.2015.12.096.

Ahmadi, M. A., Ebadi, M., Shokrollahi, A., & Majidi, S. M. J. (2013). Evolving artificial neural network and imperialist competitive algorithm for prediction oil flow rate of the reservoir. *Applied Soft Computing*, *13*(2), 1085–1098. https://doi.org/10.1016/j.asoc.2012.10.009.

Alatas, B., Akin, E., & Ozer, A. B. (2009). Chaos embedded particle swarm optimization algorithms. *Chaos, Solitons and Fractals*, *40*(4), 1715–1734. https://doi.org/10.1016/j.chaos.2007.09.063.

Atashpaz-Gargari, E., & Lucas, C. (2007). Imperialist competitive algorithm: An algorithm for optimization inspired by imperialistic competition. In *2007 IEEE congress on evolutionary computation* (pp. 4661–4667). IEEE. https://doi.org/10.1109/CEC.2007.4425083.

Awad, N. H., Ali, M. Z., Suganthan, P. N., & Reynolds, R. G. (2017). CADE: A hybridization of cultural algorithm and differential evolution for numerical optimization. *Information Sciences*, *378*, 215–241. https://doi.org/10.1016/j.ins.2016.10.039.

Becerra, R. L., & Coello, C. A. (2006). Cultured differential evolution for constrained optimization. *Computer Methods in Applied Mechanics and Engineering*, *195*(33–36), 4303–4322. https://doi.org/10.1016/j.cma.2005.09.006.

Beheshti, Z. (2020). A time-varying mirrored S-shaped transfer function for binary particle swarm optimization. *Information Sciences*, *512*, 1503–1542. https://doi.org/10.1016/j.ins.2019.10.029.

Bingol, H., & Alatas, B. (2020). Chaos based optics inspired optimization algorithms as global solution search approach. *Chaos, Solitons and Fractals*. https://doi.org/10.1016/j.chaos.2020.110434, 110434.

Bouchekara, H. R. E. H., Zellagui, M., & Abido, M. A. (2017). Optimal coordination of directional overcurrent relays using a modified electromagnetic field optimization algorithm. *Applied Soft Computing*, *54*, 267–283. https://doi.org/10.1016/j.asoc.2017.01.037.

Brest, J., Greiner, S., Boskovic, B., Mernik, M., & Zumer, V. (2006). Self-adapting control parameters in differential evolution: A comparative study on numerical benchmark problems. *IEEE Transactions on Evolutionary Computation*, *10*(6), 646–657. https://doi.org/10.1109/TEVC.2006.872133.

Cai, J., Ma, X., Li, Q., Li, L., & Peng, H. (2009). A multi-objective chaotic particle swarm optimization for environmental/economic dispatch. *Energy Conversion and Management*, *50*(5), 1318–1325. https://doi.org/10.1016/j.enconman.2009.01.013.

Cai, J., Ma, X., Li, L., & Peng, H. (2007). Chaotic particle swarm optimization for economic dispatch considering the generator constraints. *Energy Conversion and Management, 48* (2), 645–653. https://doi.org/10.1016/j.enconman.2006.05.020.

Cantú-Paz, E. (1999). *Designing efficient and accurate parallel genetic algorithms. Vol. 99017.* The University of Illinois at Urbana-Champaign.

Chu, S. C., & Tsai, P. W. (2007). Computational intelligence based on the behavior of cats. *International Journal of Innovative Computing Information and Control, 3*(1), 163–173.

Chu, S. C., Tsai, P. W., & Pan, J. S. (2006). Cat swarm optimization. In Q. Yang, & G. Webb (Eds.), *Vol. 4099. PRICAI 2006: 9th Pacific rim international conference on artificial intelligence* (pp. 854–858). Springer. https://doi.org/10.1007/11801603_94.

Chuang, L. Y., Yang, C. H., & Li, J. C. (2011). Chaotic maps based on binary particle swarm optimization for feature selection. *Applied Soft Computing, 11*(1), 239–248. https://doi.org/10.1016/j.asoc.2009.11.014.

Civioglu, P., & Besdok, E. (2013). A conceptual comparison of the cuckoo-search, particle swarm optimization, differential evolution and artificial bee colony algorithms. *Artificial Intelligence Review, 39*(4), 315–346. https://doi.org/10.1007/s10462-011-9276-0.

Clerc, M., & Kennedy, J. (2002). The particle swarm—Explosion, stability, and convergence in a multidimensional complex space. *IEEE Transactions on Evolutionary Computation, 6* (1), 58–73. https://doi.org/10.1109/4235.985692.

Coelho, L. D. (2008). A quantum particle swarm optimizer with chaotic mutation operator. *Chaos, Solitons and Fractals, 37*(5), 1409–1418. https://doi.org/10.1016/j.chaos.2006.10.028.

Coelho, L. D. (2010). Gaussian quantum-behaved particle swarm optimization approaches for constrained engineering design problems. *Expert Systems with Applications, 37*(2), 1676–1683. https://doi.org/10.1016/j.eswa.2009.06.044.

Coelho, L., Souza, R. C. T., & Mariani, V. C. (2009). Improved differential evolution approach based on cultural algorithm and diversity measure applied to solve economic load dispatch problems. *Mathematics and Computers in Simulation, 79*(10), 3136–3147. https://doi.org/10.1016/j.matcom.2009.03.005.

Coello, C. A. C., & Lechuga, M. S. (2002). MOPSO: A proposal for multiple objective particle swarm optimization. In *Vol. 2. Proceedings of the 2002 congress on computational intelligence* (pp. 1051–1056). IEEE. https://doi.org/10.1109/CEC.2002.1004388.

Coello, C. A. C., Pulido, G. T., & Lechuga, M. S. (2004). Handling multiple objectives with particle swarm optimization. *IEEE Transactions on Evolutionary Computation, 8*(3), 256–279. https://doi.org/10.1109/TEVC.2004.826067.

Corana, A., Marchesi, M., Martini, C., & Ridella, S. (1987). Minimizing multimodal functions of continuous-variables with the simulated annealing algorithm. *ACM Transactions on Mathematical Software, 13*(3), 262–280. https://doi.org/10.1145/29380.29864.

Das, S., Abraham, A., Chakraborty, U. K., & Konar, A. (2009). Differential evolution using a neighborhood-based mutation operator. *IEEE Transactions on Evolutionary Computation, 13*(3), 526–553. https://doi.org/10.1109/TEVC.2008.2009457.

Deb, K. (2001). *Multi-objective optimization using evolutionary algorithms* (1st ed.). John Wiley & Sons Ltd.

Deb, K., & Jain, H. (2014). An evolutionary many-objective optimization algorithm using reference-point-based nondominated sorting approach, part I: Solving problems with box constraints. *IEEE Transactions on Evolutionary Computation, 18*(4), 577–601. https://doi.org/10.1109/TEVC.2013.2281535.

Deb, K., Pratap, A., Agarwal, S., & Meyarivan, T. (2002). A fast and elitist multiobjective genetic algorithm: NSGA-II. *IEEE Transactions on Evolutionary Computation, 6*(2), 182–197. https://doi.org/10.1109/4235.996017.

Dorigo, M., Di Caro, G., & Gambardella, L. M. (1999). Ant algorithms for discrete optimization. *Artificial Life*, *5*(2), 137–172. https://doi.org/10.1162/106454699568728.

Dorigo, M., & Gambardella, L. M. (1997). Ant colonies for the travelling salesman problem. *Biosystems*, *43*(2), 73–81. https://doi.org/10.1016/S0303-2647(97)01708-5.

Dorigo, M., Maniezzo, V., & Colorni, A. (1996). Ant system: Optimization by a colony of cooperating agents. *IEEE Transactions on Systems Man and Cybernetics Part B—Cybernetics*, *26*(1), 29–41. https://doi.org/10.1109/3477.484436.

dos Santos Coelho, L., & Lee, C. S. (2008). Solving economic load dispatch problems in power systems using chaotic and Gaussian particle swarm optimization approaches. *International Journal of Electrical Power & Energy Systems*, *30*(5), 297–307. https://doi.org/10.1016/j.ijepes.2007.08.001.

Duman, S., Guvenc, U., Sonmez, Y., & Yorukeren, N. (2012). Optimal power flow using gravitational search algorithm. *Energy Conversion and Management*, *59*, 86–95. https://doi.org/10.1016/j.enconman.2012.02.024.

Dupanloup, I., Schneider, S., & Excoffier, L. (2002). A simulated annealing approach to define the genetic structure of populations. *Molecular Ecology*, *11*(12), 2571–2581. https://doi.org/10.1046/j.1365-294x.2002.01650.x.

Eberhart, R. C., & Shi, Y. (2000). Comparing inertia weights and constriction factors in particle swarm optimization. In *Vol. 1. Proceedings of the 2000 congress on evolutionary computation* (pp. 84–88). IEEE. https://doi.org/10.1109/CEC.2000.870279.

Gaing, Z. L. (2003). Particle swarm optimization to solving the economic dispatch considering the generator constraints. *IEEE Transactions on Power Systems*, *18*(3), 1187–1195. https://doi.org/10.1109/TPWRS.2003.814889.

Gaing, Z. L. (2004). A particle swarm optimization approach for optimum design of PID controller in AVR system. *IEEE Transactions on Energy Conversion*, *19*(2), 384–391. https://doi.org/10.1109/TEC.2003.821821.

Gandomi, A. H., Yang, X. S., & Alavi, A. H. (2013). Cuckoo search algorithm: A metaheuristic approach to solve structural optimization problems. *Engineering with Computers*, *29*(1), 17–35. https://doi.org/10.1007/s00366-011-0241-y.

Gandomi, A. H., Yun, G. J., Yang, X. S., & Talatahari, S. (2013). Chaos-enhanced accelerated particle swarm optimization. *Communications in Nonlinear Science and Numerical Simulation*, *18*(2), 327–340. https://doi.org/10.1016/j.cnsns.2012.07.017.

Ghasemi, M., Ghavidel, S., Ghanbarian, M. M., Gharibzadeh, M., & Vahed, A. A. (2014). Multi-objective optimal power flow considering the cost, emission, voltage deviation and power losses using multi-objective modified imperialist competitive algorithm. *Energy*, *78*, 276–289. https://doi.org/10.1016/j.energy.2014.10.007.

Glover, F. (1990). Tabu search—A tutorial. *Interfaces*, *20*(4), 74–94. https://doi.org/10.1287/inte.20.4.74.

Goffe, W. L., Ferrier, G. D., & Rogers, J. (1994). Global optimization of statistical functions with simulated annealing. *Journal of Econometrics*, *60*(1–2), 65–99. https://doi.org/10.1016/0304-4076(94)90038-8.

Goldberg, D. E. (1989). *Genetic algorithms in search, optimization, and machine learning* (1st ed.). Addison-Wesley.

Guo, L., Meng, Z., Sun, Y., & Wang, L. (2016). Parameter identification and sensitivity analysis of solar cell models with cat swarm optimization algorithm. *Energy Conversion and Management*, *108*, 520–528. https://doi.org/10.1016/j.enconman.2015.11.041.

Han, K. H., & Kim, J. H. (2002). Quantum-inspired evolutionary algorithm for a class of combinatorial optimization. *IEEE Transactions on Evolutionary Computation*, *6*(6), 580–593. https://doi.org/10.1109/TEVC.2002.804320.

Han, K. H., & Kim, J. H. (2004). Quantum-inspired evolutionary algorithms with a new termination criterion, H-epsilon gate, and two-phase scheme. *IEEE Transactions on Evolutionary Computation*, *8*(2), 156–169. https://doi.org/10.1109/TEVC.2004.823467.

Han, K. H., Park, K. H., Lee, C. H., & Kim, J. H. (2001). Parallel quantum-inspired genetic algorithm for combinatorial optimization problem. In *Vol. 2*. *Proceedings of the 2001 congress on evolutionary computation* (pp. 1422–1429). IEEE. https://doi.org/10.1109/CEC.2001.934358.

Holland, J. H. (1975). *Adaptation in natural and artificial systems: An introductory analysis with applications to biology, control, and artificial intelligence* (1st ed.). University of Michigan Press.

Hong, W. C. (2009). Chaotic particle swarm optimization algorithm in a support regression electric load forecasting model. *Energy Conversion and Management*, *50*(1), 105–117. https://doi.org/10.1016/j.enconman.2008.08.031.

Jayabarathi, T., Raghunathan, T., Adarsh, B. R., & Suganthan, P. N. (2016). Economic dispatch using hybrid grey wolf optimizer. *Energy*, *111*, 630–641. https://doi.org/10.1016/j.energy.2016.05.105.

Jeong, Y. W., Park, J. B., Jang, S. H., & Lee, K. Y. (2010). A new quantum-inspired binary PSO: Application to unit commitment problems for power systems. *IEEE Transactions on Power Systems*, *25*(3), 1486–1495. https://doi.org/10.1109/TPWRS.2010.2042472.

Jia, H., Peng, X., & Lang, C. (2021). Remora optimization algorithm. *Expert Systems with Applications*, *182*. https://doi.org/10.1016/j.eswa.2021.115665, 115665.

Karaboga, D., & Akay, B. (2009). A comparative study of artificial bee colony algorithm. *Applied Mathematics and Computation*, *214*(1), 108–132. https://doi.org/10.1016/j.amc.2009.03.090.

Karaboga, D., & Basturk, B. (2007). A powerful and efficient algorithm for numerical function optimization: Artificial bee colony (ABC) algorithm. *Journal of Global Optimization*, *39*(3), 459–471. https://doi.org/10.1007/s10898-007-9149-x.

Karaboga, D., & Basturk, B. (2008). On the performance of artificial bee colony (ABC) algorithm. *Applied Soft Computing*, *8*(1), 687–697. https://doi.org/10.1016/j.asoc.2007.05.007.

Karaboga, D., & Ozturk, C. (2011). A novel clustering approach: Artificial bee colony (ABC) algorithm. *Applied Soft Computing*, *11*(1), 652–657. https://doi.org/10.1016/j.asoc.2009.12.025.

Kaveh, A., Sheikholeslami, R., Talatahari, S., & Keshvari-Ilkhichi, M. (2014). Chaotic swarming of particles: A new method for size optimization of truss structures. *Advances in Engineering Software*, *67*, 136–147. https://doi.org/10.1016/j.advengsoft.2013.09.006.

Kaveh, A., & Talatahari, S. (2010). Optimum design of skeletal structures using imperialist competitive algorithm. *Computers & Structures*, *88*(21 – 22), 1220–1229. https://doi.org/10.1016/j.compstruc.2010.06.011.

Kennedy, J., & Eberhart, R. (1995). Particle swarm optimization. In *Vol. 4*. *1995 IEEE international conference on neural networks* (pp. 1942–1948). IEEE. https://doi.org/10.1109/ICNN.1995.488968.

Kim, D. H., Abraham, A., & Cho, J. H. (2007). A hybrid genetic algorithm and bacterial foraging approach for global optimization. *Information Sciences*, *177*(18), 3918–3937. https://doi.org/10.1016/j.ins.2007.04.002.

Kirkpatrick, S., Gelatt, C. D., & Vecchi, M. P. (1983). Optimization by simulated annealing. *Science*, *220*(4598), 671–680. https://doi.org/10.1126/science.220.4598.671.

Lau, T. W., Chung, C. Y., Wong, K. P., Chung, T. S., & Ho, S. L. (2009). Quantum-inspired evolutionary algorithm approach for unit commitment. *IEEE Transactions on Power Systems*, *24*(3), 1503–1512. https://doi.org/10.1109/TPWRS.2009.2021220.

Lee, K. Y., & El-Sharkawi, M. A. (2008). *Modern heuristic optimization techniques. Theory and applications to power systems*. IEEE Press – John Wiley & Sons, Inc. https://doi.org/10.1002/9780470225868.

Lin, K. C., Zhang, K. Y., Huang, Y. H., Hung, J. C., & Yen, N. (2016). Feature selection based on an improved cat swarm optimization algorithm for big data classification. *Journal of Supercomputing*, *72*(8), 3210–3221. https://doi.org/10.1007/s11227-016-1631-0.

Liu, J., & Lampinen, J. (2005). A fuzzy adaptive differential evolution algorithm. *Soft Computing*, *9*(6), 448–462. https://doi.org/10.1007/s00500-004-0363-x.

Liu, B., Wang, L., Jin, Y. H., Tang, F., & Huang, D. X. (2005). Improved particle swarm optimization combined with chaos. *Chaos, Solitons and Fractals*, *25*(5), 1261–1271. https://doi.org/10.1016/j.chaos.2004.11.095.

Lu, Y. L., Zhou, J. Z., Qin, H., Wang, Y., & Zhang, Y. C. (2011). A hybrid multi-objective cultural algorithm for short-term environmental/economic hydrothermal scheduling. *Energy Conversion and Management*, *52*(5), 2121–2134. https://doi.org/10.1016/j.enconman.2010.12.003.

Lujano-Rojas, J. M., Yusta, J. M., Artal-Sevil, J. S., & Domínguez-Navarro, J. A. (2019). Day-ahead optimal battery operation in islanded hybrid energy systems and its impact on greenhouse gas emissions. *Applied Sciences*, *9*(23), 1–30. https://doi.org/10.3390/app9235221.

Mallipeddi, R., Suganthan, P. N., Pan, Q. K., & Tasgetiren, M. F. (2011). Differential evolution algorithm with ensemble of parameters and mutation strategies. *Applied Soft Computing*, *11*(2), 1679–1696. https://doi.org/10.1016/j.asoc.2010.04.024.

Maulik, U., & Bandyopadhyay, S. (2002). Performance evaluation of some clustering algorithms validity indices. *IEEE Transactions on Pattern Analysis and Machine Intelligence*, *24*(12), 1650–1654. https://doi.org/10.1109/TPAMI.2002.1114856.

Mavrovouniotis, M., Li, C., & Yang, S. (2017). A survey of swarm intelligence for dynamic optimization: Algorithms and applications. *Swarm and Evolutionary Computation*, *33*, 1–17. https://doi.org/10.1016/j.swevo.2016.12.005.

Meng, K., Wang, H. G., Dong, Z. Y., & Wong, K. P. (2010). Quantum-inspired particle swarm optimization for valve-point economic load dispatch. *IEEE Transactions on Power Systems*, *25*(1), 215–222. https://doi.org/10.1109/TPWRS.2009.2030359.

Mirjalili, S. (2015). How effective is the grey wolf optimizer in training multi-layer perceptrons. *Applied Intelligence*, *43*(1), 150–161. https://doi.org/10.1007/s10489-014-0645-7.

Mirjalili, S., Hashim, S. Z. M., & Sardroudi, H. M. (2012). Training feedforward neural networks using hybrid particle swarm optimization and gravitational search algorithm. *Applied Mathematics and Computation*, *218*(22), 11125–11137. https://doi.org/10.1016/j.amc.2012.04.069.

Mirjalili, S., Mirjalili, S. M., & Lewis, A. (2014). Grey wolf optimizer. *Advances in Engineering Software*, *69*, 46–61. https://doi.org/10.1016/j.advengsoft.2013.12.007.

Mirjalili, S., Saremi, S., Mirjalili, S. M., & Coelho, L. D. (2016). Multi-objective grey wolf optimizer: A novel algorithm for multi-criterion optimization. *Expert Systems with Applications*, *47*, 106–119. https://doi.org/10.1016/j.eswa.2015.10.039.

Mishra, S. (2005). A hybrid least square-fuzzy bacterial foraging strategy for harmonic estimation. *IEEE Transactions on Evolutionary Computation*, *9*(1), 61–73. https://doi.org/10.1109/TEVC.2004.840144.

Narayanan, A., & Moore, M. (1996). Quantum-inspired genetic algorithms. In *Proceedings of IEEE international conference on evolutionary computation* (pp. 61–66). IEEE. https://doi.org/10.1109/ICEC.1996.542334.

Orero, S. O., & Irving, M. R. (1996). Economic dispatch of generators with prohibited operating zones: A genetic algorithm approach. *IEE Proceedings—Generation, Transmission and Distribution*, *143*(6), 529–534. https://doi.org/10.1049/ip-gtd:19960626.

Passino, K. M. (2002). Biomimicry of bacterial foraging for distributed optimization and control. *IEEE Control Systems Magazine*, *22*(3), 52–67. https://doi.org/10.1109/MCS.2002.1004010.

Pradhan, P. M., & Panda, G. (2012). Solving multiobjective problems using cat swarm optimization. *Expert Systems with Applications*, *39*(3), 2956–2964. https://doi.org/10.1016/j.eswa.2011.08.157.

Qin, A. K., & Suganthan, P. N. (2005). Self-adaptive differential evolution algorithm for numerical optimization. In *Vol. 2. 2005 IEEE congress on evolutionary computation* (pp. 1785–1791). IEEE. https://doi.org/10.1109/CEC.2005.1554904.

Qin, H., Zhou, J. Z., Lu, Y. L., Li, Y. H., & Zhang, Y. C. (2010). Multi-objective cultured differential evolution for generating optimal trade-offs in reservoir flood control operation. *Water Resources Management*, *24*(11), 2611–2632. https://doi.org/10.1007/s11269-009-9570-7.

Rahnamayan, S., Tizhoosh, H. R., & Salama, M. M. A. (2008). Opposition-based differential evolution. *IEEE Transactions on Evolutionary Computation*, *12*(1), 64–79. https://doi.org/10.1109/TEVC.2007.894200.

Rashedi, E., Nezamabadi-Pour, H., & Saryazdi, S. (2009). GSA: A gravitational search algorithm. *Information Sciences*, *179*, 2232–2248. https://doi.org/10.1016/j.ins.2009.03.004.

Reynolds, R. G. (1994). In A. V. Sebald, & L. J. Fogel (Eds.), *An introduction to cultural algorithms Proceedings of the third annual conference on evolutionary programming* (pp. 131–139). World Scientific.

Santosa, B., & Ningrum, M. K. (2009). Cat swarm optimization for clustering. In *2009 international conference of soft computing and pattern recognition* (pp. 54–59). IEEE. https://doi.org/10.1109/SoCPaR.2009.23.

Sayed, G. I., Hassanien, A. E., & Azar, A. T. (2019). Feature selection via a novel chaotic crow search algorithm. *Neural Computing and Applications*, *31*(1), 171–188. https://doi.org/10.1007/s00521-017-2988-6.

Shabani, H., Vahidi, B., & Ebrahimpour, M. (2013). A robust PID controller based on imperialist competitive algorithm for load-frequency control of power systems. *ISA Transactions*, *52*(1), 88–95. https://doi.org/10.1016/j.isatra.2012.09.008.

Sinha, N., Chakrabarti, R., & Chattopadhyay, R. K. (2003). Evolutionary programming techniques for economic load dispatch. *IEEE Transactions on Evolutionary Computation*, *7*(1), 83–94. https://doi.org/10.1109/TEVC.2002.806788.

Skinderowicz, R. (2020). Implementing a GPU-based parallel MAX-MIN ant system. *Future Generation Computer Systems*, *106*, 277–295. https://doi.org/10.1016/j.future.2020.01.011.

Soleimanpour-Moghadam, M., Nezamabadi-Pour, H., & Farsangi, M. M. (2014). A quantum inspired gravitational search algorithm for numerical function optimization. *Information Sciences*, *267*, 83–100. https://doi.org/10.1016/j.ins.2013.09.006.

Srinivasan, D., & Seow, T. H. (2003). Particle swarm inspired evolutionary algorithm (PS-EA) for multiobjective optimization problems. In *Vol. 4. The 2003 congress on evolutionary computation* (pp. 2292–2297). IEEE. https://doi.org/10.1109/CEC.2003.1299374.

Storn, R., & Price, K. (1997). Differential evolution—A simple and efficient heuristic for global optimization over continuous spaces. *Journal of Global Optimization*, *11*(4), 341–359. https://doi.org/10.1023/A:1008202821328.

Stützle, T., & Hoos, H. H. (2000). MAX–MIN ant system. *Future Generation Computer Systems*, *16*(8), 889–914. https://doi.org/10.1016/S0167-739X(00)00043-1.

Sulaiman, M. H., Mustaffa, Z., Mohamed, M. R., & Aliman, O. (2015). Using the gray wolf optimizer for solving optimal reactive power dispatch problem. *Applied Soft Computing*, *32*, 286–292. https://doi.org/10.1016/j.asoc.2015.03.041.

Sun, Y., Zhang, L. B., & Gu, X. S. (2012). A hybrid co-evolutionary cultural algorithm based on particle swarm optimization for solving global optimization problems. *Neurocomputing*, *98*, 76–89. https://doi.org/10.1016/j.neucom.2011.08.043.

Trelea, I. C. (2003). The particle swarm optimization algorithm: Convergence analysis and parameter selection. *Information Processing Letters*, *85*(6), 317–325. https://doi.org/10.1016/S0020-0190(02)00447-7.

Tsai, P. W., Pan, J. S., Chen, S. M., & Liao, B. Y. (2012). Enhanced parallel cat swarm optimization based on the Taguchi method. *Expert Systems with Applications*, *39*(7), 6309–6319. https://doi.org/10.1016/j.eswa.2011.11.117.

Tsai, P. W., Pan, J. S., Chen, S. M., Liao, B. Y., & Hao, S. P. (2008). Parallel cat swarm optimization. In *Vol. 6. Proceedings of 2008 international conference on machine learning and cybernetics* (pp. 3328–3333). IEEE. https://doi.org/10.1109/ICMLC.2008.4620980.

Vesterstrøm, J., & Thomsen, R. (2004). A comparative study of differential evolution, particle swarm optimization, and evolutionary algorithms on numerical benchmark problems. In *Vol. 2. Proceedings of the 2004 congress on evolutionary computation (CEC 2004)* (pp. 1980–1987). IEEE. https://doi.org/10.1109/CEC.2004.1331139.

Vrugt, J. A., ter Braak, C. J. F., Diks, C. G. H., Robinson, B. A., Hyman, J. M., & Higdon, D. (2009). Accelerating Markov chain Monte Carlo simulation by differential evolution with self-adaptive randomized subspace sampling. *International Journal of Nonlinear Sciences and Numerical Simulation*, *10*(3), 273–290. https://doi.org/10.1515/IJNSNS.2009.10.3.273.

Walton, S., Hassan, O., Morgan, K., & Brown, M. R. (2011). Modified cuckoo search: A new gradient free optimisation algorithm. *Chaos, Solitons and Fractals*, *44*(9), 710–718. https://doi.org/10.1016/j.chaos.2011.06.004.

Yang, X. S., & Deb, S. (2009). Cuckoo search via Lévy flights. In *2009 world congress on nature and biologically inspired computing (NaBIC)* (pp. 210–214). IEEE. https://doi.org/10.1109/NABIC.2009.5393690.

Yang, X. S., & Deb, S. (2013). Multiobjective cuckoo search for design optimization. *Computers and Operations Research*, *40*(6), 1616–1624. https://doi.org/10.1016/j.cor.2011.09.026.

Yang, X. S., & Deb, S. (2014). Cuckoo search: Recent advances and applications. *Neural Computing and Applications*, *24*(1), 169–174. https://doi.org/10.1007/s00521-013-1367-1.

Yang, B., Zhang, X. S., Yu, T., Shu, H. C., & Fang, Z. H. (2017). Grouped grey wolf optimizer for maximum power tracking of doubly-fed induction generator based wind turbine. *Energy Conversion and Management*, *133*, 427–443. https://doi.org/10.1016/j.enconman.2016.10.062.

Zhan, Z. H., Zhang, J., Li, Y., & Chung, H. S. H. (2009). Adaptive particle swarm optimization. *IEEE Transactions on Systems Man and Cybernetics Part B—Cybernetics*, *39*(6), 1362–1381. https://doi.org/10.1109/TSMCB.2009.2015956.

Zhang, J., & Sanderson, A. C. (2009). JADE: Adaptive differential evolution with optional external archive. *IEEE Transactions on Evolutionary Computation*, *13*(5), 945–958. https://doi.org/10.1109/TEVC.2009.2014613.

Zhang, R., Zhou, J. Z., Mo, L., Ouyang, S., & Liao, X. (2013). Economic environmental dispatch using an enhanced multi-objective cultural algorithm. *Electric Power Systems Research*, *99*, 18–29. https://doi.org/10.1016/j.epsr.2013.01.010.

Genetic algorithms and other heuristic techniques in power systems optimization

2

2.1 Evolution and development of genetic algorithms

The conceptual background behind a genetic algorithm (GA) is in the principles of heredity. The notion of heredity is related to the mechanisms used by nature to transmit biochemical information from one generation to another through the genes. Thus, the whole process results in the transmission of physical and behavioral characteristics from parents to offspring. Genetics studies the procedures involved in heredity, identifying genes and their relationship with specific traits. Based on this phenomenon, humanity understands the creation and transformation of biodiversity from one generation to another one as a mechanism of continuous evolution. Thus, genetic diversity could justify biodiversity as a mechanism of constant change partially guided by the combination of certain specific laws and random events (Hartwell et al., 2010).

Interpreting the process of heredity as a means of transformation and improvement, Holland (1975) formulated the first GA in 1975. In this sense, the concept of adaptation refers to the ability of any living being to systematically modify its structure to enhance its performance under specific environmental conditions. As can be observed, the idea of adaptation intrinsically refers to an optimization process. In his work, Holland established the mathematical framework of a GA, and deduced and generalized essential parameters and factors found in the behavior of biological systems and processes. Specifically, Holland introduced the concept of a schema to generalize the interaction between a group of genes, and also added genetic operators such as crossover and mutation. On the one hand, the incorporation of the schema allows us to analyze systems with interactions of high complexity. On the other hand, the genetic operators will enable us to perform bio-inspired analyses in a variety of fields, such as economy, control theory, and artificial intelligence, among others.

Later, Goldberg (1989) made a significant effort to disseminate the capabilities of GA to a broad audience. Based on his experience with people from different backgrounds, in his book, he explained the mathematical and computational tools as well as the corresponding results from the implementation of a GA to solve practical optimization problems in different fields. Thus far, most of the efforts made by Holland and Goldberg had been to develop and promote the massive implementation of GA in different fields of science and engineering; Deb (2001) extended these efforts to optimization with many variables, which allows us to analyze a more realistic situation. In his book, Deb presented a comprehensive explanation of a variety of classical approaches with practical applications. Although his book describes complex concepts and algorithms to deal with problems of multiple objectives, Deb wrote his book

Genetic Optimization Techniques for Sizing and Management of Modern Power Systems
https://doi.org/10.1016/B978-0-12-823889-9.00001-1

in a transparent manner, which has led to it being widely adopted by specialists and practitioners of many fields. The evolution of GA has taken advantage of the development of parallel computing to improve its optimization capabilities. In this regard, Cantú-Paz (1999) analyzed the problems behind the parallel implementation of GA. He studied the specific relationship between the parallel GA search power and efficiency with its parameter settings, and reported valuable conclusions about the correct design of a parallel GA in terms of population sizing, scalability, migration, and topology.

The total contribution of the authors above and many others not mentioned here for brevity has resulted in many GA-based approaches to solving problems in many knowledge fields. Fig. 2.1 shows an approximation of the number of publications related to GA between the years 1987 and 2020. It is possible to observe how interest in this technique has had an increasing trend and still could offer possibilities for the solutions to current problems.

GAs found a relevant application in the field of molecular docking, which has become a fundamental approach for the analysis of molecular recognition, estimating the binding mode and binding affinity of a structure composed of several molecules whose characteristics have been studied by scientists a priori (Roy et al., 2015). In this regard, Jones et al. (1995) proposed one of the first applications of GA to the docking problem. Expressly, a GA was incorporated to consider the full and partial flexibilities

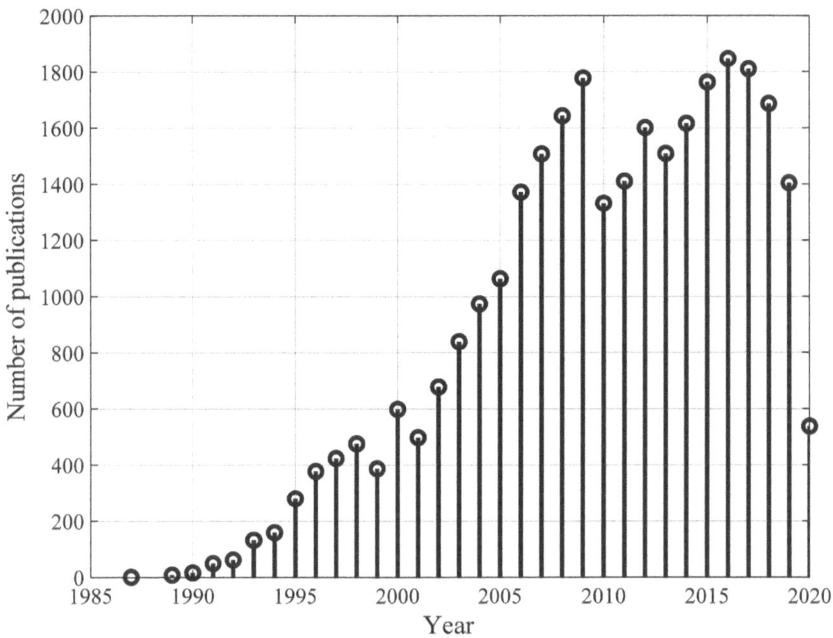

Fig. 2.1 GA literature: Number of GA-related publications per year between 1987 and 2020.

of the ligand and protein, respectively. Later, Jones et al. (1997) presented the program genetic optimization for ligand docking (GOLD). This program represents an enhanced version of the theory previously explained, improving its reliability. Morris et al. (1998) studied the docking problem over seven protein-ligand test systems using Monte Carlo simulated annealing, a traditional GA, and the Lamarckian GA. They concluded that the conventional and Lamarckian GAs could deal with problems with more degrees of freedom more efficiently and reliably than the Monte Carlo simulated annealing.

Kosakovsky et al. (2006a, 2006b) presented a genetic algorithm recombination detection (GARD) model to analyze multiple sequence alignment for recombination. Through extensive simulation, the authors demonstrated that GARD is an efficient tool that offers accurate results. Deaven and Ho (1995) developed another study based on GAs in the field of molecular geometry. Given a set of atoms, determining the configuration with the lowest energy is a combinatorial problem in which the number of settings increases exponentially with the size of the particle set. If the size of the atoms arranged is small, the scientist can solve the problem by analyzing all of the configurations. If the number of particles is modest, the scientist can use simulated annealing. Finally, if the atom set is large, the authors suggested a GA. Thus, they optimized the geometry of carbon clusters by representing them as a list of particles in the atomic Cartesian coordinates in arbitrary order. The authors established a relationship between the GA operators and parameters, and the energy and position of the atom-structure.

Another relevant field where scientists and engineers have successfully used GAs is pattern recognition and machine learning. In this regard, Siedlecki and Sklansky (1989) developed one of the first methods for feature selection based on a GA. The technique minimizes the subset of features, fulfilling some performance conditions established by the decision-maker (DM). Leardi et al. (1992) and Leardi and González (1998) developed a methodology that searches for the subset with the smallest cross-validated covariance. In this approach, the authors formulated the optimization problem as one free of constraints, where the objective function is calculated by the DM using partial least squares regression analysis. Yang and Honavar (1998) performed a multicriteria analysis of feature selection problems by incorporating the accuracy of the classification process using a neural network and the cost related to such classification. Specifically, the accuracy and cost were included by the authors in the operators of the suggested GA implementation. Oh et al. (2004) presented a hybrid GA in which a refined local search process improves the accuracy. Expressly, the user imposes the feature-subset size. Then, the GA incorporates this constraint in its operators. Huang and Wang (2006) introduced a methodology for feature selection and parameter optimization based on a GA so that the impact of kernel parameter calculation and feature selection are simultaneously considered during support vector machine training, improving the classification precision.

Determining the number of groups or clusters in a large dataset is another problem that has been treated in the technical literature using GAs. Maulik and Bandyopadhyay (2000) carried out one of the first studies in this field of knowledge. Their approach aimed to determine the cluster centers to minimize the Euclidean distances of each data point from the clusters in which the algorithm has classified them. In this regard, the implemented GA uses real coding, because continuous variables are involved in the problem. Krishna and Murty (1999) combined a GA with the K-means clustering algorithm to improve the characteristics of both, resulting in the genetic K-means algorithm. This hybrid GA uses the K-means algorithm as a substitute for some operators of a conventional GA.

Investigators have demonstrated that fuzzy theory is a crucial area of technology. GAs have also contributed to this field. Ishibuchi et al. (1995) implemented a GA to find the proper rules of a fuzzy system to enhance its performance. The reasoning behind this approach is that if the DM chooses the appropriate fuzzy rules, the system performance improves in terms of inference speed and user understanding of system operations. The authors formulated this technique as a multiobjective problem in which the GA maximizes the number of successfully classified patterns and minimizes the number of fuzzy rules. Later, the algorithm combines these objectives into a single one using weighting factors.

GAs have contributed to the solution of sequencing problems, where a determined amount of work has to be processed by a group of available machines. Solving this problem consists of finding the optimal combination of tasks, minimizing the completion time. In this regard, Reeves (1995) implemented a technique based on a GA in which the authors modified the GA-operators to create feasible individuals and improve diversity.

Optimal scheduling of power generators is another area where DMs have used GAs. There are two main problems to be solved with scheduling power generators: unit commitment (UC) and economic dispatch (ED). The UC problem consists of determining which units should be operated by the system operator at a determined time, and ED consists of determining the power to be supplied by the committed generators. Regarding the UC problem, Kazarlis et al. (1996) proposed a technique based on a GA in which a committed or de-committed unit is represented by using binaries. Concerning the ED problem, Walters and Sheble (1993), as well as Chiang (2005), developed an optimization model using a GA able to include the operational constraints related to the ramping capabilities and valve-point effects.

Coit and Smith (1996)) implemented a GA to optimize system reliability. The authors formulated the optimization problem to choose the appropriate combination of system components, considering their reliability constraints and cost. In other words, the problem consists of finding the optimal combination of elements to maximize reliability while minimizing cost, simultaneously taking into consideration some specific constraints.

The planning of water distribution systems has also incorporated GAs. In this regard, Savic and Walters (1997) included a GA on the software GANET, which is a computational tool for the assessment and rehabilitation of water distribution systems and hydraulic model calibration.

So far, we have been reviewing some relevant GA applications and tools, such as GOLD, GARD, and GANET. However, there are some basic concepts related to GAs that need to be clarified. In the next section, we are going to start by describing the general formulation of a typical optimization problem and its associated concepts.

2.2 General formulation of an optimization problem

There are several types of optimization problems, depending on the number of objectives to be optimized by the DM; researchers have classified them as mono-objective optimization (MoOO), multiobjective optimization (MuOO), and many-objective optimization (MaOO) algorithms. MoOO consists of optimizing a single objective; a widely accepted example is production cost. MuOO and MaOO refer to a problem with more than one objective to be optimized by DM. However, there is an essential difference in these terms. MuOO relates to problems with a limited number of objectives, typically two or three. In contrast, the concept of MaOO refers to a larger number of objectives, say four or higher. Researchers introduced this terminology because the increment in the number of objectives dramatically increases the mathematical complexity (Cao et al., 2017).

Eq. (2.1) shows the general formulation of a traditional optimization problem regardless of the number of objectives.

$$\begin{aligned} &\text{min or max } y_{\text{obj}}\left(\boldsymbol{x}_{\text{pop}}\right) \text{ or } \boldsymbol{y}_{\text{obj}}\left(\boldsymbol{x}_{\text{pop}}\right) \\ &\text{subject to } \boldsymbol{T}\left(\boldsymbol{x}_{\text{pop}}\right) \geq \boldsymbol{H} \\ &\qquad \boldsymbol{L}\left(\boldsymbol{x}_{\text{pop}}\right) = \boldsymbol{R} \\ &\qquad\quad \boldsymbol{x}_{\text{pop}}^{\text{min}} \leq \boldsymbol{x}_{\text{pop}} \leq \boldsymbol{x}_{\text{pop}}^{\text{max}} \end{aligned} \tag{2.1}$$

where the objective (y_{obj}) or objectives ($\boldsymbol{y}_{\text{obj}}$) to be minimized or maximized are subject to a set of inequality ($\boldsymbol{T}(\boldsymbol{x}_{\text{pop}}) \geq \boldsymbol{H}$) and equality ($\boldsymbol{L}(\boldsymbol{x}_{\text{pop}}) = \boldsymbol{R}$) constraints, as well as some bounds ($\boldsymbol{x}_{\text{pop}}^{\text{min}} \leq \boldsymbol{x}_{\text{pop}} \leq \boldsymbol{x}_{\text{pop}}^{\text{max}}$) on the decision variables.

Optimizing only one objective is probably the case with most practical applications because most of the projects aim to minimize costs or maximize economic profits. The decision variables ($\boldsymbol{x}_{\text{pop}}$) are binary, integers, or reals, so that the GA implementation needs to be accommodated by the DM for each specific formulation. In a general sense, the constraint set relates to the physical operation of the system under study. From the perspective of a GA and its biological basis, the DM can understand the possible values of decision variables ($\boldsymbol{x}_{\text{pop}}$) set as a group of individuals or a population. Consequently, the DM can understand the objective-function value (y_{obj} or $\boldsymbol{y}_{\text{obj}}$) as a measure of the skills of these individuals or their fitness. Following this parallelism, the procedure to solve an optimization problem through a GA involves the consideration of a group of living beings interacting with each other through biological processes such as natural selection and mating, as well as genetic mutation over a determined number of generations. The next section describes these biological processes during GA implementation.

2.3 General structure of a genetic algorithm

Fig. 2.2 shows the flowchart of a GA. The implementation starts by collecting information on the algorithm settings. The settings are the population size, the number of generations, the crossover rate, and the mutation rate. The population size refers to the number of individuals, the generations refer to the lifetime of the population, the crossover rate refers to the willingness of individuals to engage with each other, and the mutation rate refers to any unexpected change in the population. The knowledge and experience of the DM are crucial to set these parameters correctly, as are the computational resources available. On the one hand, the population size and the number of generations define the times that GA evaluates the objective function and, consequently, the computational time of the optimization process. On the other hand, the crossover and mutation rates represent the search characteristics of the algorithm. Usually, the crossover rate is a value close to one; this means that most of the individuals are willing to engage with each other. The mutation rate is a value close to zero, which means that sudden or unexpected changes in the characteristics of any individual rarely happen. The population size, number of generations, and the crossover and mutation rates are optimization parameters required by the algorithm at the first stage. Another essential operation is related to how the algorithm selects individuals to get ready for engagement. The GA selection operator tries to emulate the dynamic behind natural selection.

The next step consists of initializing the algorithm by randomly creating a set of individuals according to the population size previously specified. The population is initialized by the DM using a random number generator to fulfill the constraints of decision variables. According to Eq. (2.1), a group of individuals equal to the population size is created by the DM using random numbers within the bounds of decision

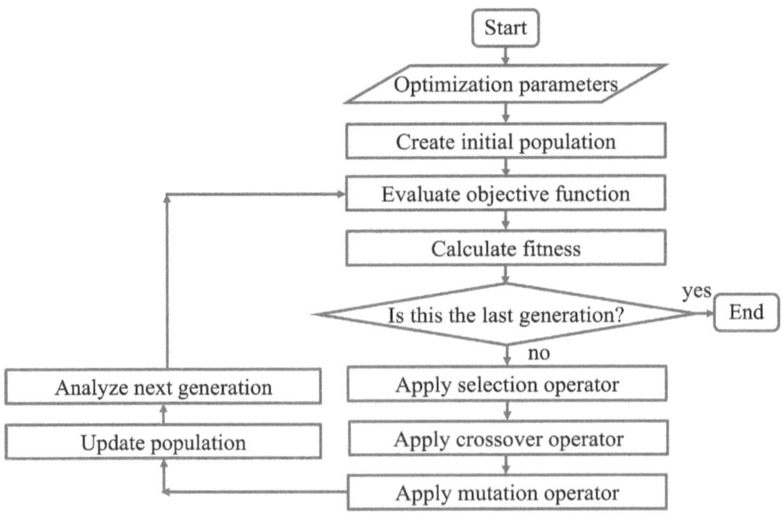

Fig. 2.2 Flowchart of a GA: General structure and implementation of a GA.

variables ($\boldsymbol{x}_{\mathbf{pop}}^{\mathbf{min}} \leq \boldsymbol{x}_{\mathbf{pop}} \leq \boldsymbol{x}_{\mathbf{pop}}^{\mathbf{max}}$). Once the DM has defined the initial population, the next step consists of calculating the values of the objective function to determine later the corresponding fitness of each individual. Researchers have defined fitness as a measure of the capabilities of an individual to survive the environmental conditions of the actual generation. As the generation is yet at its early stage, the operators related to the natural selection, crossover, and mutation are applied, defining the set of individuals, or in other words the population, of the next generation. The algorithm sequentially repeats these steps until a determined stopping criterion is achieved. The criterion used in this work consists of fulfilling the total number of generations; however, it could have been defined by the DM according to the objective function value variations.

By definition, the set of feasible solutions have to fulfill all of the constraints in Eq. (2.1). The DM can reflect the effects of these constraints on the optimization objective through weighting factors added to the objective function value. Thinking in terms of minimization, when some determined values of decision variables ($\boldsymbol{x}_{\mathbf{pop}}$) do not fulfill some of the constraints ($\boldsymbol{T}(\boldsymbol{x}_{\mathbf{pop}}) \geq \boldsymbol{H}$ and $\boldsymbol{L}(\boldsymbol{x}_{\mathbf{pop}}) = \boldsymbol{R}$), it is possible to increase the objective function value by adding the violation margins ($|\boldsymbol{T}(\boldsymbol{x}_{\mathbf{pop}}) - \boldsymbol{H}|$ and $|\boldsymbol{L}(\boldsymbol{x}_{\mathbf{pop}}) - \boldsymbol{R}|$)—the difference between the actual value of the violated constraint and the corresponding maximum one—multiplied by a penalty factor. Thus, those values of decision variables that violate some restrictions would have higher objective function costs than those feasible individuals, or in other words, a lower fitness and consequently fewer chances to appear in the next GA generation.

The traditional GA implementation could result in a performance reduction because the best individuals could become lost from one generation to another (Deb, 2001). Thus, an operator able to conserve the best individuals during each generation is usually integrated by the DM into the traditional GA implementation. Researchers know this as elitism, since the best individuals are forced by the DM to survive. Fig. 2.3 shows the implementation of elitism adopted in this work. Once the genetic operators have created the offspring, they are compared by the algorithm with their parents, and only the best individuals are selected.

Let I be the population size; the algorithm calculates the objective function for each of the offspring. After that, a set of $2 \times I$ individuals is created by the algorithm joining the parent and offspring populations. Then, the best I individuals are selected by the algorithm from this joined set.

Frequently, the algorithm needs to compare and decide what is the best individual. However, this is not a trivial issue. In the case of MoOO, the algorithm contrasts two different individuals by comparing the corresponding objective-function values. However, the case of MuOO and MaOO is more complicated because several objectives have to be compared at the same time, which does not always lead to consistent results. Thus, researchers introduced the concept of domination to overcome this theoretical problem. Thinking in terms of the minimization of two different objectives, Fig. 2.4 presents the objective function values of a population. The objective function values of dominated individuals are shown with empty circles, while the objective function values of nondominated individuals are shown with filled ones. Thus, nondominated individuals are those that better represent the minimization sense

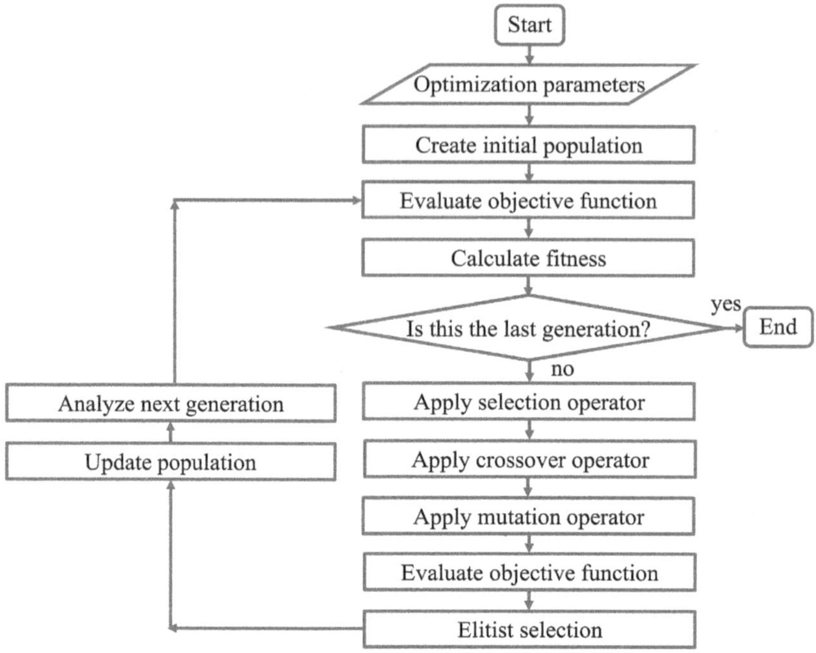

Fig. 2.3 Flowchart of an elitist GA: General structure of a GA using elitism.

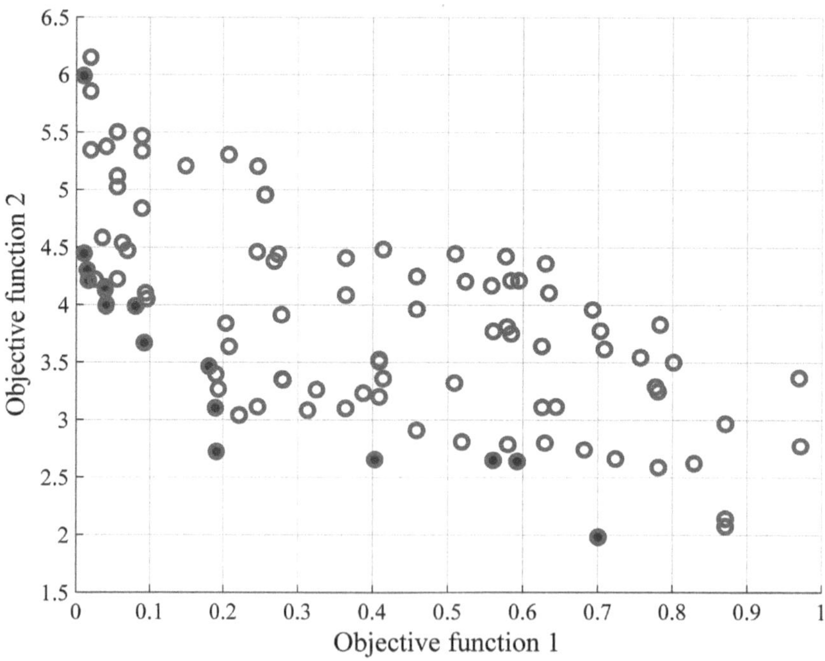

Fig. 2.4 Domination concept: Illustration of dominated and nondominated individuals.

among all of the individuals in the population. They represent an approximation to the Pareto-optimal front, which is a set of individuals ambitiously searched by all MuOO and MaOO techniques.

In MuOO and MaOO, the comparison of two individuals using the objective function values is replaced by determining which dominates the other. Accordingly, this result will define the fitness to be assigned to each of them so that those dominated individuals would have a worse fitness than those individuals less dominated. Conversely, the individuals belonging to the Pareto-optimal front, which means the nondominated ones, would have the best fitness in the population.

In a general sense, MuOO and MaOO algorithms should search for individuals as close as possible to the Pareto-optimal front, covering all of the regions of the front as well as possible. Investigators have described this reasoning using the convergence, spread, and distribution concepts (Bezerra et al., 2017). Convergence defines the extent to which the individuals are close to the Pareto-optimal front, the spread is related to the distance between the extreme individuals of the approximated Pareto-optimal front, and distribution is related to how individuals along the front are distributed.

Investigators have developed a considerable amount of approaches to solving MuOO and MaOO problems. With the creation and evolution of these approaches, scientists realized that the comparison of MuOO and MaOO is problematic because it involves processing information of a set of individuals in terms of convergence, spread, and distribution. To deal with this problem, researchers have proposed several indicators to measure the quality of a Pareto set using another one as a reference. Among the most popular indicators are the hyper-volume (HV), the inverted generational distance (IGD), and the R2 coefficients. HV considers each individual in the potential Pareto-optimal front to construct a hyper-cube using a point as a reference. Then, joining all of these hyper-cubes and normalizing the result leads to the HV value. Finally, high values of HV are desirable (Deb, 2001). Moreover, generational distance (GD) is a measure of the average distance between the potential Pareto-optimal front and the reference one. For example, for a bi-objective problem, GD is the Euclidean distance (Deb, 2001). Consequently, IGD represents an improvement of the GD indicator (Bezerra et al., 2017). Alternatively, The R2 indicator uses a set of general utility functions and their probabilities, as well as a reference set, to evaluate the performance of the potential Pareto-optimal front (Brockhoff et al., 2012).

During the early stage of MuOO development, scientist extended their knowledge about MoOO techniques to solve MuOO. Later, some approaches based on identifying nondominated individuals, other techniques based on some of the aforementioned indicators (HV, IGD, and R2), and other methods based on decomposition were incorporated. In this regard, Schaffer (1985) proposed the vector evaluated genetic algorithm (VEGA), which was the first MuOO implementation. VEGA is a direct extension of a MoOO to a MuOO since it pays attention to one objective at a time by dividing the GA population into subgroups to evaluate each target separately (Deb, 2001). Kursawe (1990) proposed the vector-optimized evolution strategy (VOES), which has some similarities with VEGA. However, VOES introduced the dominant and recessive chromosomes concepts, inspired by genetics. Hajela and

Lin (1992) developed the weighted-based genetic algorithm (WBGA) in which the fitness estimation is carried out by the algorithm using some weights combined with the search capabilities of the GA population.

Murata and Ishibuchi (1995) proposed the random weighted genetic algorithm (RWGA) following WBGA approach. The main difference lies in how the algorithm assigns fitness. Fonseca and Flemming (1993) developed the multiobjective genetic algorithm (MOGA). Using the concept of domination between two individuals and the characteristics of the Pareto front, which represents the set of individuals that are superior to the population remainder but are in equilibrium between each other (Fig. 2.4), MOGA determines the fitness to maintain the population diversity. Srinivas and Deb (1994) proposed the nondominated sorting genetic algorithm (NSGA) using the reasoning and contributions of Goldberg about the behavior of those individuals that belong to the Pareto-optimal front. Horn et al. (1994) introduced the niched Pareto genetic algorithm (NPGA), whose main difference with the previously mentioned techniques relies on how the GA selection operator is applied. Deb et al. (2002) developed an algorithm that takes advantage of elitism to solve multiobjective problems; researchers know it as the elitist nondominated sorting genetic algorithm or NSGA-II.

Many (MuOO and MaOO) elitist approaches use a different population in which the best individuals are collected and stored by the algorithm. Following this reasoning, Osyczka and Kundu (1995) introduced the distance-based Pareto genetic algorithm (DPGA), which has one population over which the algorithm applies the genetic operators, and another population in which the algorithm stores the elite individuals. Zitzler and Thiele (1998) developed the strength Pareto evolutionary algorithm (SPEA), which also uses two populations. The concept of strength is a measure of the elite influence over the standard population. The authors have incorporated this reasoning into fitness estimation. Knowles and Corne (2000) developed the Pareto archived evolution strategy (PAES). The authors created this algorithm using their experience in solving practical problems, where they analyzed the advantages of local search strategies and consequently incorporated this technique into their algorithm. Zitzler et al. (2001) presented an improved version of SPEA (SPEA2), which has an enhanced approach for fitness assignment and a more efficient mechanism to select the elite individuals. Corne et al. (2000) presented the Pareto envelope-based selection algorithm (PESA), which establishes the selection and population diversity through a scheme based on a hyper-grid. A year later, Corne et al. (2001) introduced an improved version of PESA (PESA-II), which uses a hyper-box as a selection unit for fitness estimation. Specifically, the fitness is assigned to the hyper-box to which a determined individual belongs.

Deb et al. (2003) proposed a multiobjective evolutionary algorithm based on ε-dominance (ε-MOEA). As in elitist algorithms, ε-MOEA uses two populations. These populations sequentially evolve so that the algorithm considers their corresponding offspring and parents to update the elitist and standard populations using the ε-dominance concept. The incorporation of ε-dominance allows the computational efficiency of the algorithm to be improved. Zitzler and Künzli (2004) proposed the indicator-based evolutionary algorithm (IBEA). The authors created IBEA intending to incorporate DM preferences in the optimization process. The general idea behind

IBEA consists of defining an optimization target and including it in the selection procedure. Thus, IBEA can use DM preferences to influence population diversity. Zhang and Li (2007) introduced the multiobjective evolutionary algorithm based on decomposition (MOEA/D). MOEA/D decomposes the problem into several subproblems that are solved simultaneously. MOEA/D is computationally efficient because, at the moment of solving each subproblem, it only uses information from the nearby ones, which reduces the computational burden. Using hyper-volume (S-metric), Beume et al. (2007) created the S-metric selection evolutionary multiobjective algorithm (SMS-EMOA). SMS-EMOA maximizes the dominated HV during the optimization process. Tseng and Chen (2009) proposed the multiple trajectory search (MTS). MTS divides an initially random population into the foreground and background individuals. Then, the algorithm applies local search techniques over these individuals sequentially, using a size-variable search.

Chen et al. (2015) presented a nondominated sorting and local search (NSLS) technique. NSLS performs a local search to obtain high-quality individuals and nondominated sorting to find the individuals to be used by the algorithm in the next iteration. Thus, those individuals located are involved in the nondominated sorting to improve diversity. Li et al. (2016) developed the bi-criterion evolution (BCE) approach in which the algorithm simultaneously considers the advantages of those methods based on Pareto front analysis and those based on decomposition methods or performance indicators. BCE uses two populations that share information between them; one population is used by the part based on Pareto front analysis, while the indicator-based criterion uses the other population. Thus, BCE searches for Pareto-optimal front and population diversity simultaneously. Similarly, Tian et al. (2016) developed a technique that uses an improved version of IGD indicator. Specifically, the algorithm finds those individuals that do not contribute to the evolution; the authors call them noncontributing solutions (NSs). Thus, the algorithm becomes computationally efficient. By its characteristics, researchers know it as the multiobjective evolutionary algorithm based on ICG and NS (MOEA/IGD-NS).

The complexity of the practical problems is notably growing due to technological development. As a consequence, the number of objectives functions of many fields of science and engineering is increasing. When the number of objectives is two or three, the problem can be recognized as a MuOO problem, but when the objectives are greater than or equal to four, the problem is recognized as a MaOO problem. MuOO techniques do not offer good results facing problems with more than three objectives, because the amount of nondominated individuals grows exponentially with the number of objectives. In these circumstances, the capabilities of MuOO algorithms to get close to the Pareto-optimal front are dramatically reduced. The weaknesses of MuOO to deal with a high number of objectives are overcome by MaOO using different strategies. One strategy consists of modifying the definitions used to establish the relationship among individuals. It means changing how Pareto dominance and ranking structures are applied. Some examples are the average and maximum ranking, favor relation, fuzzy Pareto dominance, etc. Another strategy consists of enhancing the procedure used to increase diversity (spread and distribution) to guide the algorithm towards the direction of the Pareto-optimal front. In other words, researchers modified some methods used to reduce the ability of MuOO algorithm to get a potential Pareto-optimal front with good spread and distribution, to improve convergence. Another policy consists of using two populations, one devoted to maintaining convergence and the other to maintain diversity.

Regarding the methods based on indicators (HV, IGD, and R2), they also have difficulties in solving MaOO problems. In the specific case of HV, the computational burden increases with the number of objectives. However, other indicators with lower computational requirements, such as R2 and IGD, can be used to replace it. Regarding the decomposition-based techniques, they tend to maintain population diversity using the weighting vectors. However, in the case of MaOO, frequently used algorithms such as MOEA/D are not able to offer good spread and distribution, but they show good convergence (Yuan, Xu, Wang, Zhang, & Yao, 2016).

Numerous MaOO algorithms have been proposed in the literature. Hughes (2003) proposed the multiple single objective Pareto sampling (MSOPS) algorithm. MSOPS carries out a parallel search, which enables it to solve problems with a large number of objectives. It can deal with limits and discontinuities in the Pareto front even when the surface cannot be traced easily by the algorithm. Later in Hughes (2007), Hughes presented an improved version of MSOPS (MSOPS-II). MSOPS-II can use an automatic target vector generation process, or it can reformulate the fitness assignment procedure for constraint handling. These changes in the fundamental methodology allow us to solve a variety of optimization problems with minimal intervention of the DM. Bringmann et al. (2011) introduced the approximation-guided evolutionary (AGE) algorithm, which solves MuOO and MaOO problems by using a formal approximation theory. Later, Wagner and Neumann (2013) developed an improved version (AGE-II) that optimizes how traditional AGE managed the archive used to store the elite individuals and their influence on the computational time. This improvement results in a balance between the accuracy and computational burden during the optimization process. Bader and Zitzler (2011) developed an algorithm that uses HV estimation (HypE). HV results in a better value for that set closest to the Pareto-optimal one; however, its calculation could be intensive for a large number of objectives. Thus, HypE calculates HV using a technique based on Monte Carlo simulation, which allows speeding up of the computing process. This characteristic makes HypE particularly suitable to solve MaOO problems.

Wang et al. (2013) used the preference-inspired coevolutionary algorithm (PICEA) to solve MaOO problems. It makes the set of DM preferences evolve jointly with a group of individuals. Yang et al. (2013) proposed a grid-based evolutionary algorithm (GrEA). The authors introduced several concepts such as grid dominance and grid difference, as well as grid ranking, grid crowding distance, and grid coordinate point distance, to establish the fitness of each individual. Thus, the algorithm can give priority to those individuals in the optimal direction, avoiding overcrowding. Deb and Jain (2014) and Jain and Deb (2014) deeply analyzed the existent MaOO algorithms and, following the reasoning behind NSGA-II, proposed the third NSGA version (NSGA-III). NSGA-III uses some reference points, which can be used adaptively by the algorithm to find the Pareto-optimal front. Li et al. (2013) introduced a shift-based density estimation (SDE) method to improve how the algorithm maintains the population diversity so that it can solve MaOO problems. Thus, SDE incorporates the population distribution and its convergence. Later, in Li, Yang, et al. (2015), they proposed the bi-goal evolution (BiGE). BiGE converts a MaOO problem into a MuOO problem of two variables in terms of individuals' proximity and diversity. The algorithm selects those potential individuals and puts them in the first stages of a classification system based on layers.

Yuan et al. (2014) proposed the ensemble fitness ranking (EFR) to incorporate a variety of fitness mechanisms and ranking methods. EFR can deal with general-purpose fitness assignment procedures so that it can balance the convergence speed and population diversity. Later, in Yuan, Xu, Wang, Zhang, and Yao (2016), the authors derived an improved version of MOEA/D and EFR techniques. They improved MOEA/D by adding a distance-based updating process (MOEA/D-DU), while in the case of EFR, they included a ranking restriction scheme (EFR-RR). Asafuddoula et al. (2015) developed a decomposition-based model that employs systematic sampling and a simple preemptive distance comparison procedure. Furthermore, optimization constraints are faced by using an adaptive epsilon formulation. Zhang et al. (2015) proposed a knee point-driven technique, which gives special preference to the knee points of those nondominated individuals. The idea behind this reasoning is that those individuals close to the knee have a large HV and consequently are good candidates to improve convergence.

Cheng et al. (2015) proposed the incorporation of directional diversity (DD) and favorable convergence (FC) operators to improve spread and distribution, as well as convergence. Expressly, DD and FC features are included in the GA selection process to achieve a trade-off between diversity and convergence. Li, Deb, et al. (2015) proposed a framework that consolidates the dominance and decomposition ideas into a unified methodology. Thus, a trade-off between diversity and convergence would be reached. Hernández and Coello (2015) presented a methodology based on R2 indicator. The authors combined the scalarizing achievement function and statistical information about the population, in terms of the position of individuals to the Pareto-optimal front, to improve the diversity of the resulting Pareto set estimation. He and Yen (2016) developed a methodology that is composed of two stages. During the first stage, several target points close to the Pareto-optimal front are searched. Then, in the second stage, a diversity improvement strategy is applied to the results of the first stage to improve spread and distribution.

Liu et al. (2017) developed an algorithm based on the continuous generation of reference points to improve spread, distribution, and convergence. Then, the best individuals are chosen using the distance between those individuals in the potential Pareto-optimal set and the reference points. Following the same trend, Cheng et al. (2016) created a methodology based on reference vector generation. The authors incorporated the angle-penalized distance as a scalarization method to balance spread, distribution, and convergence. In addition, an adaptation strategy was introduced to adjust how the reference points are distributed along with the scales of each objective. Jiang and Yang (2017) developed a version of SPEA family based on reference direction (SPEA/R). The authors took the ideas and advantages of SPEA and SPEA2 and improved them by adding a density estimator based on reference direction, an enhanced fitness assignment procedure, and an improved environmental selection strategy. Yuan, Xu, Wang, and Yao (2016) presented a technique based on an improved dominance relation to enhance the convergence of NSGA-III using a fitness assignment procedure based on decomposition, which takes advantage of the dominance definition in terms of diversity maintenance.

Concerning GA computational implementation, in terms of hardware, most algorithms to deal with MoOO, as well as MuOO and MaOO, have been intrinsically designed to be implemented by the DM in a single central processing unit (CPU). High performance computing (HPC) has been continuously evolving so that investigators can program the graphical processing unit (GPU) to reduce the computational time or improve the optimization results of a specific GA implementation. In other words, GPU can be used as an accelerator of any GA implementation. Researchers know the combination of CPU and GPU as heterogeneous architecture; programming of heterogeneous architecture requires detailed knowledge about the hardware structure. NVIDIA created the compute unified device architecture (CUDA) as a parallel computing platform for CPU-GPU architectures. CUDA can be used by programmers of C/C++ (Cheng et al., 2014), Fortran (Ruetsch & Fatica, 2013), as well as other languages. HPC can be used to process many required tasks or analyze many data elements. In this regard, task parallelism is related to the processing of several independent works that can be done at the same time. In contrast, data parallelism is related to the processing of many data elements at the same time, or in other words, sharing data processing across many cores (Cheng et al., 2014; Ruetsch & Fatica, 2013).

Due to their algorithmic characteristics, GAs are susceptible to being implemented using a data-parallel approach. Expressly, genetic operators such as selection, crossover, and mutation can be parallelized using a data-parallel method to distribute the computational burden across many processors. GA can be parallelized to reduce computational time or to improve the results in terms of convergence and quality. In this regard, a parallel genetic algorithm (PGA) can be classified into a master-slave or panmictic model, an island or coarse-grained model, or a cellular or fine-grained model (Cantú-Paz, 1999). Fig. 2.5 shows the scheme of each of these models. In the

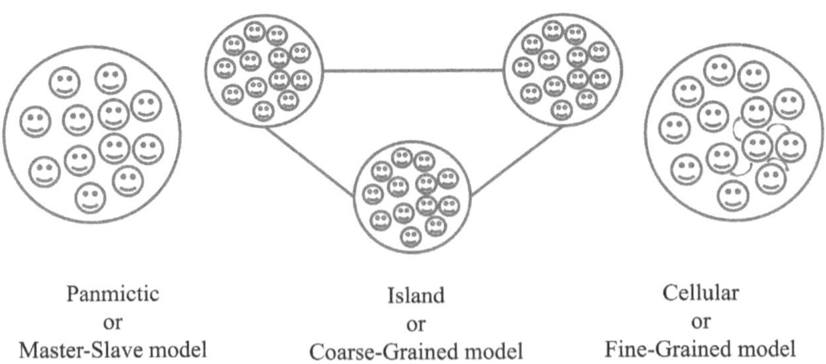

Fig. 2.5 PGA types: General idea of parallel GA classification.

master-slave or panmictic model, individuals are allowed to mate with the remainder of the population. In contrast, in a cellular or fine-grained model, individuals are allowed to mate only with the closest neighbors. Finally, on the island or coarse-grained model, the whole population is divided into several subpopulations so that individuals exclusively can be mated with other individuals within the corresponding subpopulation. Information among the subpopulations is exchanged using a migration operator (Cheng & Gen, 2019).

In a practical sense, the DM needs to optimize the operation and performance of a determined system or organization. Then, the DM carries out the problem formulation, paying attention to the priorities, which are reflected in the objective functions to be optimized. Depending on the number of objectives, the optimization process is classified (MoOO, MuOO, or MaOO), and the corresponding technique is selected among the vast family suggested in the literature. The computational implementation is done, and the advantages are quantified and evaluated. These processes can be summarized in the existence of many commercial tools for specific applications such as GOLD, GARD, and GANET. Another critical aspect is the computational infrastructure available; if an HPC architecture is accessible or affordable, the implementation of a PGA is an attractive option.

Concerning the extent of this work, special attention is paid to MoOO and MuOO because, in many practical applications for power systems, the optimization variables are one or two. On the one hand, MoOO corresponds to the minimization of operating costs. On the other hand, MuOO corresponds to the minimization of operating costs and pollutant emissions. Due to their importance and wide implementation, five MuOO techniques will be described: MOGA, NSGA, NSGA-II, SPEA, and SPEA2. Once the global picture of GA evolution has been described, the next step consists of explaining the basic concepts related to its implementation. Such concepts are described in the following sections.

2.3.1 Genetic algorithm coding

As discussed above, a GA can be formulated to deal with binary, integer, or real decision variables. Eq. (2.2) shows the representation of an individual ($i = 1, \ldots, I$) as a single vector ($\boldsymbol{x}_{\mathbf{pop}(i)}$), whose elements could be the binary, integer, or real variables to be determined.

$$\boldsymbol{x}_{\mathbf{pop}(i)} = \left[x_{\text{pop}(i,\,1)} \cdots x_{\text{pop}(i,\,v)} \cdots x_{\text{pop}(i,\,V)}\right] \tag{2.2}$$

In this context, V is the number of decision variables, and I is the number of individuals or the population size. According to Eq. (2.1), each element ($v = 1, \ldots, V$) of the individual (i) is constrained to a minimum ($\boldsymbol{x}_{\mathbf{pop}}^{\mathbf{min}}$) and maximum ($\boldsymbol{x}_{\mathbf{pop}}^{\mathbf{max}}$) value, as shown in Eq. (2.3).

$$x_{\text{pop}(v)}^{\min} \leq x_{\text{pop}(i,\,v)} \leq x_{\text{pop}(v)}^{\max} \forall v = 1, \ldots, V \tag{2.3}$$

where the factors $x^{\min}_{\text{pop}(v)}$ and $x^{\max}_{\text{pop}(v)}$ are elements of the vectors $\boldsymbol{x}^{\mathbf{min}}_{\mathbf{pop}}$ and $\boldsymbol{x}^{\mathbf{max}}_{\mathbf{pop}}$, respectively. These vectors are defined in Eqs. (2.4) and (2.5), respectively.

$$\boldsymbol{x}^{\mathbf{min}}_{\mathbf{pop}} = \left[x^{\min}_{\text{pop}(1)} \cdots x^{\min}_{\text{pop}(v)} \cdots x^{\min}_{\text{pop}(V)}\right] \tag{2.4}$$

$$\boldsymbol{x}^{\mathbf{max}}_{\mathbf{pop}} = \left[x^{\max}_{\text{pop}(1)} \cdots x^{\max}_{\text{pop}(v)} \cdots x^{\max}_{\text{pop}(V)}\right] \tag{2.5}$$

The number of elements of each individual (V) represents the dimension of the problem under study. These parameters can increase the complexity of the problem to be solved since the number of possible combinations depends exponentially on their number. A vast family of situations can be represented by using binary, integer, or real coding. For example, the connection or disconnection of a determined device at a specific hour can be modeled employing binary states. Another example could be the representation of a list of manufacturers for which it is necessary to determine the best performance. In this case, an integer number can be assigned to each manufacturer in the list, so that an integer coding can be effectively used. The case of real coding can be used to represent the settings of a parameter in an electronic device.

2.3.2 Fitness calculation

The procedure for fitness assignment strongly depends on the number of objectives to be optimized. The general formulation for the objective-function values is shown in Eq. (2.6).

$$\begin{aligned} \boldsymbol{y}_{\mathbf{obj}}\left(\boldsymbol{x}_{pop}(i)\right) &= \boldsymbol{y}_{obj}(i) \\ &= \left[y_{obj}(i1) \cdots y_{obj}(im) \cdots y_{obj}(iM)\right] \end{aligned} \tag{2.6}$$

where $m = 1, \ldots, M$ is the index of each objective to be optimized.

As stated before, MoOO ($m = 1$) consists of minimizing or maximizing a single quantity, while MuOO ($m = 2$ or 3) consists of finding out a set of points that describe the minimum or maximum values per objective. The set of optimal points per objective is known as the Pareto-optimal front; to estimate its shape, a specific procedure is required. In this regard, MoOO and MuOO require the application of different series of steps.

2.3.2.1 Fitness calculation in mono-objective optimization

In MoOO, fitness could be assigned directly according to the objective function value. However, this approach would assign a very high fitness to the best individuals and a small one for those individuals with less ability. To avoid this behavior, fitness ($y_{\text{fit}}(i)$) of a determined individual (i) is assigned linearly among all of the individuals according to Eq. (2.7).

$$y_{\mathrm{fit}}(i) = \frac{I+1-i}{\sum_{h=1}^{I}\{I+1-h\}} \forall 1 \le i \le I \tag{2.7}$$

In this way, those individuals with relatively low fitness are likely to be involved in GA functioning. The specific procedure to assign the fitness to a determined population is explained as follows:

Step 1: Calculate the factor $\sum_{h=1}^{I}\{I+1-h\}$.

Step 2: Create a table ($\mathcal{L}_{\mathrm{A}}$) with three columns. The first column contains the index $i=1, \ldots, I$; the second column contains the objective function value of each individual. The third column will be completed later.

Step 3: Sort the table created in Step 2 ($\mathcal{L}_{\mathrm{A}}$) by objective function value (column two).

Step 4: Complete the table sorted in Step 3 by filling the third column with:

$$^{(I+1-i)}\Big/\left(\sum\nolimits_{h=1}^{I}\{I+1-h\}\right)\forall i = 1,\ldots,I$$

Step 5: Sort the table filled in Step 4 by individuals (column one). The third column represents the fitness ($y_{\mathrm{fit}}(i)$) of each individual i.

2.3.2.2 *Fitness calculation in multiobjective optimization*

In the case of MoOO, the objective function value is directly used to assign the fitness of each individual. However, the case of MuOO introduces more complexity because there are M objectives under evaluation. Thus, the concept of domination was introduced; it is briefly defined in Eq. (2.8) (Deb, 2001).

$$\begin{aligned} i \le j :\Leftrightarrow \forall 1 \le m \le M : y_{\mathrm{obj}}(im) \le y_{\mathrm{obj}}(jm) \\ \wedge \exists 1 \le m \le M : y_{\mathrm{obj}}(im) < y_{\mathrm{obj}}(jm) \end{aligned} \tag{2.8}$$

It is said that i individual dominates individual j ($i \le j$) if two conditions are fulfilled. The first condition establishes that all of the objectives of individual i are not worse than all of the objectives of individual $j(y_{\mathrm{obj}}(i,m) \le y_{\mathrm{obj}}(j,m) \forall m = 1, \ldots, M)$. The second condition establishes that in at least one objective (m) the individual i has to be better than individual $j \exists 1 \le m \le M : y_{\mathrm{obj}}(i,m) < y_{\mathrm{obj}}(j,m)$ (Deb, 2001).

Based on this domination concept, there are several factors useful to assign the fitness of individuals involved in a MuOO problem. One of them is the ranking, which is defined as the number of individuals that dominate another specific individual plus one. This definition is intrinsically based on the sequential application of the concept of Eq. (2.8). Following this reasoning, the ranking ($r(i)$) of each individual (i) in a population can be determined by using the procedure explained as follows (Deb, 2001):

Step 1: Initialize the elements of the ranking variable to one $r(i) \leftarrow 1 \forall i = 1, \ldots, I$.

Step 2: Apply the procedure explained in the flowchart of Fig. 2.6.

The algorithm shown in Fig. 2.6 sequentially selects each of the individuals in the population (i) and compares it with the remainder (j), which is why both variables change within the interval [1, I]. The algorithm avoids the comparison of one individual with itself by including the condition $i \neq j$. Once this condition is verified, if a determined individual j dominates another individual i, then the corresponding ranking is increased by one. Once the algorithm is finished, the ranking vector ($r(i)$) will contain the result of the counting process for each $i = 1, \ldots, I$.

Another factor required is the number of individuals that belongs to a determined rank ($n_r(i)$). It can be determined by implementing the procedure explained as follows (Deb, 2001):

Step 1: Initialize the elements of the ranking-counter variable to zero; $n_r(i) \leftarrow 0 \forall i = 1, \ldots, I$.

Step 2: Apply the procedure explained in the flowchart of Fig. 2.7.

In the flowchart shown in Fig. 2.7, it is assumed that there are as many ranking values as individuals in the population, which is why the index for each ranking (j) varies within the interval [1, I]. For a determined ranking, the algorithm detects the individual with such a ranking ($i = r(j)$) and increments the counter variable by one ($n_r(i) \leftarrow n_r(i) + 1$).

Another critical factor is the front to which each individual belongs. The concepts of front and rank are related to each other. The procedure described as follows can be used to determine the front of a set of individuals (Deb, 2001).

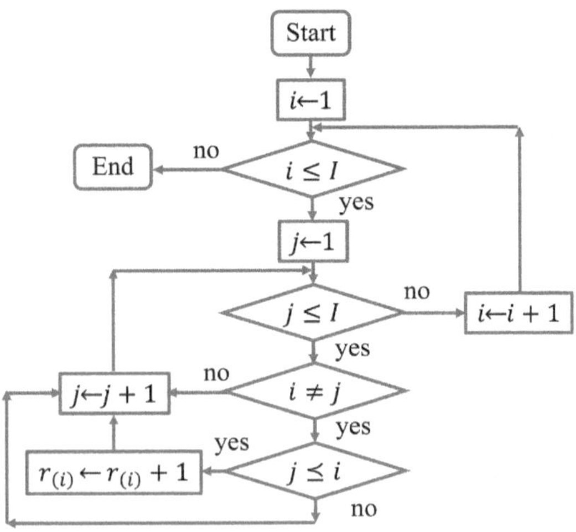

Fig. 2.6 Rank of GA population: Algorithm to determine the rank of each individual.

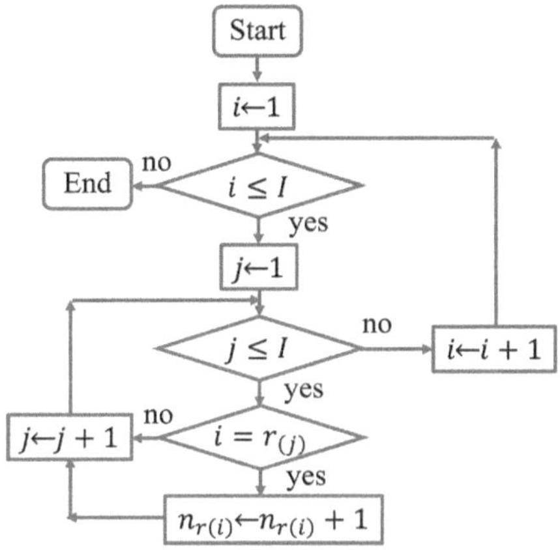

Fig. 2.7 Individuals per rank: Algorithm to determine the number of individuals in a rank.

Step 1: Create a table ($\mathcal{L}_\mathbf{B}$) with three columns. The first column stores the index of each individual ($i=1, \ldots, I$), while the second column stores the corresponding ranking ($r(i) \forall i=1, \ldots, I$). The third column will be completed later.

Step 2: Sort in ascending order the table created in Step 1 ($\mathcal{L}_\mathbf{B}$) by the rank (column 2). The resulting table ($\mathcal{L}_\mathrm{C}$) will be used in the next step.

Step 3: Apply the procedure explained in the flowchart in Fig. 2.8. As the second column of the sorted table ($\mathcal{L}_\mathrm{C}$) contains integers in ascending order, the algorithm of Fig. 2.8 uses this characteristic to determine the corresponding front. The algorithm

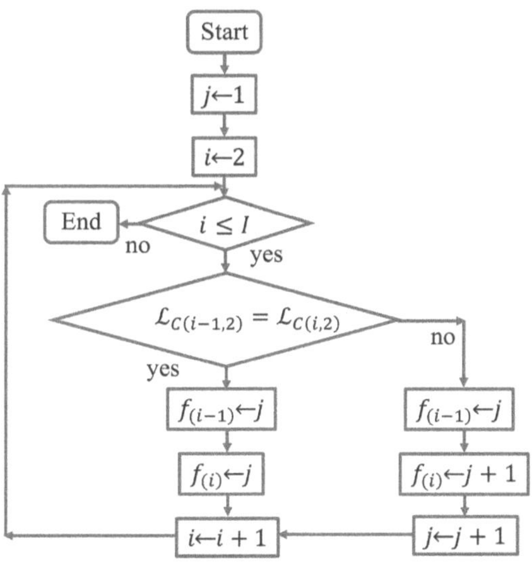

Fig. 2.8 Front of GA population: Algorithm to determine the front of each individual.

starts with the first front ($j \leftarrow 1$). Pairs of individuals are selected ($i-1$ and i) and compared ($\mathcal{L}_C(i-1,2) = \mathcal{L}_C(i,2)$); if they have the same ranking, it is recognized and stored for each individual ($f(i-1) \leftarrow j$ and $f(i) \leftarrow j$), and then, another pair of individuals are selected and compared ($i \leftarrow i+1$). On the other hand, if the ranks of both individuals are different, this fact is recognized and stored ($f(i-1) \leftarrow j$ and $f(i) \leftarrow j+1$), and another pair of individuals are selected and compared ($i \leftarrow i+1$), but considering the next front ($j \leftarrow j+1$). Once the algorithm of Fig. 2.8 is applied, the front of each individual is stored in the variable $f(i) \forall i = 1, \ldots, I$.

Step 4: The third column of $\mathcal{L}_C$ is filled with the vector $f(i) \forall i = 1, \ldots, I$. After that, the table ($\mathcal{L}_C$) is sorted by the first column. The result is a table whose first column is the index of each individual, the second column is the rank of each individual, and the third column is the front to which each individual belongs.

The number of individuals per front ($n_f(i) \forall i = 1, \ldots, I$) can be determined by using the algorithm in Fig. 2.7, but using the vector $f(i)$ instead of $r(i) \forall i = 1, \ldots, I$.

Once the methodologies for rank and front assignment have been described, it is possible to explain the specific procedure for fitness assignment suggested by MOGA, NSGA, and NSGA-II, as well as SPEA and SPEA2. It is essential to highlight that most of the concepts related to MuOO techniques previously discussed and the implementations described as follows have been adapted from Deb (2001).

Fitness assignment using MOGA starts by estimating the raw and the average fitness. Raw fitness is based on a linear relationship that assigns the value I (population size) to those individuals with the best performance and 1 to those less-promising individuals. Thus, the average fitness is the average value of the raw fitness of those individuals with the same rank. This approach drives the interest to those individuals with lower rank and reduces the attention to those individuals with poor performance. However, the relationship between individuals with the same rank has not yet been considered. In other words, the diversity of individuals with the same rank needs to be incorporated. MOGA deals with this aspect using the niche count, which uses the distance between each individual with the same rank, as well as a sharing parameter and a sharing function. Finally, the combination of the average fitness with the niche counting through a scaling factor results in scaled fitness, which is used later by some genetic operators.

The procedure of fitness assignment using MOGA for optimization problems of only two objective functions ($M = 2$) is briefly described as follows (Deb, 2001):

Step 1: Calculate the value of the sharing parameter ($\sigma_{(I)}^{\text{shr}}$) using Eq. (2.9).

$$\sigma_{(I)}^{\text{shr}} = \frac{2}{(I-1)} \tag{2.9}$$

Step 2: Determine the rank ($r(i)$) of each individual ($i = 1, \ldots, I$) and the number of individuals per rank ($n_r(i)$).

Step 3: For each individual (i) calculate the average fitness ($y_{fit}^{\text{avg}}(i) \forall i = 1, \ldots, I$) using Eq. (2.10).

$$y_{\text{fit}}^{\text{avg}}(i) = \begin{cases} I - \dfrac{1}{2}(n_r(1) - 1);\ r(i) = 1 \\ I - \displaystyle\sum_{h=1}^{r(i)-1} n_r(h) - \dfrac{1}{2}(n_r(r(i)) - 1);\ r(i) > 1 \end{cases} \tag{2.10}$$

Step 4: For all of the individuals that belong to a determined rank ($i = 1, \ldots, I$), calculate the niche count. To this end, the procedure described in Steps 4.1–4.8 can be applied.

Step 4.1: Analyze the individuals of the first rank by setting $j \leftarrow 1$.

Step 4.2: If $j \leq I$, go to Step 4.3, else go to Step 5.

Step 4.3: If $n_r(j) > 0$, go to Step 4.4, else set $j \leftarrow j+1$ and go to Step 4.2.

Step 4.4: Create a list ($\mathcal{L}_{D(h)} \forall h = 1, \ldots, n_r(j)$) of individuals with rank j. To carry out this task, the algorithm in Fig. 2.9 can be used.

Step 4.5: Calculate the normalized Euclidean distance ($\sigma_{(a,b)}^{\text{euc}}$) between all of the pairs of individuals (a, b) with rank j ($\mathcal{L}_{D(h)} \forall h = 1, \ldots, n_r(i)$) obtained in Step 4.4. The general expression for the normalized Euclidean distance is shown in Eq. (2.11).

$$\sigma_{(a,b)}^{\text{euc}} = \sqrt{\sum_{m=1}^{M} \left(\frac{y_{\text{obj}}(a,m) - y_{\text{obj}}(b,m)}{y_{\text{obj}}^{\text{max}}(m) - y_{\text{obj}}^{\text{min}}(m)} \right)^2} \tag{2.11}$$

where the vectors $\boldsymbol{y}_{\textbf{obj}}^{\textbf{min}}$ and $\boldsymbol{y}_{\textbf{obj}}^{\textbf{max}}$ are the maximum and minimum values per objective (m), respectively. Formally, they are defined in Eqs. (2.12) and (2.13), respectively.

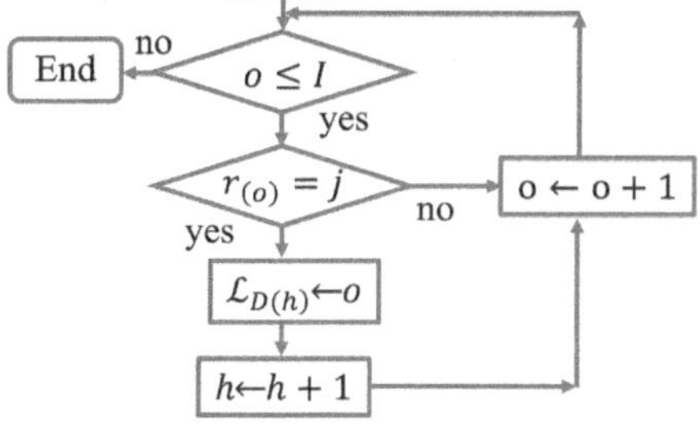

Fig. 2.9 Individuals within a rank: Algorithm to determine the individuals with rank j.

$$\mathbf{y}_{\mathbf{obj}}^{\mathbf{min}} = \left[y_{\text{obj}(1)}^{\min} \cdots y_{\text{obj}(m)}^{\min} \cdots y_{\text{obj}(M)}^{\min}\right] \tag{2.12}$$

$$\mathbf{y}_{\mathbf{obj}}^{\mathbf{max}} = \left[y_{\text{obj}(1)}^{\max} \cdots y_{\text{obj}(m)}^{\max} \cdots y_{\text{obj}(M)}^{\max}\right] \tag{2.13}$$

Step 4.6: Using the sharing parameter ($\sigma_{(I)}^{\text{shr}}$) calculated in Step 1, estimate the value of the sharing function ($g_{(a,b)}^{\text{shr}}$) for all of the pairs (a, b) of individuals with rank j. This is carried out by evaluating the results of Step 4.5 ($\sigma_{(a,b)}^{\text{euc}}$) in Eq. (2.14).

$$g_{(a,b)}^{\text{shr}} = \begin{cases} 1 - \sigma_{(a,b)}^{\text{euc}} / \sigma_{(I)}^{\text{shr}}; \sigma_{(a,b)}^{\text{euc}} \leq \sigma_{(I)}^{\text{shr}} \\ 0; \sigma_{(a,b)}^{\text{euc}} > \sigma_{(I)}^{\text{shr}}. \end{cases} \tag{2.14}$$

Step 4.7: Calculate the niche count ($n_C(j)$) for individuals with rank j using Eq. (2.15).

$$n_C\left(\mathcal{L}_{D(h)}\right) = \sum_{o=1}^{o=n_r(j)} g_{(h,o)}^{\text{shr}} \forall h = 1, \ldots, n_r(j) \tag{2.15}$$

Step 4.8: Analyze the niche count of the next rank by setting $j \leftarrow j+1$ and go to Step 4.2.

Step 5: Calculate the shared fitness ($y_{fit}^{\text{shr}}(i) \forall i = 1, \ldots, I$) by using Eq. (2.16).

$$y_{\text{fit}}^{\text{shr}}(i) = \frac{y_{\text{fit}}^{\text{avg}}(i)}{n_C(i)} \forall i = 1, \ldots, I \tag{2.16}$$

Step 6: For all of the individuals that belong to a determined rank ($t = 1, \ldots, I$), calculate the scaling factor ($\mathbf{y}_{\mathbf{fit}}^{\mathbf{scl}}$). To carry out this task, the procedure described in Steps 6.1–6.6 is implemented.

Step 6.1: Analyze the individuals of the first rank by setting $t \leftarrow 1$.

Step 6.2: If $t \leq I$, go to Step 6.3, else go to Step 7.

Step 6.3: If $n_r(t) > 0$, go to Step 6.4, else set $t \leftarrow t+1$ and go to Step 6.2.

Step 6.4: Create a list ($\mathcal{L}_{E(h)} \forall h = 1, \ldots, n_r(t)$) of individuals with rank t. To carry out this task, the algorithm in Fig. 2.9 can be used.

Step 6.5: Calculate the scaling factor ($\mathbf{y}_{\mathbf{fit}}^{\mathbf{scl}}$) using Eq. (2.17).

$$y_{\text{fit}}^{\text{scl}}\left(\mathcal{L}_{E(h)}\right) = \left(n_r(t) y_{\text{fit}}^{\text{avg}}\left(\mathcal{L}_{E(h)}\right)\right) \Big/ \left(\sum_{h=1}^{n_r(j)} y_{\text{fit}}^{\text{shr}}\left(\mathcal{L}_{E(h)}\right)\right) \forall h = 1, \ldots, n_r(t) \tag{2.17}$$

Step 6.6: Calculate the scaling factor of the next rank by setting $t \leftarrow t+1$ and go to Step 6.2.

Step 7: Calculate the scaled version of shared fitness using Eq. (2.18).

$$y_{\mathrm{fit}}^{\mathrm{shr}}(i) \leftarrow y_{\mathrm{fit}}^{\mathrm{shr}}(i) y_{\mathrm{fit}}^{\mathrm{scl}}(i) \forall i=1,\ldots,I \tag{2.18}$$

Another important technique is NSGA (nondominated sorting genetic algorithm). As its name specifies, it is based on sorting the population according to the non-domination concept. The NSGA working principle consists of dividing the population into several nondominated groups or fronts, the individuals belonging to the first front being those individuals closest to the Pareto-optimal front; consequently, those individuals belonging to subsequent fronts are farther from the optimal one. Then, the fitness is assigned so that those individuals within the first front have the highest. Consequently, those individuals belonging to subsequent fronts would have lower fitness. This procedure guarantees that the algorithm goes in the optimal direction. However, the diversity among the individuals has to be forced by a specific procedure. To this end, NSGA modifies the assigned fitness according to the number of neighboring individuals using the niching theory. Fitness assignment of NSGA could be implemented by following the steps (Deb, 2001):

Step 1: Select a value for the sharing parameter (σ_{shr}).

Step 2: Determine the front of each individual ($f(i) \forall i=1, \ldots, I$) and the number of individuals per front ($n_f(i) \forall i=1, \ldots, I$), as previously explained.

Step 3: Determine the number of fronts (W) in the population as $W=\max(f(i) \forall i=1,\ldots,I)$.

Step 4: Initialize the minimum shared fitness ($y_{\mathrm{min}}^{\mathrm{shr}}$) to a value slightly higher than I.

Step 5: For all of the individuals that belong to a determined front ($w=1, \ldots, W$), calculate the niche count, assigned fitness, and shared fitness. To this end, the procedure described in Steps 5.1–5.10 can be used.

Step 5.1: Analyze all of the individuals in the first front by assigning $w \leftarrow 1$.

Step 5.2: If $w \leq W$, go to Step 5.3, else stop.

Step 5.3: Create a list ($\mathcal{L}_{\mathrm{F}}$) of individuals that belong to front w, using an approach similar to that explained in Fig. 2.9, but changing the rank ($r(i) \forall i=1, \ldots, I$) by the front ($f(i) \forall i=1, \ldots, I$).

Step 5.4: Calculate the normalized Euclidean distance between all of the individuals within a front ($f(h) \forall h=1, \ldots, n_f(w)$) using Eq. (2.11).

Step 5.5: Using the sharing parameter (σ_{shr}) selected in Step 1, calculate the value of the sharing function ($g_{(a,b)}^{\mathrm{shr}}$) for all of the pairs (a, b) of individuals in the front w. This is carried out by evaluating the results of Step 5.4 ($\sigma_{(a,b)}^{\mathrm{euc}}$) in Eq. (2.19).

$$g_{(a,b)}^{\mathrm{shr}} = \begin{cases} 1-\left(\sigma_{(a,b)}^{\mathrm{euc}} / \sigma_{(I)}^{\mathrm{shr}}\right)^2; \sigma_{(a,b)}^{\mathrm{euc}} \leq \sigma_{(I)}^{\mathrm{shr}} \\ 0; \sigma_{(a,b)}^{\mathrm{euc}} > \sigma_{(I)}^{\mathrm{shr}} \end{cases} \tag{2.19}$$

Step 5.6: Calculate the niche count ($n_C(w)$) for individuals in the front w using Eq. (2.20).

$$n_C\left(\mathcal{L}_{F(h)}\right) = \sum_{o=1}^{n_f(w)} g_{(h,o)}^{\text{shr}} \forall h = 1, \ldots, n_f(w) \tag{2.20}$$

Step 5.7: Calculate the assigned fitness ($\boldsymbol{y_{\text{fit}}^{\text{asg}}}$) of the elements in the list using Eq. (2.21).

$$y_{\text{fit}}^{\text{asg}}\left(\mathcal{L}_{F(h)}\right) \leftarrow y_{\text{min}}^{\text{shr}} \tag{2.21}$$

Step 5.8: Calculate the shared fitness ($\boldsymbol{y_{\text{fit}}^{\text{shr}}}$) using Eq. (2.22).

$$y_{\text{fit}}^{\text{shr}}\left(\mathcal{L}_{F(h)}\right) = \frac{y_{\text{fit}}^{\text{asg}}\left(\mathcal{L}_{F(h)}\right)}{n_C\left(\mathcal{L}_{F(h)}\right)} \forall h = 1, \ldots, n_f(w) \tag{2.22}$$

Step 5.9: Calculate the minimum shared fitness as $y_{\text{min}}^{\text{shr}} \leftarrow \min\left\{y_{\text{fit}}^{\text{shr}}\left(\mathcal{L}_{F(h)}\right) \forall h = 1, \ldots, n_f(w)\right\}$.

Step 5.10: Analyze the next front by setting $w \leftarrow w+1$ and going to Step 5.2.

Regarding the NSGA-II fitness assignment, it is also based on nondomination sorting. However, its strong capabilities rely on elitism.

Besides the population, SPEA uses an archive that stores the best individuals, or in other words, the elite, during GA evolution. Fitness assignment is performed for the individuals of both the population ($i = 1, \ldots, I$) and the archive ($q = 1, \ldots, Q$); the fitness assigned to the individuals in the archive ($y_{\text{fit}}^{\text{ach}}(q) \forall q = 1, \ldots, Q$) is known as the strength, and depends on the number of individuals in the archive (q) that dominate individuals in the population (i). Thus, an individual in the archive that dominates more individuals in the population would have a better, or smaller, fitness. The procedure shown in Fig. 2.10 can be used to evaluate the fitness of an archive with Q individuals.

Once the fitness ($y_{\text{fit}}^{\text{ach}}(q)$) of the individuals ($q = 1, \ldots, Q$) in the archive has been estimated, the next step is assigning fitness ($y_{\text{fit}}^{\text{pop}}(i)$) to the individuals ($i = 1, \ldots, I$) in the population. Fitness of a determined individual (i) in the population is influenced by the total strength of those individuals ($q = 1, \ldots, Q$) that dominate i plus one. In other words, fitness can be estimated as the addition of the fitness of all those individuals in the archive that dominate i plus one. The addition of one is done to make those individuals in the population have a worse fitness than those in the archive. For SPEA, an individual with lower fitness is better (Deb, 2001). The fitness of individuals in the population can be assigned using the procedure shown in Fig. 2.11.

When compared with SPEA, SPEA2 introduces important changes in the manner fitness is assigned. First of all, a new population has to be created by joining or merging the standard and elite populations. Strength ($y_{\text{fit}}^{\text{str}}(e)$) is assigned to all individuals of the joint population ($e = 1, \ldots, E = I + Q$); the strength is calculated by the number of individuals that a specific individual dominates. It allows us to avoid assigning the same fitness in those situations in which individuals of the standard population are

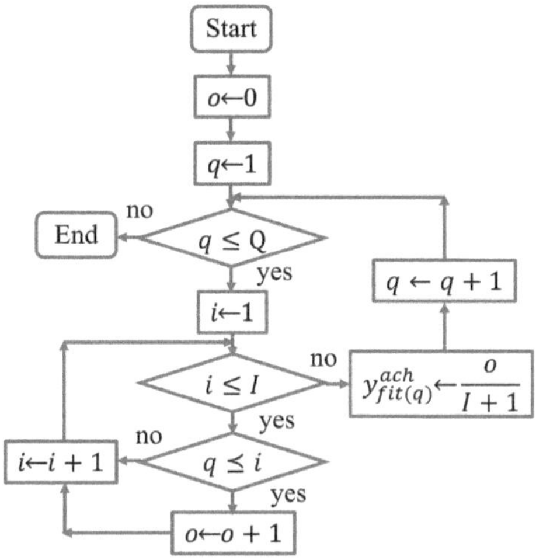

Fig. 2.10 Strength calculation (SPEA): Algorithm to determine the fitness (strength) of individuals in the archive.

dominated by the same number of individuals in the elite population (Zitzler et al., 2001). Fitness can be assigned using the algorithm in Fig. 2.12.

Once the strength ($y_{\text{fit}}^{\text{str}}(e)$) of each individual ($e = 1, \ldots E = I + Q$) in the joint population has been estimated, the raw fitness ($y_{\text{fit}}^{\text{raw}}(e)$) of a particular individual (e) is estimated from the strength of those individuals that dominate him or her. In this regard, a low value of raw fitness corresponds to a nondominated individual, while a high value of raw fitness corresponds to an individual dominated by many other individuals. Raw fitness can be calculated using the algorithm in Fig. 2.13.

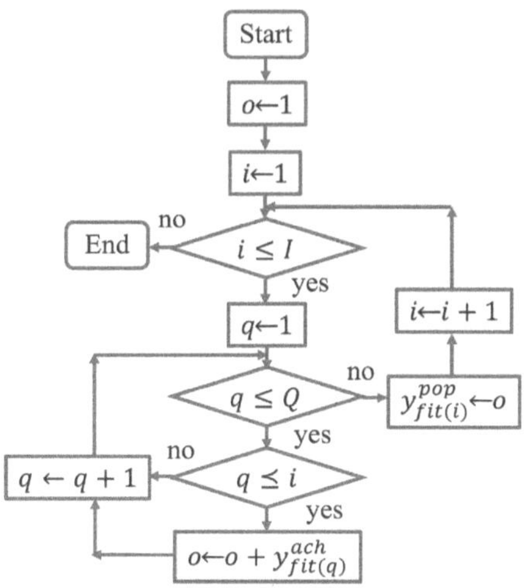

Fig. 2.11 Fitness calculation (SPEA): Algorithm to determine the fitness of individuals in the population.

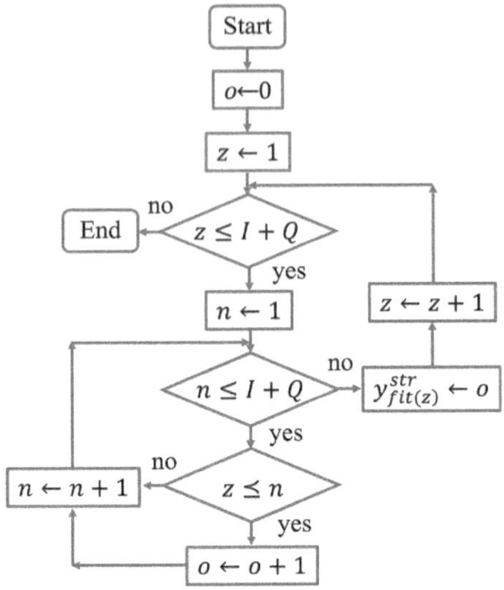

Fig. 2.12 Strength calculation (SPEA2): Algorithm to determine the fitness strength of individuals in the archive and population.

Distance between individuals within the population can also be defined using the distance sought ($\sigma^{\text{sgh}}_{(z,k)}$) concept, which identifies the closest kth individual from another individual z. The distance sought ($\sigma^{\text{sgh}}_{(z,k)}$) can be estimated by using the procedure explained as follows:

Step 1: Set $z \leftarrow 1$.

Step 2: If $z \leq I+Q$ go to Step 3, else Stop.

Step 3: Create a list ($\mathcal{L}_{\text{G}}$) with $I+Q$ elements.

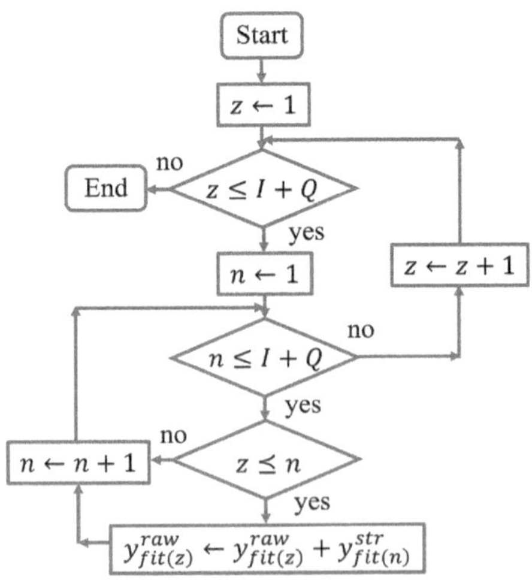

Fig. 2.13 Raw fitness calculation (SPEA2): Algorithm for the raw fitness of individuals in the archive and population.

Step 4: Set $n \leftarrow 1$.

Step 5: If $n \leq I+Q$ go to Step 6, else go to Step 8.

Step 6: Calculate the normalized Euclidean distance ($\sigma^{\text{euc}}_{(z,n)}$) between individuals z and n, and store the result in the element z of the list created in Step 3 ($\mathcal{L}_{G(z)}$).

Step 7: Set $n \leftarrow n+1$ and go to Step 5.

Step 8: Sort in ascending order the elements of the list created in Step 3 and completed in Step 6 ($\boldsymbol{\mathcal{L}}_{\mathbf{G}}$).

Step 9: Select the element $k=\sqrt{I+Q}$ of the sorted list of Step 8. The result will be the distance sought ($\sigma^{\text{sgh}}_{(z,k)}$) of the individual z.

Step 10: Set $z \leftarrow z+1$ and go to Step 2.

After calculating the distance sought ($\sigma^{\text{sgh}}_{(e,k)}$) of each individual ($e$) in the joint population, the density ($y^{\text{den}}_{\text{fit}}(e)$) is calculated using Eq. (2.23).

$$y^{\text{den}}_{\text{fit}}(e) = \frac{1}{\sigma^{\text{sgh}}_{(e,\,k)} + 2} \forall e = 1,\ldots,E \tag{2.23}$$

Finally, the fitness of each individual ($y_{\text{fit}}(e)$) is calculated as the addition between the raw fitness ($y^{\text{raw}}_{\text{fit}}(e)$) and the density ($y^{\text{den}}_{\text{fit}}(e)$). This idea is shown in Eq. (2.24).

$$y_{\text{fit}}(e) = y^{\text{raw}}_{\text{fit}}(e) + y^{\text{den}}_{\text{fit}}(e) \forall e = 1,\ldots,E = I+Q \tag{2.24}$$

2.3.3 Genetic operators

In a previous section, genetic operators were briefly defined, and the ideas that motivate their existence were explained. In the next sections, the authors describe some classical and commonly used techniques to simulate the natural selection, crossover, and mutation processes.

2.3.3.1 Selection operator

In this section, some of the classical selection operators such as proportional and tournament selection will be explained (Deb, 2001). In the proportional selection, the number of selected individuals tends to be proportional to their fitness. Its implementation is inspired by the working principle of a roulette wheel. First, the selection probability ($y^{\text{prob}}_{\text{fit}}(i)$) of a determined individual (i) is calculated using Eq. (2.25):

$$y^{\text{prob}}_{\text{fit}}(i) = {}^{y_{\text{fit}}(i)} \Big/ \left(\sum\nolimits_{h=1}^{I} y_{\text{fit}}(h) \right) \tag{2.25}$$

while its cumulative probability ($y^{\text{cum}}_{\text{fit}}(i) \forall i = 1, \ldots, I$) is calculated according to Eq. (2.26).

$$y_{\text{fit}}^{\text{cum}}(i)=\sum_{h=1}^{i}y_{\text{fit}}^{\text{prob}}(h)\forall i=1,\ldots,I \tag{2.26}$$

Once the cumulative probability ($\mathbf{y_{fit}^{cum}}$) is known, the selection process can be implemented as follows:

Step 1: Set $i\leftarrow 1$.

Step 2: If $i\leq I$ go to Step 3, else stop.

Step 3: Generate a random number (*RAND*) in the interval [0, 1].

Step 4: Apply the algorithm in Fig. 2.14 and go to Step 5.

Step 5: Set $i\leftarrow i+1$ and go to Step 2.

Another frequently used method is tournament selection. In this approach, two different individuals are forced to compete, and the corresponding winner is that with better fitness. Thus, the mating pool is composed of the favorite individuals. This approach is repeated systematically so that each individual would compete in two tournaments. Then, the best individual would win twice, while the worse would be discarded. At the end of the process, a specific individual could appear once or twice in the mating pool. This approach is particularly important in NSGA-II. This algorithm uses a more sophisticated version of the tournament selection approach known as crowding tournament selection (CTS). CTS uses the crowding distance (CD) and the rank of each individual as the metric for comparison during each tournament. First of all, a table of individuals per front is required, which can be built by using the procedure explained as follows (Deb, 2001):

Step 1: Create a table ($\boldsymbol{\mathcal{L}}_{\mathbf{H}}$) to store a list of individuals in each front. The table has to have $\max(f(i)\forall i=1,\ldots,I)$ rows and $\max(n_f(i)\forall i=1,\ldots,I)$ columns.

Step 2: Complete the table created in Step 1 ($\boldsymbol{\mathcal{L}}_{\mathbf{H}}$) by applying the procedure in Fig. 2.15.

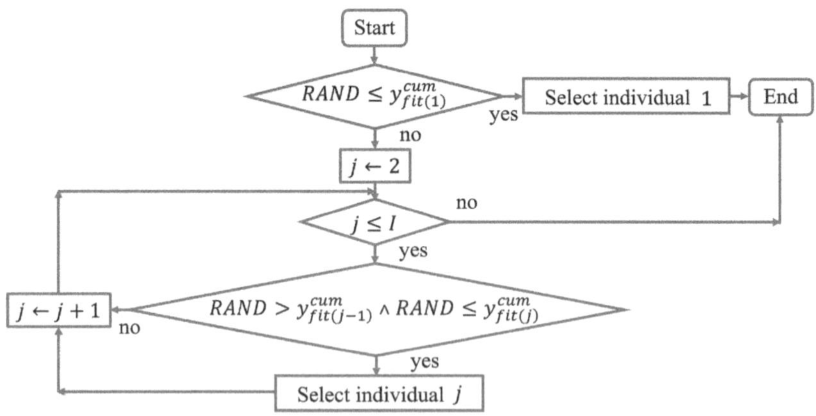

Fig. 2.14 Roulette wheel selection: Algorithm for proportional selection.

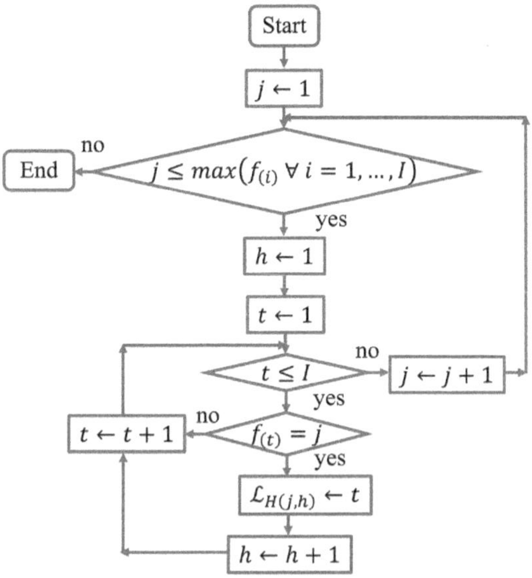

Fig. 2.15 Individuals within a front: Algorithm to create a table of individuals per front.

Considering the individuals belonging to a specific front (j), CD can be calculated by implementing the method described as follows (Deb, 2001):

Step 1: Create a table ($\boldsymbol{\mathcal{L}}_\mathbf{I}$) of individuals in the front j. This can be obtained by reading the row j and the columns 1, …, $n_f(j)$ in the table $\boldsymbol{\mathcal{L}}_\mathbf{H}$ of Fig. 2.15.

Step 2: Create a table ($\boldsymbol{\mathcal{L}}_\mathbf{J}$) to store the objective function values of individuals in the front j, which can be done using Eq. (2.27).

$$\boldsymbol{\mathcal{L}}_\mathbf{J} = \mathbf{y}_\mathbf{obj}\left(\mathcal{L}_{I(j,h)}\right) \forall h = 1, \ldots, n_f(j) \tag{2.27}$$

Step 3: Sort each column of the table $\boldsymbol{\mathcal{L}}_\mathbf{J}$ and store the indices in another table ($\boldsymbol{\mathcal{L}}_\mathbf{K}$).

Step 4: Create a table ($\boldsymbol{\mathcal{L}}_\mathbf{L}$) to store CD. Initialize $\boldsymbol{\mathcal{L}}_\mathbf{L}$ to zero ($\mathcal{L}_{L(h)} = 0 \forall h = 1, \ldots, n_f(j)$).

Step 5: Calculate CD for the individuals in front j. To this end, follow the procedure explained in Steps 5.1–5.5.

Step 5.1: Set $m \leftarrow 1$.

Step 5.2: If $m \leq M$ go to Step 5.3, else stop.

Step 5.3: Set CD of boundary individuals to infinity. It is $\mathcal{L}_L(\mathcal{L}_K(1, m)) \leftarrow \infty$ and $\mathcal{L}_L\left(\mathcal{L}_K\left(n_f(j), m\right)\right) \leftarrow \infty$.

Step 5.4: Calculate CD of nonboundary individuals by applying Eq. (2.28) for $o = 2, \ldots, n_f(j) - 1$.

$$\begin{aligned}\mathcal{L}_L(\mathcal{L}_K(o, m)) \leftarrow\ & \mathcal{L}_L(\mathcal{L}_K(o, m)) \\ & + \frac{\mathcal{L}_J(\mathcal{L}_K(o+1, m), m) - \mathcal{L}_J(\mathcal{L}_K(o-1, m), m)}{y_\text{obj}^\text{max}(m) - y_\text{obj}^\text{min}(m)}\end{aligned} \tag{2.28}$$

Step 5.5: Set $m \leftarrow m+1$ and go to Step 5.2.

The procedure aforementioned is repeated for each front $j=1, \ldots, \max(f(i)\,\forall i=1,\ldots,I)$ in the population. After that, CTS is performed. During CTS, two different individuals are compared; the winner will be that with a lower rank. If the compared individuals have the same rank, the winner will be that with the higher CD.

Baker (1985) proposed another broadly adopted method known as stochastic universal sampling (SUS). SUS uses a group of equidistant numbers generated from a single random one. Thus, an individual corresponding to each number in the equidistant set is selected by using the cumulative probability values (Deb, 2001).

2.3.3.2 Crossover operator

Once the mating pool has been obtained from the application of the selection operator, the next step in GA implementation consists of applying the crossover operator. The implementation of this operator depends on the GA codification used. The basic idea behind the crossover operator is illustrated in Fig. 2.16 for the case of a binary-coded GA or an integer-coded GA.

According to Fig. 2.16, a specific point in all individuals is selected, and one section of each individual is sequentially exchanged at a time, obtaining a pair of new individuals (Deb, 2001). In the case of a real-coded GA, the crossover procedure can be done by a simulation of the process previously explained; it is known as the simulated binary crossover (SBX) operator (Deb & Agrawal, 1995).

As previously explained, DM specifies the crossover rate (p_c). However, in the case of SBX, an additional parameter has to be specified: the distribution index (n_c). The application of the crossover operator is constrained to a random number (*RAND*). If *RAND* is lower than the crossover rate, the crossover operation is performed over a pair of individuals; if not, another pair of individuals is selected, and it is verified whether the crossover operator should be applied or not. The crossover simulation can be done by following the procedure described as follows (Deb, 2000):

Step 1: Set $v \leftarrow 1$.

Step 2: If $v \leq V$, go to Step 3, else stop.

Step 3: Generate a random number (*RAND*) in the interval [0, 1].

Step 4: If $RAND \leq {}^{1}/_{2}$, go to Step 5, else set $v \leftarrow v+1$ and go to Step 2.

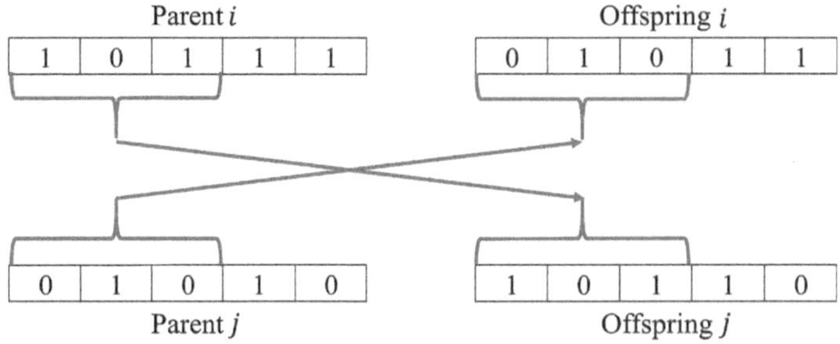

Fig. 2.16 Crossover operator: Illustration of single-point crossover.

Step 5: If $x_{\text{pop}}(i,v) > x_{\text{pop}}(j,v)$ set $x_1 \leftarrow x_{\text{pop}}(j,v)$ and $x_2 \leftarrow x_{\text{pop}}(i,v)$, else $x_1 \leftarrow x_{\text{pop}}(i,v)$ and $x_2 \leftarrow x_{\text{pop}}(j,v)$.

Step 6: Calculate the parameters α and β using Eqs. (2.29) and (2.30).

$$\beta = 1 + \frac{2}{x_2 - x_1} \min\left[\left(x_1 - x_{\text{pop}}^{\min}(v)\right), \left(x_{\text{pop}}^{\max}(v) - x_2\right)\right] \tag{2.29}$$

$$\alpha = 2 - \beta^{-(n_C - 1)} \tag{2.30}$$

Step 7: Generate a random number (*RAND*) in the interval [0, 1] and calculate the parameter δ using Eq. (2.31).

$$\delta = \begin{cases} (\alpha \times RAND)^{1/(n_C+1)}; RAND \leq {}^1/_\alpha \\ \left(\dfrac{1}{2 - \alpha \times RAND}\right)^{1/(n_C+1)}; RAND > {}^1/_\alpha \end{cases} \tag{2.31}$$

Step 8: Assign the offspring using Eqs. (2.32) and (2.33).

$$x_{\text{pop}}(i, v) \leftarrow \frac{1}{2}\left[(x_1 + x_2) - \delta |x_1 - x_2|\right] \tag{2.32}$$

$$x_{\text{pop}}(j, v) \leftarrow \frac{1}{2}\left[(x_1 + x_2) + \delta |x_1 - x_2|\right] \tag{2.33}$$

Step 9: Assign $v \leftarrow v + 1$ and go to Step 2.

2.3.3.3 Mutation operator

Implementation of the mutation operator also depends on the GA coding and on the mutation rate (p_m) selected by the DM. Fig. 2.17 illustrates the implementation of this operator for a binary-coded or integer-coded GA. It consists of randomly selecting an individual (i) in the population and choosing an element (v) on this individual to change its value. If $x_{\text{pop}}(i,v) = 1$ then it is changed to 0 ($x_{\text{pop}}(i,v) \leftarrow 0$). In contrast, if $x_{\text{pop}}(i,v) = 0$ then it is changed to 1 ($x_{\text{pop}}(i,v) \leftarrow 1$).

In the case of a real-coded GA, the methodology to be applied is different; it is known as polynomial mutation and the DM needs to provide a distribution index

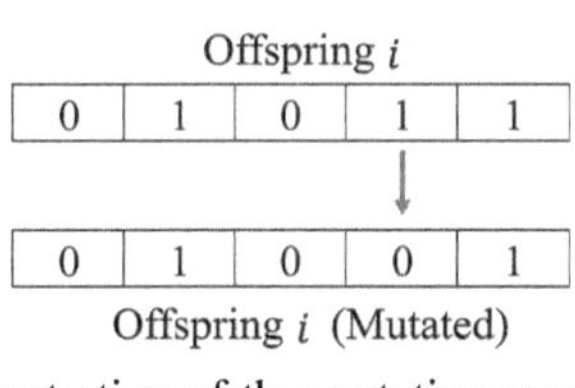

Fig. 2.17 Mutation operator: Illustration of the mutation operator.

(n_m) in addition to the mutation rate (p_m). Implementation of polynomial mutation operator is briefly explained as follows (Deb, 2000):

Step 1: Set $v \leftarrow 1$.

Step 2: If $v \leq V$, go to Step 3, else stop.

Step 3: Calculate the parameter φ as shown in Eq. (2.34).

$$\varphi = \frac{\min\left[\left(x_{\text{pop}}(i,v) - x_{\text{pop}}^{\min}(v)\right), \left(x_{\text{pop}}^{\max}(v) - x_{\text{pop}}(i,v)\right)\right]}{x_{\text{pop}}^{\max}(v) - x_{\text{pop}}^{\min}(v)} \tag{2.34}$$

Step 4: Generate a random number (*RAND*) in the interval [0, 1].

Step 5: If $RAND < p_m$ go to Step 6, else set $v \leftarrow v+1$ and go to Step 2.

Step 6: Generate a random number (*RAND*) in the interval [0, 1].

Step 7: Using the random number generated in Step 6 (*RAND*), calculate θ using Eq. (2.35).

$$\theta = \begin{cases} \left[2 \times RAND + (1 - 2 \times RAND)(1-\varphi)^{(n_m+1)}\right]^{1/(n_m+1)} - 1; RAND \leq {}^1/_2 \\ 1 - \left[2(1-RAND) + 2\left(RAND - {}^1/_2\right)(1-\varphi)^{(n_m+1)}\right]^{1/(n_m+1)}; RAND > {}^1/_2 \end{cases} \tag{2.35}$$

Step 8: Calculate the mutated individual by applying Eq. (2.36).

$$x_{\text{pop}}(i,v) \leftarrow x_{\text{pop}}(i,v) + \theta\left(x_{\text{pop}}^{\max}(v) - x_{\text{pop}}^{\min}(v)\right) \tag{2.36}$$

Step 9: Set $v \leftarrow v+1$ and go to Step 2.

In addition to the techniques mentioned here for natural selection, crossover, and mutation, other classical approaches can be found in Herreda et al. (1998).

2.4 Basic implementation of a genetic algorithm

This section describes the basic concepts behind GA implementation. In a general sense, a GA implementation follows the guidelines explained in Figs. 2.2 and 2.3 for nonelitist and elitist approaches, respectively. However, some methods have specific details that should be highlighted. In the case of MoOO, the elitist and nonelitist approaches can be directly implemented by following the flowcharts given in Figs. 2.2 and 2.3, respectively. Regarding MuOO, this is true only for the case of nonelitist methods such as MOGA and NSGA. The elitist approaches such as NSGA-II, SPEA, and SPEA2 have important details that need to be explained.

NSGA-II is initialized with a population of individuals generated randomly, while the offspring population is assumed to be empty. Under these conditions, the joint population is built by merging the parent and offspring populations into one. The rank of each individual of the joint population and the front to which it belongs is determined.

Then, the individuals to be included in the population for the next generation have to be identified. To this end, the algorithm described next can be used (Deb, 2001).

Step 1: Calculate the cumulative values of $n_f(i)\forall i=1, \ldots, I$ using Eq. (2.37).

$$n_f^{\text{cum}}(i) = \sum_{h=1}^{i} n_f(h)\forall i = 1,\ldots,I \tag{2.37}$$

Step 2: Calculate the fronts (u) to be included in the elite population. To this end, the algorithm in Fig. 2.18 can be used.

Step 3: Create a list ($\mathcal{L}_{\mathbf{M}}$) with elements.

Step 4: If $n_f^{\text{cum}}(u)=I$, store all individuals in the fronts $1, 2, \ldots, u$ in the list $\mathcal{L}_{\mathbf{M}}$, else go to Step 5.

Step 5: Store all individuals in the fronts 1, 2, …, $u-1$ in the list $\mathcal{L}_{\mathbf{M}}$.

Step 6: Create a table ($\mathcal{L}_{\mathbf{N}}$) with $n_f(u)$ elements, calculate CD for the individuals in the front u and store the results in the table $\mathcal{L}_{\mathbf{N}}$.

Step 7: Sort the elements in the table according to CD in ascending order.

Step 8: Fulfill the elements $n_f(u-1)+1, \ldots, I$ with the first individuals of the sorted table from Step 7.

Once the population for the next generation has been determined, the corresponding offspring is determined by applying CTS, crossover, and mutation operators.

SPEA starts by creating a population with randomly generated individuals, while the archive remains empty. Then, the nondominated individuals in the population are included in the archive, extending its size. After that, the nondominated individuals in the extended archive are found, discarding the dominated ones, which changes the structure of the archive again. If the number of elements in the archive is higher than a predefined limit (Q) established by the DM, a clustering algorithm is applied to reduce the size of the archive to Q. In contrast, if the number of individuals in the

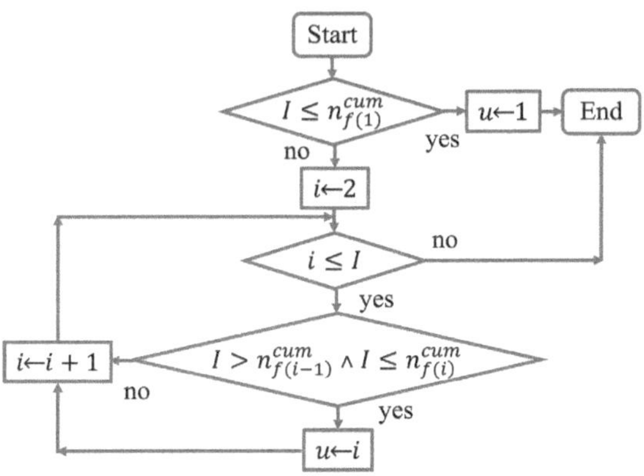

Fig. 2.18 Front selection (NSGA-II): Algorithm to determine the front to be included in the elite population.

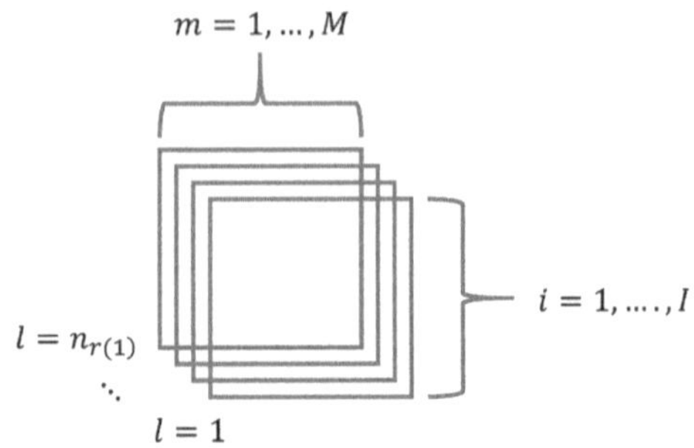

Fig. 2.19 Cluster array (SPEA): Illustration of a cluster representation.

archive do not exceed the limit (Q), it will remain untouched. It can be observed that the archive size is variable through GA evolution (Deb, 2001).

Now, our attention will be centered on the implementation of the clustering algorithm. Fig. 2.19 shows the structure ($\mathcal{L}_O$) used in this work to represent the clusters of individuals. The table $\mathcal{L}_O$ has three indices, the first for each individual (i), the second for each objective function, and the third for the cluster (ℓ). In addition to the table $\mathcal{L}_O$, another list ($\mathcal{L}_P$) is required to store the number of individuals included in each cluster. Thus, both structures ($\mathcal{L}_O$ and $\mathcal{L}_P$) can be used to find a determined individual in a specific cluster.

The procedure required to reduce the number of individuals in the archive through clustering is briefly explained as follows (Deb, 2001):

Step 1: Fill the elements of the table $\mathcal{L}_O(i, m, \ell)$ with all individuals with rank one of the joint population. Each index ℓ corresponds to one individual of the joint population with rank one, that is why the table $\mathcal{L}_O$ only have $n_r(1)$ elements for the index ℓ.

Step 2: As each cluster (ℓ) has only one individual, each element in the list $\mathcal{L}_P$ is set to one. ($\mathcal{L}_O(\ell) \leftarrow 1 \forall \ell = 1, \ldots, n_r(1)$).

Step 3: Create a table ($\mathcal{L}_Q$) with three columns. For each pair of clusters (a, b), calculate the average Euclidean distance among all the individuals in both clusters. The first two columns are the pair of individuals, and the third one will be the average Euclidean distance.

Step 4: Sort the table $\mathcal{L}_Q$ according to the third column (average Euclidean distance) in ascending order.

Step 5: Using the sorted table from Step 4, select the pair of clusters with the smallest average Euclidean distance.

Step 6: Join or merge the pair of clusters selected (a, b) of Step 5 into one. Take the individuals in the table $\mathcal{L}_O(i, m, a)$ and $\mathcal{L}_O(i, m, b)$ and join them into one table.

Step 7: Assign the joint table from Step 6 to the position of cluster $a(\mathcal{L}_O(i, m, a) \leftarrow \mathcal{L}_O(i, m, a) \cup \mathcal{L}_O(i, m, b))$ and delete the elements in cluster $b(\mathcal{L}_O(i, m, b) \leftarrow \emptyset)$.

This procedure sequentially reduces the number of clusters; it has to be applied until the number of clusters becomes equal to Q. Once the number of clusters is equal to Q, it is necessary to choose only one individual per cluster. In those clusters with more than one individual, the centroid of the cluster is calculated. Then, the Euclidean distance between all of the individuals in the cluster and the centroid is estimated. Finally, the individual closest to the centroid is selected, while the remaining individuals are discarded. The procedure just explained results in an archive with Q individuals. Thus, fitness is assigned to the population and archive individuals. Both population and archive are joined into one set, and tournament selection is applied to obtain the mating pool. Finally, crossover and mutation operators are applied to obtain the population of the next generation.

SPEA2 starts by initializing the population using random individuals, while the archive is assumed to be empty. Then, the fitness of all individuals in the population and the archive are calculated. Join the population and archive into a single set and find the nondominated individuals. If the number of nondominated individuals is equal to Q, this will be the archive for the next generation. If the number of nondominated individuals is lower than Q, individuals in subsequent ranks are selected to complete the population. The selection process can be implemented by sorting the joint population according to the fitness values in ascending order and selecting the first Q individuals. Finally, if the number of nondominated individuals is higher than the allowed size (Q), a truncation procedure is applied. The truncation process is based on evaluating Eq. (2.38) (Zitzler et al., 2001).

$$\begin{aligned} &\forall 1 \le k \le n_r(1) : \sigma_{(ik)}^{\text{sgh}} = \sigma_{(jk)}^{\text{sgh}} \\ &\vee \\ &\exists 1 \le h < n_r(1) : \left[\left(\forall 1 \le o < h : \sigma_{(io)}^{\text{sgh}} = \sigma_{(jo)}^{\text{sgh}} \right) \wedge \sigma_{(ih)}^{\text{sgh}} < \sigma_{(jh)}^{\text{sgh}} \right] \end{aligned} \tag{2.38}$$

The procedure consists of discarding all of the individuals for which the condition of Eq. (2.38) is fulfilled. As can be observed, the basic concept behind Eq. (2.38) is the distance sought ($\sigma_{(i,k)}^{\text{sgh}}$), calculated for a wide number of intervals (k). For a determined individual i, distance sought can be determined using the algorithm shown in Fig. 2.20.

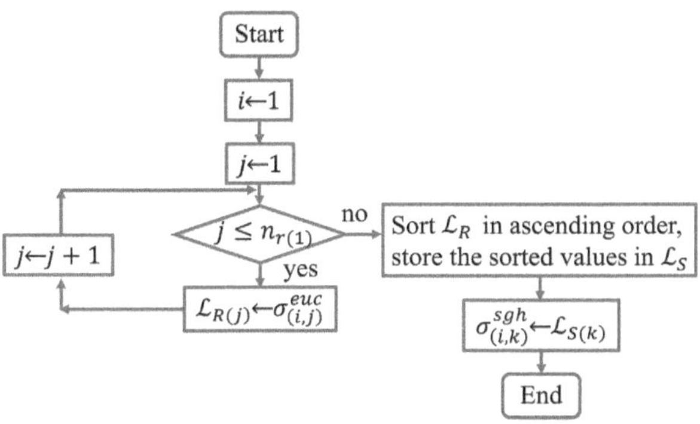

Fig. 2.20 Distance sought calculation: Algorithm to calculate the distance sought.

Eq. (2.38) discards all the individuals with identical distance sought, or those with equal distance sought until a specific value (o) and smaller distance sought for higher ones (Zitzler et al., 2004). Essentially, the truncation procedure consists of sequentially deleting that individual with the minimum distance to another individual. This idea has been implemented in this work by selecting the individual with the smallest distance sought in one step ($k=1$). This reasoning is an important simplification of the original formulation. It can be observed that the archive size is fixed through evolution of the GA, which is an important difference between SPEA and its improved version SPEA2. Once the archive has been fulfilled, tournament selection is applied over the archive, obtaining the mating pool. Thus, crossover and mutation operators are applied, resulting in the population of the next generation.

So far, the authors have presented the basic concepts behind a GA's working principles. In the next sections, some illustrative cases are discussed.

2.5 Case studies

In the previous sections, the basic concepts related to GA operation for MoOO and MuOO have been briefly presented. In this section, all of these concepts are illustrated through some general-purpose examples. The first case studies a binary-coded GA for MoOO; the second case studies a real-coded GA for MoOO, and the third case studies a real-coded GA for MuOO. The computer used to implement these examples was provided with i7-3630QM CPU at 2.40GHz, 8GB of RAM, and 64-bit operating system. In addition, the computing language used was MATLAB®.

2.5.1 Feature selection

In the introduction, the application and importance of GA to solve feature selection problems was discussed. Thus, this section analysis a feature-selection problem taken from Leardi et al. (1992). The optimization problem consists of finding out the combination of variables that minimizes the squared error related to the cross-validated variance with leave-one-out technique, using multiple linear regression. The data used in this example is shown in Table 2.1, where 11 variables with 16 objects describe the output variable or the response.

Binary-coded GA with elitism was implemented considering the proportional selection and the uniform-random mutation operator. Regarding the settings, the population size and number of generations were set as 100 and 250, respectively. The crossover rate was set to 0.8, and the mutation rate was set to 0.1.

Fig. 2.21 shows GA evolution minimizing squared error. The DM should select variables 1, 7, 10, and 11. These results are similar to those reported in the original reference (Leardi et al., 1992). The analysis of the data provided in Table 2.1 can be simplified using the four variables above obtained from the GA optimization process. About the computational time, the whole process took 2.09min.

Table 2.1 Data for feature selection.

Object	1	2	3	4	5	6	7	8	9	10	11	Response
1	5	0	15	−1	0	75	−5	0	0	−15	25	123
2	5	4	15	1	20	75	5	60	4	15	25	406
3	15	4	15	−1	60	225	−15	60	−4	−15	225	1400
4	5	0	160	1	0	800	5	0	0	160	25	352
5	15	0	160	−1	0	2400	−15	0	0	−160	225	1740
6	5	4	160	−1	20	800	−5	640	−4	−160	25	600
7	15	4	160	1	60	2400	15	640	4	160	225	1820
8	25	0	15	−1	0	375	−25	0	0	−15	625	1300
9	25	4	15	1	100	375	25	60	4	15	625	2510
10	25	0	160	1	0	4000	25	0	0	160	625	1950
11	25	4	160	−1	100	4000	−25	640	−4	−160	625	1270
12	2	0	15	1	0	30	2	0	0	15	4	179
13	7	0	15	−1	0	105	−7	0	0	−15	49	420
14	7	0	15	−1	0	105	−7	0	0	−15	49	662
15	2	0	160	−1	0	320	−2	0	0	−160	4	166
16	7	0	160	1	0	1120	7	0	0	160	49	1110

Information of feature-selection case study.

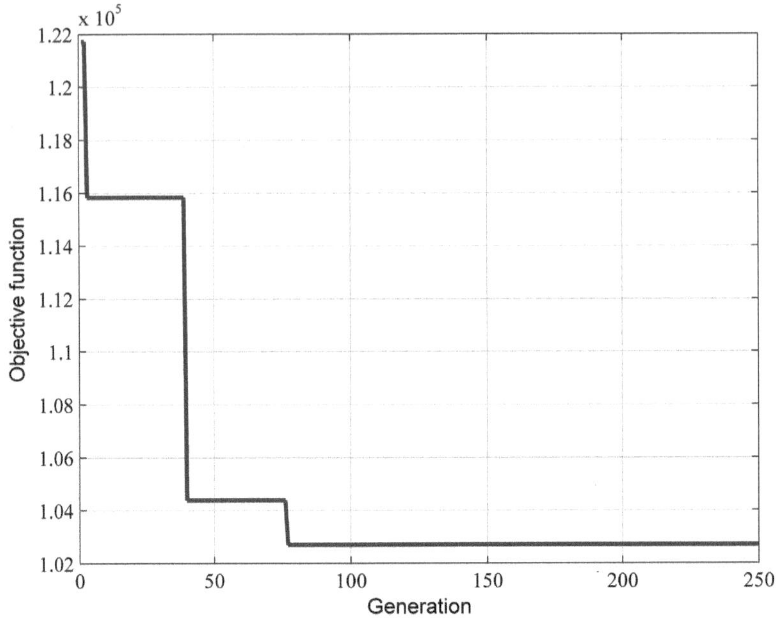

Fig. 2.21 Feature Selection example: Evolution of GA for the feature selection example.

2.5.2 Test functions for mono-objective optimization

The performance of real-coded GA for MoOO will be discussed using the functions suggested by the organizers of the congress on evolutionary computation (CEC) of the Institute of Electrical and Electronics Engineers (IEEE) of the year 2005 (IEEE CEC-2005). The CEC-2005 test suit has 25 functions with different characteristics. Tables 2.2–2.4 show the name of each function; their specific description will not be reported in this document, the reader can find this information in Suganthan et al. (2005), while the implementation in MATLAB® can be found in Suganthan (2020). Thus, real-coded GA described in the previous sections was implemented in MATLAB® to optimize CEC-2005 test functions with a dimension of 10 ($V=10$).

GA implementation was carried out using an elitist approach, adopting proportionate selection, SBX, and uniform-random mutation. The population size was set to 1500 individuals; the number of generations was set to 500, the crossover rate and the crossover distribution index were set to 0.95 and 1, respectively. The mutation rate was set to 0.01. In a general sense, GA requires a large population size and many generations to exploit its optimization potential, which is why each of these parameters was set to a high value.

The GA evolution for each problem is shown in Figs. 2.22–2.25, while Tables 2.5–2.8 show the optimization results, the optimal objective function per problem, and the results reported in Ballester et al. (2005) using an improved version of GA known as SPC-PNX.

Table 2.2 Unimodal functions.

Number	Function name
1	Shifted Sphere Function
2	Shifted Schwefel's Problem 1.2
3	Shifted Rotated High Conditioned Elliptic Function
4	Shifted Schwefel's Problem 1.2 with Noise in Fitness
5	Schwefel's Problem 2.6 with Global Optimum on Bounds

Description of the CEC-2005 test functions (unimodal functions).

Table 2.3 Basic and expanded functions.

Number	Function name
6	Shifted Rosenbrock's Function
7	Shifted Rotated Griewank's Function without Bounds
8	Shifted Rotated Ackley's Function with Global Optimum on Bounds
9	Shifted Rastrigin's Function
10	Shifted Rotated Rastrigin's Function
11	Shifted Rotated Weierstrass Function
12	Schwefel's Problem 2.13
13	Expanded Extended Griewank's plus Rosenbrock's Function (F8F2)
14	Shifted Rotated Expanded Schaffer's F6

Description of the CEC-2005 test functions (basic and expanded functions).

Table 2.4 Hybrid composition functions.

Number	Function name
15	Hybrid Composition Function
16	Rotated Hybrid Composition Function
17	Rotated Hybrid Composition Function with Noise in Fitness
18	Rotated Hybrid Composition Function
19	Rotated Hybrid Composition Function with a Narrow Basin for the Global Optimum
20	Rotated Hybrid Composition Function with the Global Optimum on the Bounds
21	Rotated Hybrid Composition Function
22	Rotated Hybrid Composition Function with High Condition Number Matrix
23	Noncontinuous Rotated Hybrid Composition Function
24	Rotated Hybrid Composition Function
25	Rotated Hybrid Composition Function without Bounds

Description of the CEC-2005 test functions (hybrid composition functions).

Table 2.5 Results unimodal functions.

Function	GA	Optimal	Error	Error (Ballester et al., 2005)
1	−450	−450	0	0
2	−449.82	−450	0.18	0
3	99,545.19	−450	99,995.19	[774.57, 352,160]
4	−448.71	−450	1.29	0
5	−307.24	−310	2.76	0

Optimization results for unimodal functions.

In a general sense, the IEEE CEC-2005 test suite posed challenging problems with a high computational burden; in this regard, the computational time was between 11.05 and 152.42 min. According to the obtained results, the implemented GA offers modest results for unimodal (Table 2.5), single (Table 2.6), and expanded (Table 2.7) functions. However, this is not the case for the hybrid composition functions (Table 2.8) in which large errors were observed.

The original report of the IEEE CEC-2005 test suite not only presents the mathematical formulation of each function but also describes some guides to assess the performance of a determined algorithm. The IEEE CEC-2005 organizers recommend evaluating the GA 25 times and selecting the 1st, 7th, 13th, 19th, and 25th, as well as the mean and standard deviation. For the number of variables selected ($V=10$), the maximum number of objective function evaluations allowed was 100,000. In Tables 2.5–2.8 the smallest (1st) and maximum (25th) errors were added for a qualitative comparison. It is a qualitative comparison because the general settings used here and in (Ballester et al., 2005) are not the same, and the random initialization was not carried out using the same numbers. For these reasons, it is not possible to carry out a quantitative comparison.

2.5.3 Test functions for multiobjective optimization

The performance of MuOO previously explained is going to be illustrated using some of the test problems proposed by Zitzler, Deb, and Thiele (ZDT) (Zitzler et al., 2000). Five of the six problems of this test suite have been selected; the first problem (ZDT-1) was solved using MOGA, the second problem (ZDT-2) was solved using NSGA, the third problem (ZDT-3) was solved using NSGA-II, the fourth problem (ZDT-4) was solved using SPEA, while the sixth problem (ZDT-6) was solved using SPEA2. In a general sense, ZDT problems have two objective functions ($M=2$) to be minimized. The first objective function only depends on the first variable ($v=1$) of the corresponding individual ($\boldsymbol{x}_{\mathbf{pop}}(i)$) under evaluation, while the second function can be obtained by evaluating two functions: y_{aux}^1 and y_{aux}^2. Furthermore, the Pareto-optimal front is reached when the function y_{aux}^1 is equal to one.

GA settings were similar for all techniques. The population size was set to 100 individuals ($I=100$), and the generations to 450, while the crossover and mutation rates were set to 0.95 ($p_C=0.95$) and 0.01 ($p_m=0.01$), respectively. The archive sizes in

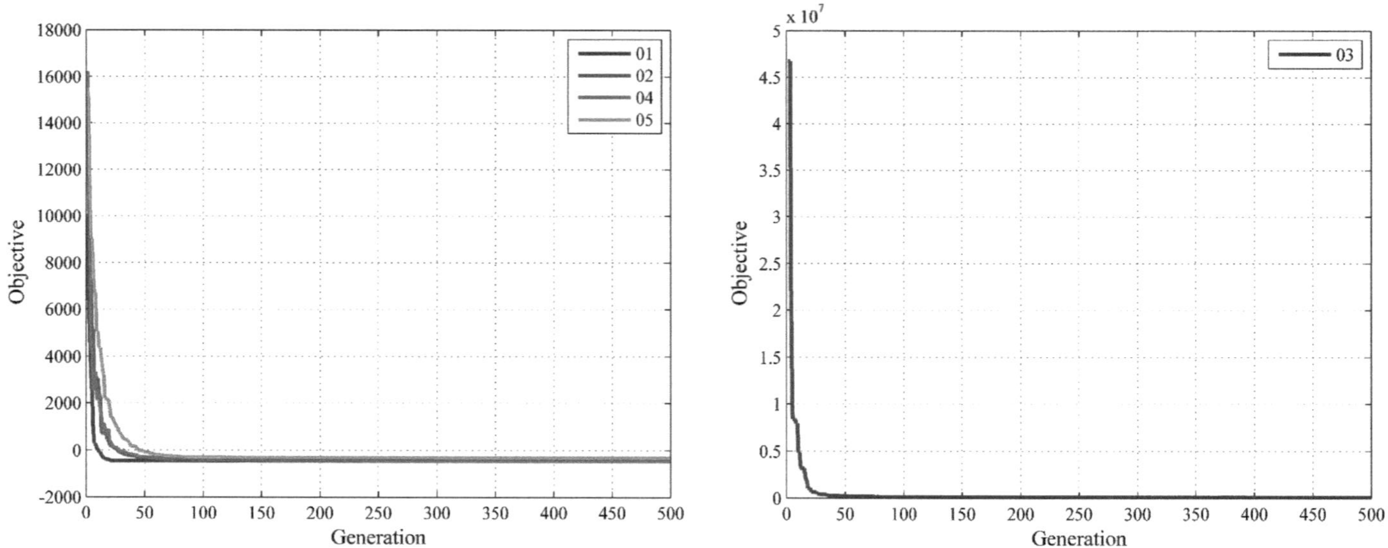

Fig. 2.22 GA convergence of problems: Evolution of GA for CEC problems 1–5.

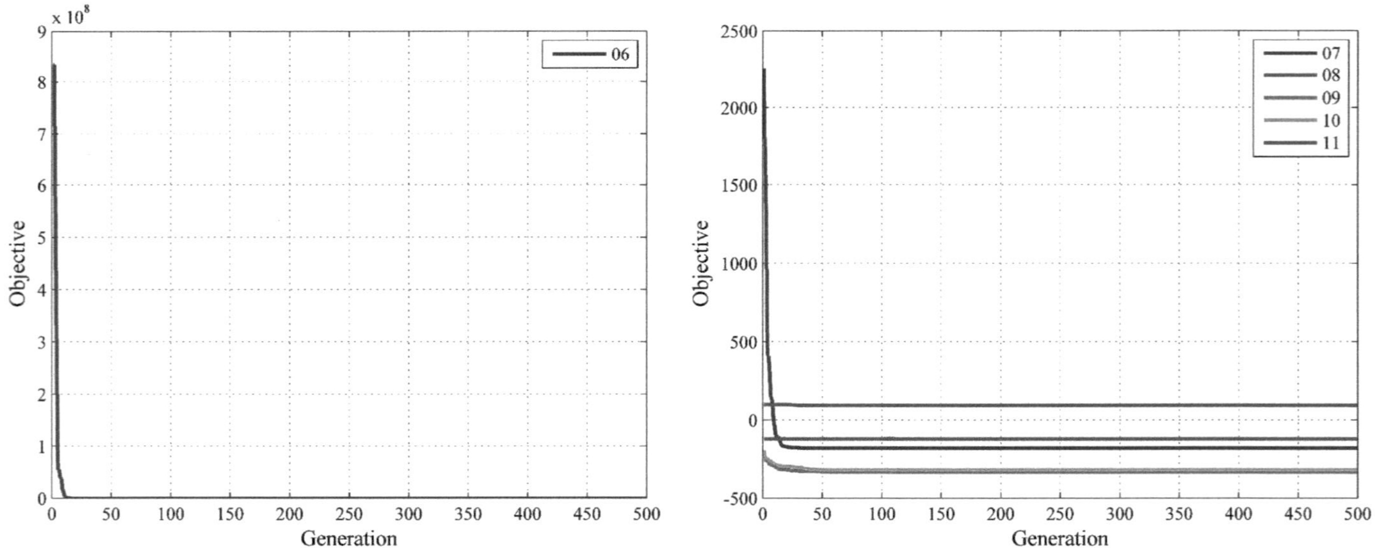

Fig. 2.23 GA convergence of problems: Evolution of GA for CEC problems 6–11.

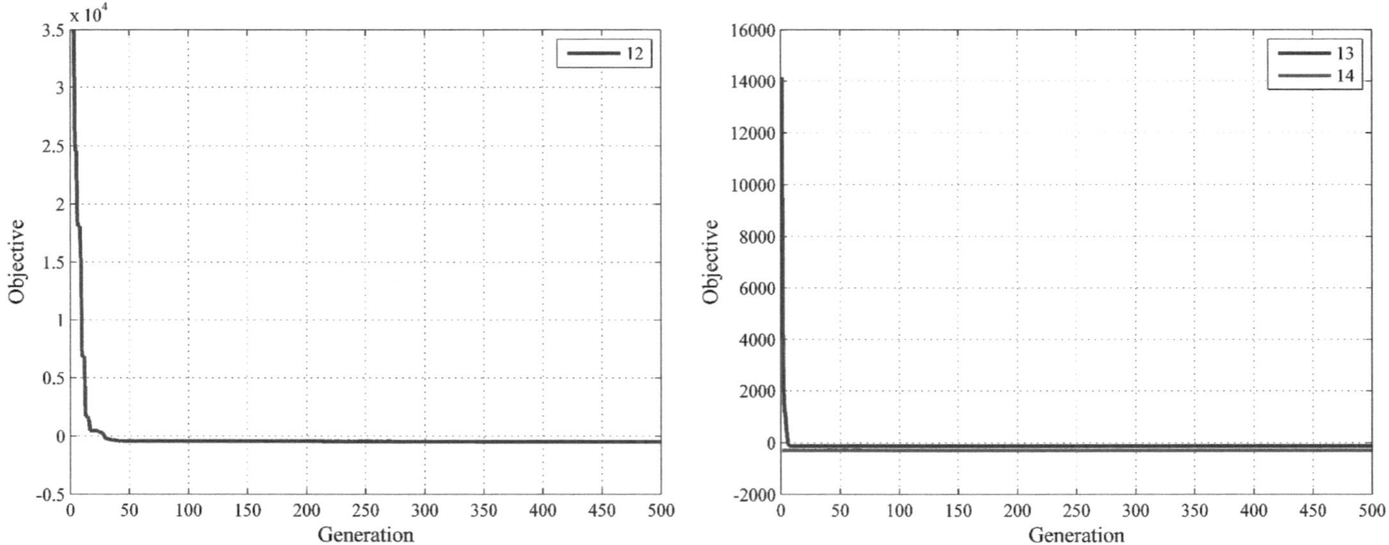

Fig. 2.24 GA convergence of problems: Evolution of GA for CEC problems 12–14.

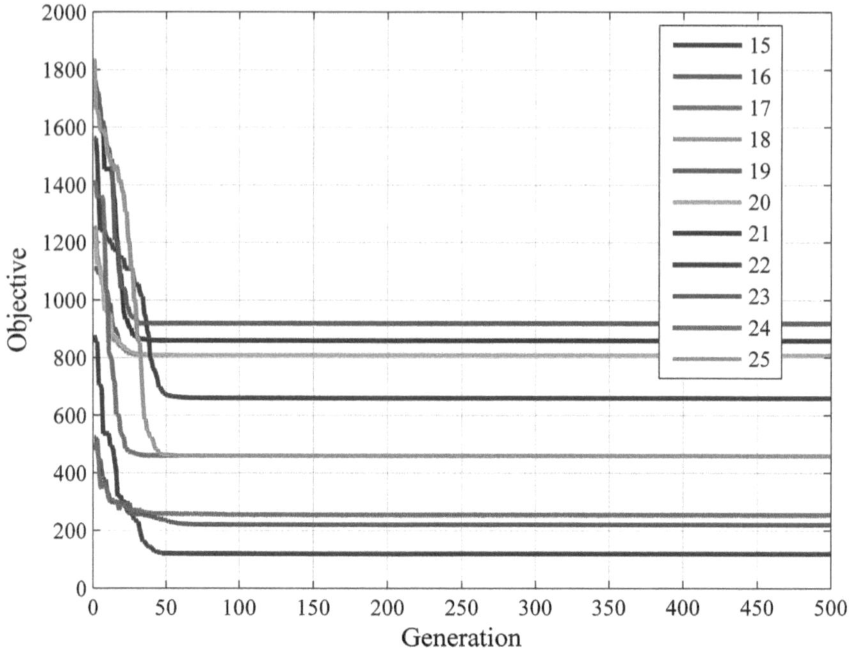

Fig. 2.25 GA convergence of problems: Evolution of GA for CEC problems 15–25.

SPEA and SPEA2 algorithms were set to 20 individuals ($Q=20$). Regarding genetic operators, SUS was adopted in all problems; its implementation in MATLAB® was taken from Chipperfield et al. (1993). SBX-operator was also implemented in MATLAB®, considering a distribution index equal to 15 ($n_c=15$). The polynomial mutation was used in problems ZDT-1, ZDT-3, and ZDT-6, considering a distribution index equal to 20 ($n_m=20$), while the uniform-random mutation was used in problems ZDT-2 and ZDT-4.

The formulation of the ZDT-1 problem is presented in Eq. (2.39). In this problem, the decision variable has 30 elements ($V=30$) limited in the interval [0, 1].

Table 2.6 Results single functions.

Function	GA	Optimal	Error	Error (Ballester et al., 2005)
6	394.82	390	4.82	[0.02, 148.30]
7	−179.89	−180	0.11	[0.01, 0.25]
8	−119.82	−140	20.18	[20.81, 21.07]
9	−330	−330	0	[0.99, 11.94]
10	−318.06	−330	11.94	[1.99, 27.39]
11	94.1	90	4.1	[0.00, 4.49]
12	−456.32	−460	3.68	[3.91, 1558.30]

Optimization results for single functions.

Table 2.7 Results expanded functions.

Function	GA	Optimal	Error	Error (Ballester et al., 2005)
13	−129.97	−130	0.03	[0.35, 1.32]
14	−297.34	−300	2.66	[1.39, 3.61]

Optimization results for expanded functions.

$$y_{\mathrm{obj}}(i,1) = x_{\mathrm{pop}}(i,1) y_{\mathrm{aux}}^{1} = 1 + 9\left\{\sum_{v=2}^{V} x_{\mathrm{pop}}(i,v) \Big/ (V-1)\right\} y_{\mathrm{aux}}^{2}$$

$$= 1 - \sqrt{x_{\mathrm{pop}}(i,1) \Big/ y_{\mathrm{aux}}^{1}}\; y_{\mathrm{obj}}(i,2) = y_{\mathrm{aux}}^{1} y_{\mathrm{aux}}^{2} \tag{2.39}$$

The Pareto front obtained from MOGA implementation is shown in Fig. 2.26. In this case, MOGA got the Pareto-optimal front, which is convex. The computational time spent was 1.28 min.

The formulation of the ZDT-2 problem is presented in Eq. (2.40). It has a similar structure to the ZDT-1 problem, except for the function y_{aux}^{2}, which is modified, obtaining a nonconvex Pareto-optimal front. In this case, the characteristics of decision variables are the same as the ZDT-1 problem.

$$y_{\mathrm{obj}}(i,1) = x_{\mathrm{pop}}(i,1) y_{\mathrm{aux}}^{1} = 1 + 9\left\{\sum_{v=2}^{V} x_{\mathrm{pop}}(i,v) \Big/ (V-1)\right\} y_{\mathrm{aux}}^{2}$$

$$= 1 - \left(x_{\mathrm{pop}}(i,1) \Big/ y_{\mathrm{aux}}^{1}\right)^{2} y_{\mathrm{obj}}(i,2) = y_{\mathrm{aux}}^{1} y_{\mathrm{aux}}^{2} \tag{2.40}$$

Fig. 2.27 shows the Pareto-optimal front for this case. The computational time spent was 1.04 min.

Table 2.8 Results hybrid composition functions.

Function	GA	Optimal	Error	Error (Ballester et al., 2005)
15	120	120	0	[63.16, 425.41]
16	220.59	120	100.59	[91.14, 139.53]
17	253.93	120	133.93	[98.92, 151.44]
18	810	10	800	[300.00, 901.46]
19	810	10	800	[300.00, 800.00]
20	810	10	800	[300.00, 800.00]
21	860	360	500	[300.00, 1173.60]
22	660	360	300	[300.01, 777.37]
23	919.47	360	559.47	[559.47, 970.50]
24	460	260	200	200
25	460	260	200	[405.57, 406.41]

Optimization results for hybrid composition functions.

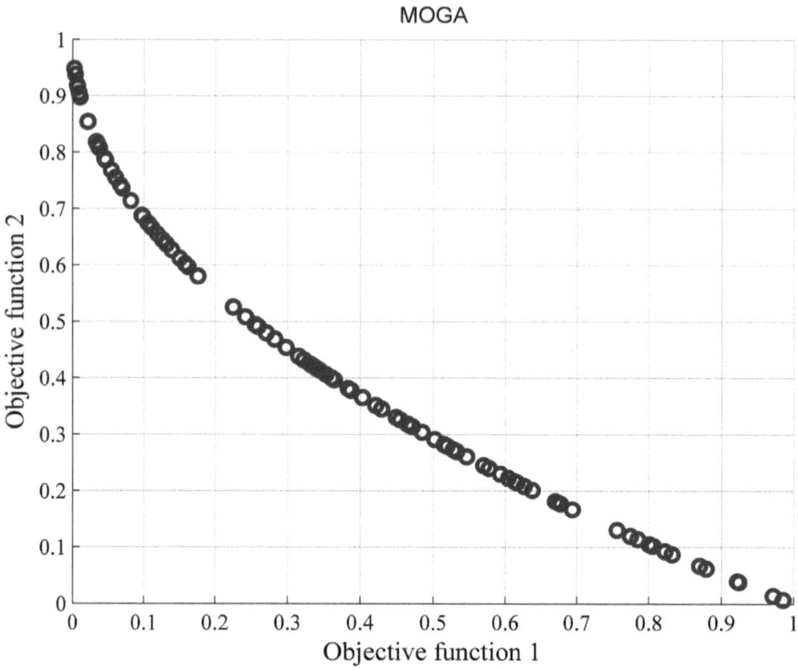

Fig. 2.26 Pareto front (MOGA ZDT-1): Evolution of GA for ZDT-1 problem.

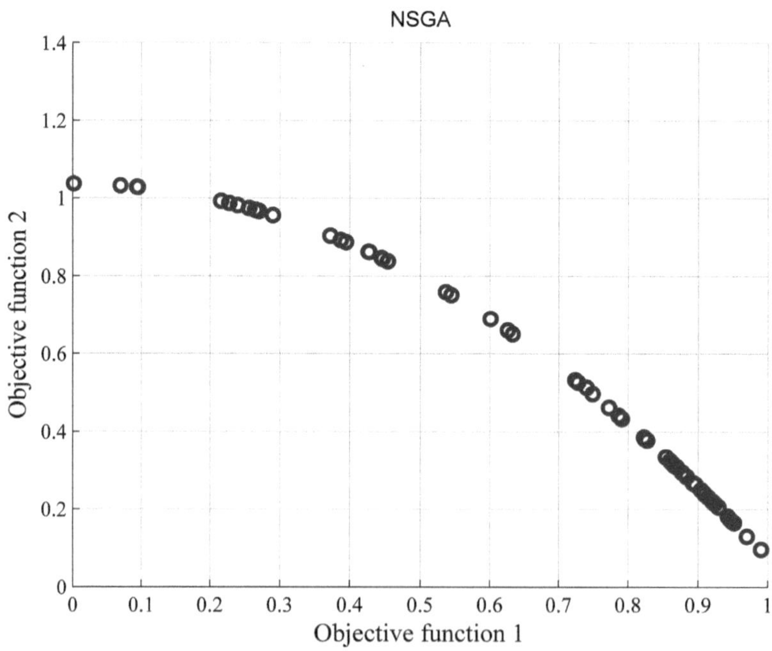

Fig. 2.27 Pareto front (NSGA ZDT-2): Evolution of GA for ZDT-2 problem.

The formulation of the ZDT-3 problem is shown in Eq. (2.41), where the function y^2_{aux} is modified by including the sinusoidal function, which makes the Pareto-optimal front discontinuous. As in the previous cases, $V=30$, and the decision variables are in the interval [0, 1].

$$y_{obj}(i,1)=x_{pop}(i,1)y^1_{aux}=1+9\left\{\sum\nolimits_{v=2}^{V} {x_{pop}(i,v)}\Big/_{(V-1)}\right\}y^2_{aux}$$

$$=1-\sqrt{{x_{pop}(i,1)}\Big/_{y^1_{aux}}}-\left({x_{pop}(i,1)}\Big/_{y^1_{aux}}\right)\sin\left(10\pi y_{obj}(i,1)\right)y_{obj}(i,2)$$

$$=y^1_{aux}y^2_{aux} \tag{2.41}$$

Fig. 2.28 shows the Pareto-optimal obtained using NSGA-II. The computational time spent was 1.97 min.

The structure of the ZDT-4 problem is presented in Eq. (2.42). This problem has been specifically designed to test the ability of GA to solve problems whose objective function is multimodal. This problem has 21^9 local Pareto-optimal fronts. The decision variable has 10 elements ($V=10$); the first element ($v=1$) is constrained in the interval [0, 1], while the remaining variables are limited in the interval [−5, 5].

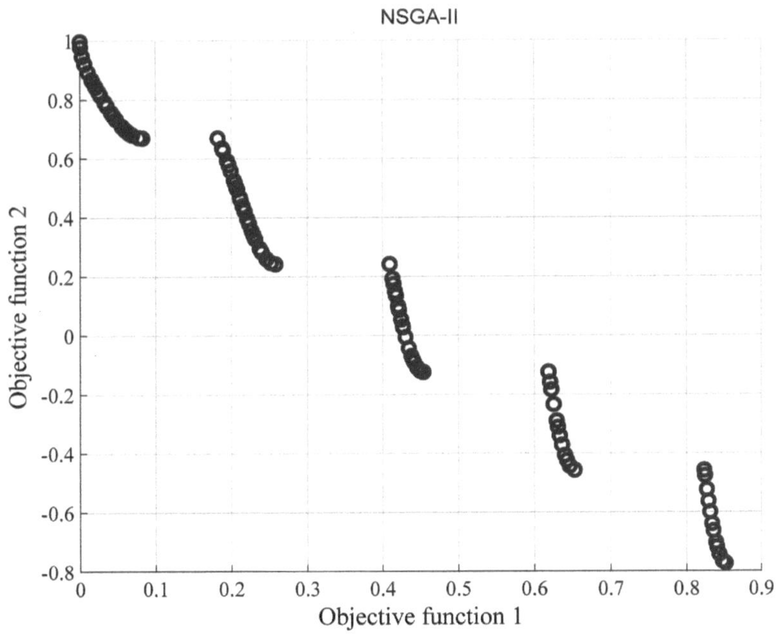

Fig. 2.28 Pareto front (NSGA-II ZDT-3): Evolution of GA for ZDT-3 problem.

$$y_{obj}(i,1) = x_{pop}(i,1)y_{aux}^1 = 1 + 10(V-1) + \sum_{v=2}^{V}\left\{\left(x_{pop}(i,v)\right)^2 - 10\cos\left(4\pi x_{pop}(i,v)\right)\right\}y_{aux}^2 = 1 - \sqrt{x_{pop}(i,1)/y_{aux}^1}\, y_{obj}(i,2) = y_{aux}^1 y_{aux}^2 \tag{2.42}$$

Fig. 2.29 shows the Pareto front obtained from the implementation of SPEA. This Pareto front is close to the optimal one due to the difficulty previously explained. The computational time spent was 45.68 min. Probably, this time consumption is related to the clustering algorithm and how this procedure deletes individuals from the archive. On the one hand, each individual is sequentially discarded from the archive, one at the time. On the other hand, as the algorithm evolves, the Pareto front gets close to the optimal one; in this situation, the number of individuals to be deleted increases and, consequently, so does the computational time.

The mathematical formulation of the ZDT-6 problem is shown in Eq. (2.43). This problem incorporates important difficulties related to the nonuniform behavior of the search space.

$$\begin{aligned} y_{obj}(i,1) &= 1 - \exp\left(-4x_{pop}(i,1)\right)\left(\sin\left(6\pi \times x_{pop}(i,1)\right)\right)^6 y_{aux}^1 \\ &= 1 + 9\left(\sum_{v=2}^{V} x_{pop}(i,v)/(V-1)\right)y_{aux}^2 \\ &= 1 - \left(x_{pop}(i,1)/y_{aux}^1\right)^2 y_{obj}(i,2) = y_{aux}^1 y_{aux}^2 \end{aligned} \tag{2.43}$$

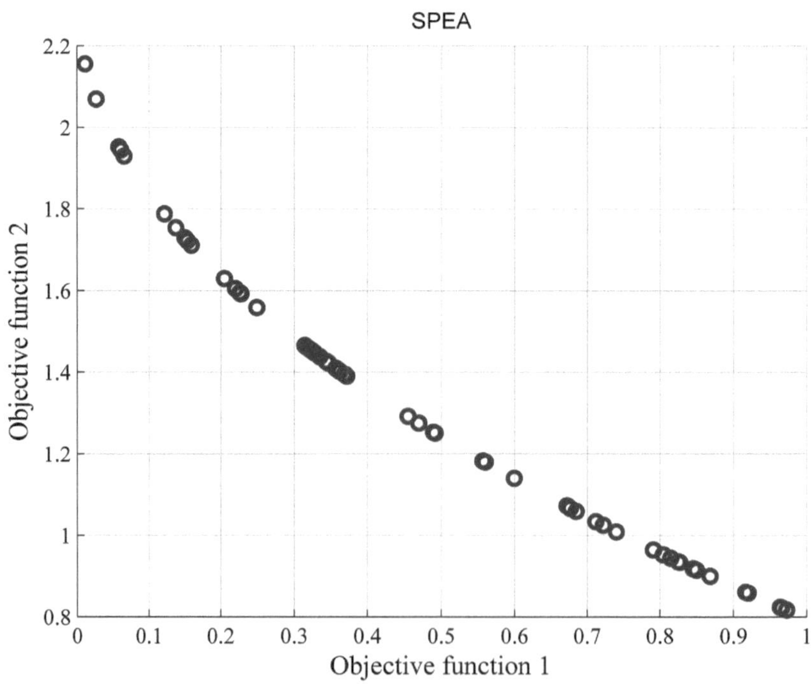

Fig. 2.29 Pareto front (SPEA ZDT-4): Evolution of GA for ZDT-4 problem.

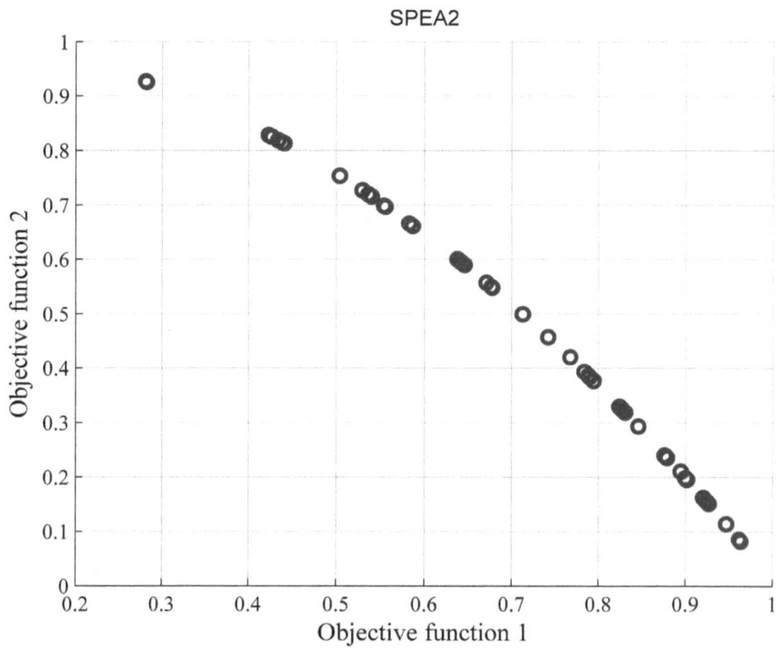

Fig. 2.30 Pareto front (SPEA2 ZDT-6): Evolution of GA for ZDT-6 problem.

Fig. 2.30 shows the Pareto-optimal front for this problem obtained from the implementation of SPEA2. The computational time spent was 8.2 min.

2.6 Other heuristic techniques

In a general sense, heuristic techniques have important industrial applications related to wind farm design and siting, parameter tuning of control devices, generation scheduling, and control of wind energy conversion systems. The authors have paid special attention to GA. However, according to the information presented in Chapter 1, this is not the only heuristic technique available in the literature. Heuristic techniques can be classified according to the field that inspires them. They can be classified as biology-based algorithms, physics-based algorithms, and geography-based algorithms. In the biology-based category are those algorithms inspired by the evolutionary processes and swarm behavior. Some examples are GA, evolutionary programming, evolutionary strategy, differential evolution, harmony search algorithm, artificial immune system, particle swarm optimization, bacteria foraging optimization, cuckoo search algorithm, artificial bee colony algorithm, ant colony optimization, coral reef optimization, teaching-learning based optimization, firefly algorithm, shuffled frog leaping algorithm, and pigeon inspired optimization. In the physics-based category are the simulated annealing algorithm, gravitational search algorithm, chaotic optimization algorithm, and intelligent water drops algorithm. In the geography-based category are the tabu search algorithm and the imperialistic competition algorithm (Behera et al., 2015).

In the specific case of GA, it has been incorporated in many commercial tools to be applied by power systems engineers; a representative example is etap software (etap, 2020), which uses a GA for optimal capacitor placement in electricity distribution systems. Another example developed by some of the authors is iHOGA software (iHOGA, 2020) for sizing and management of energy systems based on renewable energies.

2.7 Conclusions

The authors have presented a comprehensive picture of the basic concepts, rudiments, and classical methods for the implementation of a GA. Historical evolution and some popular applications were discussed, the general structure of a GA was described, including the nonelitist and elitist approaches, and the main differences between MoOO, MuOO, and MaOO algorithms were explained. In addition, the role and benefits of HPC on the implementation of a GA were discussed. Giving special attention to MuOO algorithms, popular techniques such as MOGA, NSGA, NSGA-II, SPEA, and SPEA2 were described. Thus, the reader can understand the evolution of these methodologies through time. Classical GA implementation was illustrated using three general-purpose case studies. The first case was based on the feature selection problem, and was included to demonstrate the operation of a binary- or integer-coded GA. The second case was the IEEE CEC-2005 test suite, included to evaluate the performance of the classical real-coded GA for MoOO. The third case was the ZDT test problems used to assess the performance of MuOO. All of these tests account for 31 optimization problems, providing a complete overview of GA capabilities and limitations. According to the obtained results, GA is able to solve a vast family of problems, experiencing some difficulties with hybrid composition functions (IEEE CEC-2005 test suite) and multimodal two objectives functions (ZDT-4).

References

Asafuddoula, M., Ray, T., & Sarker, R. (2015). A decomposition-based evolutionary algorithm for many objective optimization. *IEEE Transactions on Evolutionary Computation*, *19*(3), 445–460. https://doi.org/10.1109/TEVC.2014.2339823.

Bader, J., & Zitzler, E. (2011). HypE: An algorithm for fast hypervolume-based many-objective optimization. *Evolutionary Computation*, *19*(1), 45–76. https://doi.org/10.1162/EVCO_a_00009.

Baker, J. (1985). In J. Grefenstette (Ed.), *Adaptive selection methods for genetic algorithms* (pp. 101–111). L. Erlbaum Associates Inc.

Ballester, P., Stephenson, J., Carter, J., & Gallagher, K. (2005). *Real-parameter optimization performance study on the CEC-2005 benchmark with SPC-PNX. Vol. 1* (pp. 498–505). IEEE. https://doi.org/10.1109/CEC.2005.1554724.

Behera, S., Sahoo, S., & Pati, B. (2015). A review on optimization algorithms and application to wind energy integration to grid. *Renewable and Sustainable Energy Reviews*, *48*, 214–227. https://doi.org/10.1016/j.rser.2015.03.066.

Beume, N., Naujoks, B., & Emmerich, M. (2007). SMS-EMOA: Multiobjective selection based on dominated hypervolume. *European Journal of Operational Research, 181*(3), 1653–1669. https://doi.org/10.1016/j.ejor.2006.08.008.

Bezerra, L., López-Ibáñez, M., & Stützle, T. (2017). In H. Trautmann, et al. (Eds.), *An empirical assessment of the properties of inverted generational distance on multi- and many-objective optimization*. Cham: Springer. https://doi.org/10.1007/978-3-319-54157-0_3.

Bringmann, K., Friedrich, T., Neumann, F., & Wagner, M. (2011). In T. Walsh (Ed.), *Approximation-guided evolutionary multi-objective optimization* (pp. 1198–1203). AAAI Press. https://www.ijcai.org/Proceedings/11/Papers/204.pdf.

Brockhoff, D., Wagner, T., & Trautmann, H. (2012). In T. Soule (Ed.), *On the properties of the R2 indicator* (pp. 465–472). Association for Computing Machinery. https://doi.org/10.1145/2330163.2330230.

Cantú-Paz, E. (1999). Designing efficient and accurate parallel genetic algorithms. In *Technical report no. 99017* The University of Illinois at Urbana-Champaign.

Cao, B., Zhao, J., Lv, Z., Liu, X., Yang, S., Kang, X., & Kang, K. (2017). Distributed parallel particle swarm optimization for multi-objective and many-objective large-scale optimization. *IEEE Access, 5*, 8214–8221. https://doi.org/10.1109/ACCESS.2017.2702561.

Chen, B., Zeng, W., Lin, Y., & Zhang, D. (2015). A new local search-based multiobjective optimization algorithm. *IEEE Transactions on Evolutionary Computation, 19*(1), 50–73. https://doi.org/10.1109/TEVC.2014.2301794.

Cheng, J. R., & Gen, M. (2019). Accelerating genetic algorithms with GPU computing: A selective overview. *Computers & Industrial Engineering, 128*, 514–525. https://doi.org/10.1016/j.cie.2018.12.067.

Cheng, J., Grossman, M., & McKercher, T. (2014). *Professional CUDA C programming* (1st ed.). (Wrox) John Wiley & Sons. https://www.wiley.com/en-us/Professional+CUDA+C+Programming-p-9781118739327.

Cheng, R., Jin, Y., Olhofer, M., & Sendhoff, B. (2016). A reference vector guided evolutionary algorithm for many-objective optimization. *IEEE Transactions on Evolutionary Computation, 20*(5), 773–791. https://doi.org/10.1109/TEVC.2016.2519378.

Cheng, J., Yen, G., & Zhang, G. (2015). A many-objective evolutionary algorithm with enhanced mating and environmental selections. *IEEE Transactions on Evolutionary Computation, 19*(4), 592–605. https://doi.org/10.1109/TEVC.2015.2424921.

Chiang, C. (2005). Improved genetic algorithm for power economic dispatch of units with valve-point effects and multiple fuels. *IEEE Transactions on Power Systems, 20*(4), 1690–1699. https://doi.org/10.1109/TPWRS.2005.857924.

Chipperfield, A., Fleming, P., Pohlheim, H., & Fonseca, C. (1993). *Genetic algorithm toolbox for use with MATLAB*. Department of Automatic Control and Systems Engineering University of Sheffield.

Coit, D., & Smith, A. (1996). Reliability optimization of series-parallel systems using a genetic algorithm. *IEEE Transactions on Reliability, 45*(2), 254–260. https://doi.org/10.1109/24.510811.

Corne, D., Jerram, N., Knowles, J., & Oates, M. (2001). In L. Spector, E. Goodman, A. Wu, W. Langdon, & H.-M. Voigt (Eds.), *PESA-II: Region-based selection in evolutionary multiobjective optimization* (pp. 283–290). Morgan Kaufmann Publishers Inc. https://doi.org/10.5555/2955239.2955289.

Corne, D., Knowles, J., & Oates, M. (2000). In M. Schoenauer, et al. (Eds.), *The Pareto envelope-based selection algorithm for multiobjective optimization* (pp. 839–848). Springer. https://doi.org/10.1007/3-540-45356-3_82.

Deaven, D., & Ho, K. (1995). Molecular geometry optimization with a genetic algorithm. *Physical Review Letters*, *75*(2), 288–291. https://doi.org/10.1103/PhysRevLett.75.288.

Deb, K. (2000). An efficient constraint handling method for genetic algorithms. *Computer Methods in Applied Mechanics and Engineering*, *186*, 311–338. https://doi.org/10.1016/S0045-7825(99)00389-8.

Deb, K. (2001). *Multi-objective optimization using evolutionary algorithms* (1st ed.). John Wiley & Sons Ltd. https://www.wiley.com/en-us/Multi+Objective+Optimization+using+Evolutionary+Algorithms-p-9780471873396.

Deb, K., & Agrawal, R. (1995). Simulated binary crossover for continuous search space. *Complex Systems*, *9*(8), 115–148. https://www.complex-systems.com/abstracts/v09_i02_a02/.

Deb, K., & Jain, H. (2014). An evolutionary many-objective optimization algorithm using reference-point-based nondominated sorting approach, part I: Solving problems with box constraints. *IEEE Transactions on Evolutionary Computation*, *18*(4), 577–601. https://doi.org/10.1109/TEVC.2013.2281535.

Deb, K., Mohan, M., & Mishra, S. (2003). In C. Fonseca, P. Fleming, E. Zitzler, L. Thiele, & K. Deb (Eds.), *Towards a quick computation of well-spread Pareto-optimal solutions* (pp. 222–236). Springer. https://doi.org/10.1007/3-540-36970-8_16.

Deb, K., Pratap, A., Agarwal, S., & Meyarivan, T. (2002). A fast and elitist multiobjective genetic algorithm: NSGA-II. *IEEE Transactions on Evolutionary Computation*, *6*(2), 182–197. https://doi.org/10.1109/4235.996017.

etap. (2020). https://etap.com/.

Fonseca, C., & Flemming, P. (1993). In S. Forrest (Ed.), *Genetic algorithms for multiobjective optimization: Formulation, discussion and generalization* (pp. 416–423). Morgan Kaufmann Publishers Inc.

Goldberg, D. (1989). *Genetic algorithms in search, optimization, and machine learning* (1st ed.). Addison-Wesley.

Hajela, P., & Lin, C.-Y. (1992). Genetic search strategies in multi-criterion optimal design. *Structural Optimization*, *4*(2), 99–107. https://doi.org/10.1007/BF01759923.

Hartwell, L., Hood, L., Goldberg, M., Reynolds, A., & Silver, L. (2010). *Genetics: From genes to genomes* (4th ed.). McGraw-Hill Education.

He, Z., & Yen, G. (2016). Many-objective evolutionary algorithm: Objective space reduction and diversity improvement. *IEEE Transactions on Evolutionary Computation*, *20*(1), 145–160. https://doi.org/10.1109/TEVC.2015.2433266.

Hernández, R., & Coello, C. (2015). In S. Silva (Ed.), *Improved metaheuristic based on the R2 indicator for many-objective optimization* (pp. 679–686). Association for Computing Machinery. https://doi.org/10.1145/2739480.2754776.

Herreda, F., Lozano, M., & Verdegay, J. (1998). Tackling real-coded genetic algorithms: Operators and tools for behavioural analysis. *Artificial Intelligence Review*, *12*, 265–319. https://doi.org/10.1023/A:1006504901164.

Holland, J. (1975). *Adaptation in natural and artificial systems: An introductory analysis with applications to biology, control, and artificial intelligence* (1st ed.). University of Michigan Press.

Horn, J., Nafpliotis, N., & Goldberg, D. (1994). *A niched Pareto genetic algorithm for multiobjective optimization* (pp. 82–87). IEEE. https://doi.org/10.1109/ICEC.1994.350037.

Huang, C.-L., & Wang, C.-J. (2006). A GA-based feature selection and parameters optimization for support vector machines. *Expert Systems with Applications*, *31*(2), 231–240. https://doi.org/10.1016/j.eswa.2005.09.024.

Hughes, E. (2003). *Multiple single objective Pareto sampling* (pp. 2678–2684). IEEE. https://doi.org/10.1109/CEC.2003.1299427.

Hughes, E. (2007). *MSOPS-II: A general-purpose many-objective optimiser* (pp. 3944–3951). IEEE. https://doi.org/10.1109/CEC.2007.4424985.

iHOGA. (2020). https://ihoga.unizar.es/en/.

Ishibuchi, H., Nozaki, K., Yamamoto, N., & Tanaka, H. (1995). Selecting fuzzy if-then rules for classification problems using genetic algorithms. *IEEE Transactions on Fuzzy Systems*, *3* (3), 260–270. https://doi.org/10.1109/91.413232.

Jain, H., & Deb, K. (2014). An evolutionary many-objective optimization algorithm using reference-point based nondominated sorting approach, part II: Handling constraints and extending to an adaptive approach. *IEEE Transactions on Evolutionary Computation*, *18*(4), 602–622. https://doi.org/10.1109/TEVC.2013.2281534.

Jiang, S., & Yang, S. (2017). A strength Pareto evolutionary algorithm based on reference direction for multiobjective and many-objective optimization. *IEEE Transactions on Evolutionary Computation*, *21*(3), 329–346. https://doi.org/10.1109/TEVC.2016.2592479.

Jones, G., Willett, P., & Glen, R. (1995). Molecular recognition of receptor sites using a genetic algorithm with a description of desolvation. *Journal of Molecular Biology*, *245*(1), 43–53. https://doi.org/10.1016/S0022-2836(95)80037-9.

Jones, G., Willett, P., Glen, R., Leach, A., & Taylor, R. (1997). Development and validation of a genetic algorithm for flexible docking. *Journal of Molecular Biology*, *267*(3), 727–748. https://doi.org/10.1006/jmbi.1996.0897.

Kazarlis, S., Bakirtzis, A., & Pedridis, V. (1996). A genetic algorithm solution to the unit commitment problem. *IEEE Transactions on Power Systems*, *11*(1), 83–90. https://doi.org/10.1109/59.485989.

Knowles, J., & Corne, D. (2000). Approximating the non-dominated front using the Pareto archived evolution strategy. *Evolutionary Computation*, *8*(2), 149–172. https://doi.org/10.1162/106365600568167.

Kosakovsky, S., Posada, D., Gravenor, M., Woelk, C., & Frost, S. (2006a). Automated phylogenetic detection of recombination using a genetic algorithm. *Molecular Biology and Evolution*, *23*(10), 1891–1901. https://doi.org/10.1093/molbev/msl051.

Kosakovsky, S., Posada, D., Gravenor, M., Woelk, C., & Frost, S. (2006b). GARD: A genetic algorithm for recombination detection. *Bioinformatics*, *22*(24), 3096–3098. https://doi.org/10.1093/bioinformatics/btl474.

Krishna, K., & Murty, M. (1999). Genetic K-means algorithm. *IEEE Transactions on Systems, Man, and Cybernetics. Part B, Cybernetics*, *29*(3), 433–439. https://doi.org/10.1109/3477.764879.

Kursawe, F. (1990). In H. Schwefel, & R. Männer (Eds.), *Vol. 496. A variant of evolution strategies for vector optimization* (pp. 193–197). Springer.

Leardi, R., Boggia, R., & Terrile, M. (1992). Genetic algorithms as a strategy for feature selection. *Journal of Chemometrics*, *6*(5), 267–281. https://doi.org/10.1002/cem.1180060506.

Leardi, R., & González, A. (1998). Genetic algorithms applied to feature selection in PLS regression: How and when to use them. *Chemometrics and Intelligent Laboratory Systems*, *41*(2), 195–207. https://doi.org/10.1016/S0169-7439(98)00051-3.

Li, K., Deb, K., Zhang, Q., & Kwong, S. (2015). An evolutionary many-objective optimization algorithm based on dominance and decomposition. *IEEE Transactions on Evolutionary Computation*, *19*(5), 694–716. https://doi.org/10.1109/TEVC.2014.2373386.

Li, M., Yang, S., & Liu, X. (2013). Shifted-based density estimation for Pareto-based algorithms in many-objective optimization. *IEEE Transactions on Evolutionary Computation*, *18*(3), 348–365. https://doi.org/10.1109/TEVC.2013.2262178.

Li, M., Yang, S., & Liu, X. (2015). Bi-goal evolution for many-objective optimization problems. *Artificial Intelligence*, *228*, 45–65. https://doi.org/10.1016/j.artint.2015.06.007.

Li, M., Yang, S., & Liu, X. (2016). Pareto or non-Pareto: Bi-criterion evolution in multi-objective optimization. *IEEE Transactions on Evolutionary Computation*, *20*(5), 645–665. https://doi.org/10.1109/TEVC.2015.2504730.

Liu, Y., Gong, D., Sun, X., & Zhang, Y. (2017). Many-objective evolutionary optimization based on reference points. *Applied Soft Computing*, *50*, 344–355. https://doi.org/10.1016/j.asoc.2016.11.009.

Maulik, U., & Bandyopadhyay, S. (2000). Genetic algorithm-based clustering technique. *Pattern Recognition*, *33*(9), 1455–1465. https://doi.org/10.1016/S0031-3203(99)00137-5.

Morris, G., Goodsell, D., Halliday, R., Huey, R., Hart, W., Belew, R., & Olson, A. (1998). Automated docking using a Lamarckian genetic algorithm and an empirical binding free energy function. *Journal of Computational Chemistry*, *19*(14), 1639–1662. https://doi.org/10.1002/(SICI)1096-987X(19981115)19:14<1639::AID-JCC10>3.0.CO;2-B.

Murata, T., & Ishibuchi, H. (1995). *MOGA: Multi-objective genetic algorithms* (pp. 289–294). IEEE. https://doi.org/10.1109/ICEC.1995.489161.

Oh, I.-S., Lee, J.-S., & Moon, B.-R. (2004). Hybrid genetic algorithms for feature selection. *IEEE Transactions on Pattern Analysis and Machine Intelligence*, *26*(11), 1424–1437. https://doi.org/10.1109/TPAMI.2004.105.

Osyczka, A., & Kundu, S. (1995). A new method to solve generalized multicriteria optimization problems using the simple genetic algorithm. *Structural Optimization*, *10*(2), 94–99. https://doi.org/10.1007/BF01743536.

Reeves, C. (1995). A genetic algorithm for flowshop sequencing. *Computers & Operations Research*, *22*(1), 5–13. https://doi.org/10.1016/0305-0548(93)E0014-K.

Roy, K., Kar, S., & Das, R. (2015). *Understanding the basics of QSAR for applications in pharmaceutical sciences and risk assessment* (1st ed.). Academic Press.

Ruetsch, G., & Fatica, M. (2013). *CUDA Fortran for scientists and engineers* (1st ed.). Morgan Kaufmann. https://www.elsevier.com/books/cuda-fortran-for-scientists-and-engineers/ruetsch/978-0-12-416970-8.

Savic, D., & Walters, G. (1997). Genetic algorithms for least-cost design of water distribution networks. *Journal of Water Resources Planning and Management. American Society of Civil Engineers*, *123*(2), 67–77. https://doi.org/10.1061/(ASCE)0733-9496(1997)123:2(67).

Schaffer, J. (1985). *Some experiments in machine learning using vector evaluated genetic algorithms* (1st ed.). Vanderbilt University.

Siedlecki, W., & Sklansky, J. (1989). A note on genetic algorithms for large-scale feature-selection. *Pattern Recognition Letters*, *10*(5), 335–347. https://doi.org/10.1016/0167-8655(89)90037-8.

Srinivas, N., & Deb, K. (1994). Multiobjective optimization using nondominated sorting in genetic algorithms. *Evolutionary Computation*, *2*(3), 221–248. https://doi.org/10.1162/evco.1994.2.3.221.

Suganthan, P. (2020). *Personal repository*. https://github.com/P-N-Suganthan.

Suganthan, P., Hansen, N., Liang, J., Deb, K., Chen, Y.-P., Auger, A., & Tiwari, S. (2005). Problem definitions and evaluation criteria for the CEC 2005 special session on real-parameter optimization. In *2005 IEEE congress on evolutionary computation*. Edinburgh, Scotland, UK: Nanyang Technical University and Kanpur Genetic Algorithms Laboratory.

Tian, Y., Zhang, X., Cheng, R., & Jin, Y. (2016). A multi-objective evolutionary algorithm based on an enhanced inverted generational distance metric. In *2016 IEEE Congress on Evolutionary Computation (CEC), Vancouver, BC, Canada*IEEE. https://doi.org/10.1109/CEC.2016.7748352.

Tseng, L., & Chen, C. (2009). *Multiple trajectory search for unconstrained/constrained multi-objective optimization* (pp. 1951–1958). IEEE. https://doi.org/10.1109/CEC.2009.4983179.

Wagner, M., & Neumann, F. (2013). In C. Blum (Ed.), *A fast approximation-guided evolutionary multi-objective algorithm* (pp. 687–694). Association for Computing Machinery. https://doi.org/10.1145/2463372.2463448.

Walters, D., & Sheble, G. (1993). Genetic algorithm solution of economic-dispatch with valve point loading. *IEEE Transactions on Power Systems*, *8*(3), 1325–1332. https://doi.org/10.1109/59.260861.

Wang, R., Purshouse, R., & Fleming, P. (2013). Preference-inspired coevolutionary algorithms for many-objective optimization. *IEEE Transactions on Evolutionary Computation*, *17*(4), 474–494. https://doi.org/10.1109/TEVC.2012.2204264.

Yang, J., & Honavar, V. (1998). Feature subset selection using a genetic algorithm. *IEEE Intelligent Systems & Their Applications*, *13*(2), 44–49. https://doi.org/10.1109/5254.671091.

Yang, S., Li, M., Liu, X., & Zheng, J. (2013). A grid-based evolutionary algorithm for many-objective optimization. *IEEE Transactions on Evolutionary Computation*, *17*(5), 721–736. https://doi.org/10.1109/TEVC.2012.2227145.

Yuan, Y., Xu, H., & Wang, B. (2014). In C. Igel (Ed.), *Evolutionary many-objective optimization using ensemble fitness ranking* (pp. 669–676). Association for Computing Machinery. https://doi.org/10.1145/2576768.2598345.

Yuan, Y., Xu, H., Wang, B., & Yao, X. (2016). A new dominance relation-based evolutionary algorithm for many-objective optimization. *IEEE Transactions on Evolutionary Computation*, *20*(1), 16–37. https://doi.org/10.1109/TEVC.2015.2420112.

Yuan, Y., Xu, H., Wang, B., Zhang, B., & Yao, X. (2016). Balancing convergence and diversity in decomposition-based many-objective optimizers. *IEEE Transactions on Evolutionary Computation*, *20*(2), 180–198. https://doi.org/10.1109/TEVC.2015.2443001.

Zhang, Q., & Li, H. (2007). MOEA/D: A multiobjective evolutionary algorithm based on decomposition. *IEEE Transactions on Evolutionary Computation*, *11*(6), 712–713. https://doi.org/10.1109/TEVC.2007.892759.

Zhang, X., Tian, Y., & Jin, Y. (2015). A knee point-driven evolutionary algorithm for many-objective optimization. *IEEE Transactions on Evolutionary Computation*, *19*(6), 761–776. https://doi.org/10.1109/TEVC.2014.2378512.

Zitzler, E., Deb, K., & Thiele, L. (2000). Comparison of multiobjective evolutionary algorithms: Empirical results. *Evolutionary Computation*, *8*(2), 173–195. https://doi.org/10.1162/106365600568202.

Zitzler, E., & Künzli, S. (2004). In X. Yao, et al. (Eds.), *Indicator-based selection in multi-objective search* Springer. https://doi.org/10.1007/978-3-540-30217-9_84.

Zitzler, E., Laumanns, M., & Bleuler, S. (2004). In X. Gandibleux, M. Sevaux, K. Sörensen, & V. T'kindt (Eds.), *Vol. 535. A tutorial on evolutionary multiobjective optimization* (pp. 3–37). Springer. https://doi.org/10.1007/978-3-642-17144-4_1.

Zitzler, E., Laumanns, M., & Thiele, L. (2001). *SPEA2: Improving the strength Pareto evolutionary algorithm.* Computer Engineering and Communication Networks Lab and Swiss Federal Institute of Technology. TIK-Report 103 https://pdfs.semanticscholar.org/6672/8d01f9ebd0446ab346a855a44d2b138fd82d.pdf.

Zitzler, E., & Thiele, L. (1998). *An evolutionary algorithm for multiobjective optimization: The strength Pareto approach.* Computer Engineering and Communication Networks Lab and Swiss Federal Institute of Technology. https://sop.tik.ee.ethz.ch/publicationListFiles/zt1998a.pdf.

Estimation of natural resources for renewable energy systems

3

3.1 Development of renewable energies

Building a sustainable society requires the efficient use of locally available natural resources. In this regard, wind, solar, wave, tidal, hydropower, biomass, and geothermal have been widely discussed by policymakers due to their importance to sustainable development. However, the incorporation of renewable energies is not the only step in transforming into a sustainable society. Other topics related to increasing energy savings and energy conversion efficiency are also important. Thus, creating a system based on renewable energies depends on natural resources and technological improvements (Lund, 2007).

Renewable energies allow us to mitigate greenhouse gas emissions and, consequently, global warming by reducing fossil fuel power generation. In a general sense, renewable energies have applications in different fields. As an example, wind energy has applications in power generation and water pumps. Solar energy has applications in solar home systems, solar dryers, solar cookers, photovoltaic (PV) and thermal power generation, and water heaters. Hydropower has a relevant application in power production. Biomass energy has applications in heat and power generation, pyrolysis, gasification, and digestion. Geothermal energy has applications in urban heating, and power generation (Panwar et al., 2011).

Full integration of renewable energies would require essential changes in the generation mix and electricity infrastructure. If policymakers and practitioners fully explore wind, hydro, and solar resources worldwide, humanity would reduce the world demand by approximately 30%, incrementing the land use by only 0.41%. However, the variability introduced by these resources is a technical problem, which requires solutions. One option would be to include the interconnection of dispersed generators. Other options could be installing hydropower units, adopting demand response programs, installing energy storage devices, over-sizing some generation units, and developing appropriate weather forecasting techniques. However, the most crucial obstacle to the complete adoption of renewable energies is not related to techno-economic aspects; it is related to social and political perspectives (Delucchi & Jacobson, 2011; Jacobson & Delucchi, 2011). Appropriate governmental policies can produce the energy mix transformation by promoting the different sources' industrial development, considering production, distribution, and consumption; in other words, the government must create a mechanism to encourage the adoption of renewable power sources, considering local legislation and international treaties. The USA, Canada, Germany, Spain, Australia, China, and France, among many others, are some representative examples where several instruments, such as feed-in tariff (FIT) and renewable portfolio standard, have shown positive results (Solangi et al., 2011).

Genetic Optimization Techniques for Sizing and Management of Modern Power Systems
https://doi.org/10.1016/B978-0-12-823889-9.00004-7

Another angle in the integration of renewable energies is improving the communication infrastructure and power electronics. In this regard, the smart grid concept plays a vital role in integrating scattered resources that neighboring customers can easily use, increasing system efficiency (Ellabban et al., 2014).

Successful adoption of wind energy strongly depends on the geographical characteristics and wind speed profile, or in other words, the relationship between the wind flow and the topology of the location in which the wind farm is intended to be installed by the policymaker. Such a relationship defines the effects of wind availability, wind gust, and installation height on the wind farm's annual energy production (Herbert et al., 2007).

Wind energy has experienced notable growth in the European Union. Specifically, Germany, Spain, the UK, and France have increased their installed capacities up to 31,026, 18,602, 7894, and 6108 MW, respectively (Borawski et al., 2020). Worldwide, wind energy increased its capacity from around 17.2 GW in 2000 to 563.27 GW in 2018 (US Energy Information Administration, 2020b).

Solar is another energy source that has experienced rapid expansion. A representative example is China, where the installed capacity has increased from approximately 1 GW in the year 2010 to 131 GW in 2017. Such an increase is directly related to energy policy and promotion based on FIT (Li & Huang, 2020). However, the growth of solar energy is a global tendency. In the year 2000, worldwide installed capacity was around 0.98 GW, which increased to 485.36 GW in 2018 (US Energy Information Administration, 2020b).

Worldwide, available wave energy is estimated at 1 TW, approximately. Thus, the exploitation of this energy amount would be beneficial in reducing electricity demand from fossil fuel sources (Falnes, 2007). This idea has been developed through the years by policymakers. The result is an increment in the installed capacity of tidal and wave generation from 0.477 GW in the year 2000 to 1.842 GW in the year 2018 (US Energy Information Administration, 2020b).

People have used biomass for a long time in developing areas, and recently it has gained the attention of developed countries seeking to identify appropriate sources for power production. However, using biomass as a power source strongly depends on replanting significant areas in due course. In general, practitioners burn woody, herbaceous plants and grasses for direct energy production; however exceptions to this exist, depending on the form in which the specific process requires energy. For example, scientists have used biomass during the combustion process to produce steam and power using turbo-generators. Furthermore, some specific species are better for producing gaseous or liquid fuels (McKendry, 2002). In the year 2000, the worldwide consumption of biomass and waste resources totaled 32.6 GW. While in the year 2018, it reached 127.37 GW (US Energy Information Administration, 2020b).

Scientists have exploited geothermal energy in the forms of natural steam and hot water. Due to its availability, geothermal energy is reliable. It has been used for a long time to supply the heat demand in large municipal districts and as a primary source for power plants. Thus, the industry dedicated to this source is highly developed (Barbier, 2002). During the year 2000, the worldwide geothermal capacity was 8.32 GW, increasing to 13.26 GW in 2018 (US Energy Information Administration, 2020b).

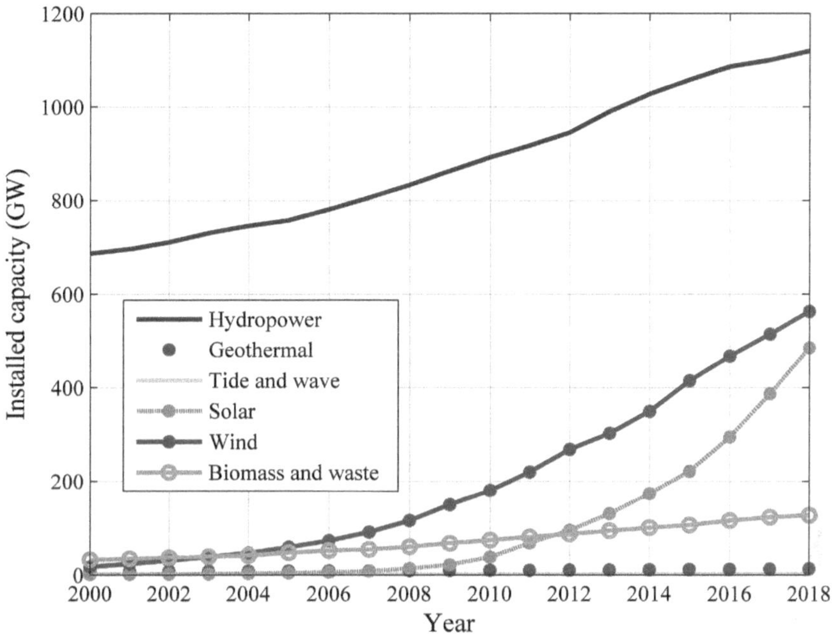

Fig. 3.1 Adoption of renewable power sources: Capacity of renewable energies between the years 2000 and 2018.

To summarize, Fig. 3.1 (US Energy Information Administration, 2020b) shows the evolution of wind, solar, wave and tide, hydropower, biomass, and waste energy production, as well as geothermal, for the period between the years 2000 and 2018. Paying particular attention to the last years of this interval, that is between 2015 and 2018, it can be seen that these technologies have grown at a rate of 49.242, 88.392, 0.0613, 19.813, 6.898, and 0.4827 GW per year, respectively. Thus, wind and solar energies are, by far, the technologies with the highest growth.

According to data from the year 2017, the US had a leading role in generating nuclear, renewables, and other types of energies, with 20 quadrillion BTU production. In addition, the US had a significant role in producing natural gas, with 28 quadrillion BTU. On the other hand, China played a crucial part in primary energy production (total energy production), with 112 quadrillions BTU, and in coal production, with 81 quadrillion BTU (total energy production from coal). Saudi Arabia had a remarkable position on energy production from petroleum and other liquids with 25 quadrillions BTU (US Energy Information Administration, 2020c).

Looking ahead, in the period between 2020 and 2050, for the US's specific case, experts predict that energy consumption will increase slower than the gross domestic product because of the improvements in energy efficiency. Furthermore, the generation mix will continue changing, specifically regarding wind and solar energies, because federal tax credits incentivize these technologies to fulfill ambitious goals. The reduced increment in the energy demand complements the changes in the generation mix. Consequently, the adoption of coal and nuclear-based units reduces with respect to those based on natural gas. The United States will produce energy at high rates, which, combined with a slow increment in the energy demand, results in an increment of US exports of crude oil, petroleum products, and natural gas (US Energy Information Administration, 2020a).

Moreover, China is planning to expand its renewable generation capacity to 20% by 2030, focusing on wind energy. However, even when the installed capacity of wind generation in China is higher than in the US, the energy produced is lower. Wind power curtailment, turbine quality, and electricity grid expansion could improve this situation (Lu et al., 2016).

During the early stages of renewable energy implementation, identifying natural resources is a crucial step that requires the acquisition of a large dataset, which in some cases is not easy. Thus, approximation techniques based on neural networks (NNs) have become popular to deal with the lack of information. This chapter describes a method using the measure-correlate-predict (MCP) technique based on an NN to estimate renewable resources. The NN presented here is trained using a genetic algorithm (GA), which is the central topic. In other words, this chapter explains how to solve the optimization problem of training an NN using a GA. Specifically, we consider two types of NNs: the feedforward neural network (FFNN) and the recurrent neural network (RNN).

3.2 Natural resources and renewable energy systems

Expressing natural resources using the typical meteorological year (TMY) (Wilcox & Marion, 2008) is a crucial step for dimensioning any renewable energy system properly. However, this is not an easy step since it requires monitoring and collecting information over a long time. Meteorologists and other scientists do not collect the data to promote renewable energy systems deployment specifically; instead, it is measured to describe the environmental conditions of places located relatively close to the metropolitan areas in flat terrain or low elevation areas. Thus, most of the information publicly available in regional wind atlases provides a general perspective to engineers and policymakers.

Due to their relevance to renewable energies, we pay special attention to wind and solar energies. In wind energy, scientists describe the potential using seven wind power density classes, where those places with a class lower than two have a marginal resource. In comparison, those with higher classifications are candidate locations for wind energy applications (AWS Scientific, Inc, 1997). In solar energy, solar atlases frequently specify useful estimations of global, direct, and diffuse radiation, cloud cover, temperature, heating and cooling, and wind speed. Meteorologists usually present these results as hourly and daily statistics.

In many cases, using this information requires processing to make it suitable for the application of interest. A representative example is a PV panel, which practitioners install on a tilted plane to increase its energy production. Thus, solar radiation and temperature for solar PV applications and concentrating technology are provided monthly, including the mean, minimum, and maximum values for one and two axis tracking (Renné et al., 2008). Another option consists of interpolating data over the area of interest using as starting point information of meteorological stations located nearby. Finally, measuring the wind and solar resources directly in the area of interest allows us to consider the effects of the terrain on the wind speed and direction and any obstacle to solar radiation. In a general sense, the approach to be adopted depends on the information available.

We discuss essential topics related to measuring environmental variables, renewable resource databases, and some estimation techniques in the next sections.

3.2.1 Measurement of environmental variables

The data quality required strongly depends on the stage of project development. For the prefeasibility study, long-term information already available is used to generalize the technology to be adopted. As can be expected, the uncertainty at this level is high. During the feasibility study, engineers combine hourly times series representing a preliminary approximation to TMY with optimization models to improve power production and acquisition and maintenance costs estimation. During the due diligence and acceptance period, a final analysis of TMY, including minute by minute and daily variations, are incorporated, obtaining an accurate estimate of the energy system performance. Once the system is functioning, the continuous operation and maintenance process provides high-quality information with low uncertainty about the operating strategy and power production. Measuring the long-term time series of renewable resources, estimating energy production under real-time conditions, and investigating the whole project's profitability are the feasibility study's main steps (Sengupta et al., 2017).

In this process, the variability involved in resource estimation plays a key role. It has two primary components, one temporal and another spatial. Temporal variability is related to the variations throughout the seasons, while spatial variability refers to the unique characteristics of the region under study.

For a solar resource, these variability components can influence its estimation in different manners (Sengupta et al., 2017). First, the measuring instruments could undergo an instantaneous failure or functional problem, creating a time series gap. In this case, scientists could complete the time series using statistical techniques based on a Markov process or similar. However, the incorporation of synthetically generated data increases the associated uncertainty (Sengupta et al., 2017). Second, if information from some meteorological stations nearby is available, scientists could extrapolate to estimate the specific site's resource. In these circumstances, it is possible that the spatial variability may introduce a palpable error; whether it does would depend on the orography of the region under analysis (Sengupta et al., 2017). Third, the yearly variations in the resource introduce uncertainty that impacts the project's profitability due to the long-term effects on power generation and power system stability (Sengupta et al., 2017).

The spatial smoothing effect can reduce the variability of a fleet of PV power plants. In other words, different places located close to each other experience a high correlation degree in solar resources. Consequently, the resources increase and decrease at a similar rate at the same time. On the other hand, places located far from each other would have uncorrelated solar resources. Finally, between these two extreme behaviors, scientists could observe a partial correlation. In addition, the regional cloud speed would significantly affect correlation and uncertainty (Sengupta et al., 2017).

Temperature, pressure, and relative humidity are among the frequently considered variables for local measurement. Scientists use the time series of temperature and pressure to estimate the air density, which influences the wind power generation from the perspective of wind energy. Ambient temperature can be measured using wire bobbins, thermocouples, thermistors, and temperature detectors based on platinum resistance. Usually, the probes for temperature and relative humidity are installed together in an aspired radiation shield at 2 m or 10 m. Similarly, technicians install the barometer at the same height as the thermometer (Lundquist et al., 2019).

Radiometers can identify a determined frequency band. The shortwave and longwave bands characterize solar radiation. Shortwave is related to solar radiation energy, while the longwave band is associated with the radiation from the atmosphere and the Earth. In addition to these bands, scientists can measure ultraviolet bands and others used by some solar PV cells and technologies. Radiometers employ various technologies. Photometers can measure shortwave radiation, obtaining the photopic irradiance. The broadband pyranometer measures the solar irradiance on a horizontal plane, receiving the global horizontal irradiance or diffuse irradiance, depending on the use of a shading ball. The tilted pyranometer can measure solar irradiance in an inclined plane, allowing us to obtain the global or diffuse irradiances in the PV panel's specific plane. The narrowband pyranometer captures the irradiance within a determined spectral band. Thus, scientists measure the ultraviolet irradiance, the photosynthetically active radiation, and the spectral band of PV cells. The pyrogeometer measures the irradiance in the infrared band, allowing us to report the downwelling or upwelling infrared radiation. The pyrheliometer enables us to determine the direct normal irradiance by measuring the irradiance at approximately 6 degree with respect to the sun.

Using a PV cell is another manner to estimate the irradiance and power production from a PV generator. Finally, the spectral radiometer offers us an estimation of the spectrally resolved irradiance using the radiation in different spectral bands. In addition to these technologies, the position in which the scientists install the instruments offers us relevant information. Global horizontal irradiance can be measured by orienting the devices in a horizontal position (facing up) to detect the shortwave and longwave radiation jointly (Lundquist et al., 2019). Similarly, the shortwave radiation and the longwave radiation, reflected and emitted by the surface, respectively, can be detected by positioning the instrument in a horizontal position (facing down). If technicians install the measuring device in a tilted position with a fixed angle, they can determine the PV cell's irradiance. Furthermore, technicians can evaluate the direct normal irradiance using a sun-tracking system with a limited field of view. Using a similar disposition of instruments but with the sun blocked out and global horizontal irradiance allows us to estimate the diffuse radiation (Lundquist et al., 2019).

Other environmental variables of interest are related to the surface fluxes of heat, moisture, and momentum. Scientists use this information to define the surface boundary conditions required in atmospheric simulations. In this regard, scientists use sonic anemometer thermometers to measure wind vectors in three dimensions, temperature and surface heat, and momentum. Sonic anemometers can detect the behavior of turbulent structures in the range of centimeters. A sonic anemometer measures the time taken by a pulse of sound to go from one transducer to another. This time depends on the air temperature, moisture, and pressure. However, the temperature measurement is not precisely the ambient temperature, but depends on temperature and moisture according to a specific function. A scintillometer estimates the variations in the intensity of a beam of electromagnetic radiation. This estimation depends on the air's refractive index variations and, consequently, some parameters related to the temperature, heat flux, and momentum flux (Lundquist et al., 2019).

Wind speed can be measured considering different heights by installing several anemometers in a tower. Some types of anemometers frequently used by technicians are the cup, propeller, and sonic anemometers. In addition, Investigators have proposed alternative technologies to characterize the wind speed profile. These include five-hole pressure probes, pulsed Doppler lidar (the Technical University of Denmark SpinnerLidar and OpenLidar), sodar, scanning radar, radar wind profilers, acoustic tomography, particle image velocimetry, and the Sandia wake imaging system (Lundquist et al., 2019).

Cup, propeller, and sonic anemometers have their performance reduced by environmental conditions such as rain and freezing. On the one hand, the impact of rain on the anemometers' operation is not clear, and technicians could mitigate it by incorporating directional wicks. On the other hand, freezing reduces the ability of the anemometer to measure wind speed. Technicians can minimize the adverse effects of freezing by using heating coils. However, this alternative could influence the measuring process of other variables such as heat fluxes, specifically in the case of sonic anemometers. In a general sense, cup anemometers have been widely accepted for wind energy applications because many wind engineers have used them in numerous projects, and they are economically affordable (Lundquist et al., 2019).

Ambient temperature is an essential variable because it influences atmospheric stability and, consequently, the energy conversion process and wind turbine loads. Meteorologists can measure the temperature profile using a radio acoustic sounding system (RASS), a microwave radiometer, and a radiosonde (Lundquist et al., 2019).

RASS uses an acoustic signal emitter and radar to detect the signal perturbed by the atmospheric refractive index changes. Another option is the microwave radiometer, which uses atmospheric brightness temperature data combined with radiative equations and an NN. Finally, radiosondes are attached to a balloon with radio communication. It communicates with a base located on the ground, which receives the radiosonde's information. Thus, temperature, pressure, humidity, and wind can be measured (Lundquist et al., 2019).

From a wind engineering perspective, the turbulence concept is directly related to measuring horizontal wind speed variance over a period of 10 min using a cup anemometer. However, the idea of turbulence can change if an atmospheric scientist or micro-meteorologist is consulted (Lundquist et al., 2019).

With the growing interest in offshore wind generation, meteorologists have proposed floating lidar systems to measure wind speed and direction. This application of lidar requires a communication system, power generation, and the implementation of a motion compensation system (Lundquist et al., 2019).

Among many others, a representative example of a meteorological station built for research of solar PV cells performance assessment is the National Institute of Standards and Technology weather station installed at Gaithersburg, Maryland, on August 01, 2014. This station can measure the direct normal irradiance, diffuse horizontal irradiance, global horizontal irradiance, infrared irradiance, spectral irradiance curves, snow depth, wind speed and direction, humidity, precipitation, barometric pressure, hail count, and ambient temperature (Boyd, 2016). The station uses a data acquisition system that performs measurements in 1 s and the conditions of the solar tracker and other elements. The system records the current–voltage curve with a maximum power point tracker and the corresponding load for the reference modules every minute. Images of the entire sky are also taken every 8 s (Boyd, 2016).

3.2.2 Renewable resource databases

Investigation and exploration of natural resources are relevant in several fields such as weather forecasting, aviation, agriculture, hydrological cycle studies, and renewable energies, among others. Expertise in this area allows us to improve the operation of related equipment and its efficiency. This idea becomes particularly important in relation to those resources under depletion, which will need to be substituted by policymakers in the short-term.

These circumstances have motivated many governmental agencies and research centers to discover the natural phenomena interactions at local, regional, and global scales. From these efforts, investigators have contributed with many proposals and models in the technical literature.

In this regard, Rajagopalan and Lall (1999) developed a method using a nonparametric approximation of a multivariate Markov process. For the environmental variables (solar radiation, wind speed, precipitation, and temperature), the approach recognizes the *k*th nearest neighbors (k-NN) within the historical data using the weighted Euclidean distance. Then, resampling one element from the k-NN set using a kernel function results in the simulation values. The probability mass function defines the kernel function. This technique is an alternative to those methods that require a specific probability density function and those based on Gaussian theory and linear analysis.

Yan et al. (2002) statistically analyzed the most extended daily temperature time series in Europe and China. As a result, they observed some relevant changes in the extreme temperature. They concluded that the coldest winter temperature has increased at three times the rate of the recent annual mean warming, which is around 2 and 3 °C per century.

Zhang and Zheng (2004) evaluated several planetary boundary layer parameterizations, including time series of average surface wind speed and temperature. According to the reported results, the diurnal cycles of surface wind speed are in phase with the surface temperature. However, wind directions are out of phase between the fifth and sixth hours. The simulation of these variables strongly depends on how the associated physical processes are parameterized. These results represent a significant contribution to solving those problems related to the prediction of environmental variables and wind energy.

Feng et al. (2004) carried out an evaluation of 726 meteorological stations in China between 1951 and 2000. The assessment consisted of implementing quality-control methods to detect typos and incorrect measurements. They reported the results as a database of 10 meteorological variables, which are useful to understand the weather's behavior at local and regional scales, as well as improve the accuracy of decision making in energy-related projects.

Using another type of model, Murtugudde and Busalacchi (1999) intensely studied the ocean's dynamics. Analyzing the tropical Indian Ocean's thermodynamic behavior, they investigated how the sea surface temperature can influence the atmospheric circulation. To this end, the authors combined the ocean general circulation model with the atmospheric mixed layer model. The work complements the findings previously reported by the authors (Murtugudde et al., 1996), where they show the benefits of incorporating information from Africa, Arabia, and northwest Australia during the analysis of surface heat flux, as well as the Indonesian Throughflow on the study of seasonal and interannual variations of the Indian Ocean.

Meteorologists have widely accepted reanalysis as a technique for climate analysis. It combines numerical models with meteorological measurements to explain the weather's behavior at a worldwide scale in the recent past. The method defines some geographical grids in which technicians install the meteorological instrumentation. The collected information is then processed by a data assimilation system and a forecast model, resulting in a database that describes the atmosphere and the oceans' characteristics (European Centre for Medium-Range Weather Forecasts (ECMWF), 2020; National Centers for Environmental Information National Oceanic and Atmospheric Administration (NOAA), 2020).

In this regard, Robertson and Mechoso (2000) used the reanalysis database of the National Center for Environmental Prediction (NCEP) and the National Center for Atmospheric Research (NCAR) with information from the years 1958–97. They carried out a study of the circulation variability in southern South America on an interannual and decadal basis. In addition, they analyzed the relationship between the sea surface temperature and river flows to explain the precipitation parameter.

Maurer et al. (2002) created a model using information from the US, Canada, and Mexico for the period between 1950 and 2000 with a step of 3 h and a resolution of 0.125 degree. The methodology applied differs from the reanalysis because information related to precipitation is obtained directly from measurements, while land surface water and energy budgets balance at every step.

Sheffield et al. (2006) created a global meteorological database of 50 years using NCEP and NCAR databases. To correct the error introduced by the reanalysis process, they used precipitation, air temperature, and radiation measurements. With similar purposes, they incorporated information from the World Meteorological Organization Solid Precipitation Measurement Intercomparison. In addition, they carried out a spatial disaggregation of precipitation using statistical downscaling in the space of 1 degree. Then, they performed a temporal disaggregation utilizing the Tropical Rainfall Measuring Mission information in steps of 3 h.

As stated before, many research centers around the world have been working hard on climate reanalysis. In this regard, the European Centre for Medium-Range Weather Forecasts (ECMWF) and other research centers have developed several reanalyses, including the ERA-40, ERA-Interim, ERA-20C, and ERA5.

ERA-40 (Uppala et al., 2005) used meteorological information from 1957 to 2002, obtained by ECMWF using instruments supported by satellites complemented with other technologies such as ocean-buoy and aircraft. This version included significant operational improvements since 1990 when the scientific community already had access to other databases such as NCEP-NCAR. ERA-Interim (Dee et al., 2011) was a reanalysis model developed as the successor to ERA-40. This version overcomes several difficulties of ERA-40, including some related to the hydrological cycle and the quality of the stratospheric circulation, among others. ERA-20C (Poli et al., 2016) is another version that performs reanalysis using data from 1900 to 2010, assimilating surface pressure and marine wind measurements. This model improved the estimation of climate indices using feedback information. ERA5 (Hersbach et al., 2020) is built by ECMWF using refined information on the global atmosphere, land surface, and ocean waves. It improves ERA-Interim's reanalysis capabilities by incorporating a recently developed integrated forecasting system, dynamic models, and data assimilation. Besides, ERA5 has a higher resolution than ERA-Interim and uncertainty estimations.

Vidal et al. (2010) presented a reanalysis model using 50 years of information with high resolution in France. They used a methodology similar to that applied by ECMWF combined with the measurements of the Météo-France database and the *Systéme d' Analyze Fournissant des Renseignements Atmosphériques à la Neige* (SAFRAN) system. In addition, they analyzed the variables' sensitivity with the number of data-points, finding that the precipitation and air temperature are the most robust.

Similarly, the Japan Meteorological Agency (JMA) developed JRA-25 and JRA-55. JRA-25 (Onogi et al., 2007) was the first reanalysis developed in Asia. It used information from 1979 to 2004 obtained from several measurement technologies such as geostationary satellites. JRA-25 offered useful information about precipitation forecast, as well as wind profiles around tropical cyclones. JRA-55 (Kobayashi et al., 2015) is the improved version of JRA-25 reanalysis. Scientists designed JRA-55 to estimate multidecadal variations and to perform climate change studies. JRA-55 mitigated some problems related to the bias error of JRA-25.

Following these efforts, the National Aeronautics and Space Administration (NASA) presented the modern-era retrospective analysis for research and applications (MERRA) and its improved version MERRA2. NASA developed MERRA (Rienecker et al., 2011) to report Earth Observing System satellite measurements used in many applications such as climate studies and the hydrological cycle. MERRA has provided significant contributions to climatology, specifically on those studies related to precipitation and water vapor. MERRA2 (Molod et al., 2015) offers refined information in all resolutions and applications when compared to MERRA.

NCEP and the Department of Energy (DOE) have made essential efforts to improve retrospective analysis. An example is the NCEP-DOE Atmospheric Model Intercomparison Project (AMIP-II) reanalysis (Kanamitsu et al., 2002). Investigators developed it to support national and international projects in different fields, including analyzing, interpreting, modeling, and synthesizing the World Ocean Circulation Experiment and the International Satellite Land Surface Climatology. Another example is the NCEP Climate Forecast System (CFSv2) (Saha et al., 2014), which has unique weather prediction capabilities in seasonal and subseasonal periods. Specifically, CFSv2 has shown outstanding results in the forecast of global sea surface temperature and the weather forecast horizon's increment.

The US National Oceanic and Atmospheric Administration (NOAA) worked on the Twentieth Century Reanalysis (20CR) (Compo et al., 2011) project, which was intended to create a database of global atmospheric circulation over the 20th century. The 20CR reanalysis works with spatial and temporal resolutions of 6h and 2°C, respectively. It assimilates information related to the surface pressure and sea-surface temperature, among other data, providing information whose uncertainty is already estimated so that scientists on climate model simulations can use it.

WorldClim 2 (Fick & Hijmans, 2017) presented a database of interpolated temperature data, precipitation, solar radiation, vapor pressure, and wind speed with a 1 km^2 resolution. Investigators created the dataset using information from many meteorological stations (between 9000 and 60,000 stations) and satellites, combined with an interpolation technique that considered geographical features such as elevation and distance. Its creators carried out an interpolation process considering 23 areas defined according to the weather stations' locations. Simultaneously, scientists used satellite information to improve the prediction in those areas with a low number of stations, improving forecasting accuracy.

Pfenninger and Staffell (2016) and Staffell and Pfenninger (2016) studied the integration of solar PV energy and wind power using MERRA and MERRA2 reanalyzes regarding renewable energy applications. In the case of solar PV integration, they used reanalysis products and satellite information to simulate TMY and PV power production in several European countries. Then, they estimated the uncertainty degree using data from already installed PV systems, as well as information reported by the transmission system operators. As in previous studies, the authors identified and corrected the bias error of reanalysis, obtaining high-quality results. The same investigators carried out a similar analysis for wind energy, using MERRA and MERRA2 with European countries' data. For wind resources, they observed an overestimation of about 50% in northwest Europe and an underestimation of 30% in the Mediterranean. As a result, the authors reported the corresponding correcting factor to mitigate such errors.

Following this trend, during the Energy System Potentials for Renewable Energy Sources (ENSPRESO) project (Ruiz et al., 2016), researchers used data from several European countries from 2010 to 2050 to create an environmental database at national and regional levels. The authors reached essential conclusions about the energy potential of Europe. On the one hand, the available wind and solar resources can supply three times the European energy demand of 2016. On the other hand, the European population can use biomass, agriculture, forestry, and waste resources to provide at least 10%, 1.5%, and 1%, respectively, of the primary energy demand.

Meteorologists and computing scientists have joined efforts to offer these results in a user-friendly manner through the World Wide Web. Some representative examples are the US National Renewable Energy Laboratory, the EU Joint Research Centre, NCEP-NCAR, ECMWF, Météo-France, JMA, NASA, DOE, and NOAA, among others.

3.2.3 Estimation of environmental variables using empirical models

Measurements of solar radiation in the place of interest and neighboring locations are sometimes unavailable. However, it is possible to estimate the solar resource using empirical radiation models during sunshine and cloudy hours. In this regard, meteorological centers (NOAA, NASA, etc.) provide values of radiation integrated over a determined time period, which could be an hour or a day, and the averaging period, which is frequently a month. Another essential piece of information often reported is the measuring position. In most cases, it is carried out over the horizontal plane, although technicians sometimes consider surfaces with a fixed tilt. Thus, scientists and practitioners have accepted two widely used formats: the monthly average daily global radiation on a horizontal plane and the hourly global radiation on a horizontal plane (Duffie & Beckman, 1991).

Generating an hourly time series of solar radiation from average values requires implementation of a complete empirical model that considers the sun position and the geographical location of the PV generator and its installation. The general procedure involves estimating the solar radiation incident on a horizontal plane, which depends on the specific day of the year, the zenith angle, and the solar constant. The zenith angle is geometrically related to the latitude, hour angle, and declination. Declination is a variable that has been empirically estimated by scientists and depends on the day of the year (Duffie & Beckman, 1991). Then, mathematically integrating the incident solar radiation over a horizontal plane for an hour and a day allows us to obtain useful information to define the corresponding clearness index. Clearness index, defined on hourly, daily, and monthly bases, empirically relates to the hours of bright sunshine. The solar radiation is then decomposed into direct, diffuse, and reflected (albedo) components, using empirical relations. Technicians use the obtained results to estimate the radiation over the tilted surface using geometrical properties (Duffie & Beckman, 1991).

Many researchers have made significant efforts to determine the empirical relation between the clearness index and bright sunshine hours. Akinoğlu and Ecevit (1990) presented one of the earliest works in which they proposed a quadratic form to estimate global solar radiation from sunshine duration. They carried out a comparative analysis considering four models, using data from 100 locations worldwide. The results revealed the capabilities of the quadratic model, which has a low average percentage error. Samuel (1991) applied a graphical method to calculate the monthly mean daily global radiation in Sri Lanka from a modified Ångström equation, using sunshine duration data.

Supit and Van Kappel (1998) developed a method based on several European countries' measurements, which uses information related to the mean daytime cloud cover, as well as the maximum and minimum temperature. The objective of this work was to propose an alternative method for those based on sunshine duration. Meza and Varas (2000) compared the models based on the number of sunshine hours with those based on maximum and minimum temperature. Using data from Chile, they concluded that models based on temperature offer more accurate results than cloud cover. Yang et al. (2001) proposed a formulation that incorporates information related to the direct and diffuse radiations obtained from the spectral model. The performance of the proposed formulation was verified using data from Japan, showing good results.

Almorox and Hontoria (2004) compared different empirical relations to calculate solar radiation from sunshine duration using data from several meteorological stations in Spain. Specifically, they considered the linear, quadratic, third degree, logarithmic, and exponential models. According to the results, they concluded that third-degree models offer the best results. However, adjusting the linear model is more comfortable, so that practitioners prefer it.

Chen et al. (2004) compared several models based on temperature and sunshine duration, using data from China. According to the results, those models based on temperature do not offer favorable results. In contrast, those models based on sunshine duration revealed better performance. Finally, they proposed a model that combines the clearness index with the maximum and minimum temperature difference, using a logarithmic function. Similarly, Trnka et al. (2005) compared several models under different conditions using Central European data. According to the results, the model based on sunshine duration offered the most accurate results. In addition, they recommended the relation based on cloud cover and temperature for solar radiation when no information about sunshine duration is available. Moreover, they suggested using information related to daily precipitation to estimate global solar radiation when cloud cover information is not available. Similarly, they recommended using the function based on temperature when daily precipitation data is not available. Finally, they suggest an error propagation analysis because the root mean square error (RMSE) is between 10% and 20%.

Yang et al. (2006) proposed a model using sunshine duration, surface pressure, precipitable water, ozone thickness, and the Ångström turbidity coefficient. The model considers the effects of intrinsic processes in the atmosphere, which improves its accuracy. The model was tested and calibrated using information from Japan, China, the US, and Saudi Arabia. Yorukoglu and Celik (2006) explained some relevant statistical processes related to global radiation estimation, including some data-interpretation issues. They analyzed linear, quadratic, cubic, logarithmic, and exponential models. Bulut and Buyukalaca (2007) proposed a simplified approach, which uses a trigonometric function that only depends on the day of the year. They tested the model using data from Turkey, finding good results.

REST2 (Gueymard, 2008) is a model developed and tested using detailed radiometric instrumentation for direct irradiance, diffuse irradiance, and global irradiance.

It improves the previously developed Code for Physical Computation of Radiation, 2 bands (CPCR2) model. In addition, Gueymard discussed the impact of turbidity on the irradiance. Almorox et al. (2011) calibrated seven models previously proposed in the literature, observing good agreement from those models based on temperature. Furthermore, they contributed with an additional model using temperature and vapor pressure, which offers better performance than its predecessors.

Besharat et al. (2013) analyzed several models using data from Iran, observing good results for all of them. The best approximation is obtained from a model based on sunshine duration. Hassan et al. (2018) paid particular attention to those models only based on the day number, which means those models that only require information about the specific day to estimate global solar radiation. They tested the proposed approach using data from Egypt, observing relatively good results since the models are straightforward; the coefficient of determination estimated was at least 82%. Despotovic et al. (2015) used data from 924 locations worldwide to test and evaluate 101 models. To this end, they considered 10 error indicators and proposed a global performance indicator to rank all of the models using only one parameter. Nwokolo (2017) classified and compared 732 models taken from recently published literature. The researcher collected this information to assess natural resources in Africa.

Recently, the interest of investigators on NNs for estimating global solar radiation has been growing. In this regard, Zhang et al. (2017) presented a comparison of different models considering the time scale and estimation type. They also incorporated models based on NNs, obtaining a novel framework for model comparison. Similarly, Fan et al. (2019) performed a comparative work incorporating 12 empirical models and 12 models based on machine learning algorithms. To classify them, they used six error indicators, a global performance index, and the computational burden. According to the results from a case study in China, they found better results on those models based on machine learning. The cubic model, the modified linear-logarithmic model, and the power model offered the best results among the empirical models. Among the machine learning-based models, the adaptive neuro-fuzzy inference system (ANFIS), extreme learning machine (ELM), least-square support vector machine (LSSVM), and multivariate adaptive regression spline (MARS) had the best performance.

Another essential process during the estimation of solar radiation is determining the global component portions, which correspond to the direct radiation, diffuse radiation, and reflected radiation (albedo). In this regard, Liu and Jordan (1960) contributed techniques to estimate the diffuse radiation in clear days on an hourly and daily basis considering different clouding conditions. They designed their statistical model using data from the US and Canada. Orgill and Hollands (1977) presented one of the first experimental works to estimate diffuse radiation from global radiation. This effort resulted in a formulation that could be included in a computational simulation program.

Erbs et al. (1982) developed and tested a model to estimate the diffuse radiation from the clearness index using experimental data from the US and Australia. In addition, the authors discussed the incorporation of their proposed model in TRNSYS software. Graham and Hollands (1990) developed a complete algorithm to generate hourly solar radiation time series using the monthly radiation data. They based their work on a stochastic disaggregation procedure whose parameters were supposed to be independent of the place of interest. The model results are useful to estimate solar resources in remote locations.

Wong and Chow (2001) assessed a family of seven models to estimate global radiation, direct radiation, and diffuse radiation. They carried out the evaluation using data from Hong Kong. De Miguel et al. (2001) evaluated several models to estimate diffuse solar radiation using meteorological data of the North Mediterranean Belt zone. They proposed two empirical piecewise correlations using the information mentioned above. Jacovides et al. (2006) carried out an extensive analysis of Cyprus's meteorological data to develop an empirical formulation to estimate the diffuse radiation. They presented a comparative study using 10 models frequently used in the literature and several error indicators. According to the observed results, they concluded that their model estimates diffuse radiation reasonably, and they observed that the comparative models seem to be independent of the location of interest.

El-Sebaii et al. (2010) developed an empirical model to estimate the diffuse solar radiation from sunshine hours, ambient temperature, relative humidity, and cloud cover, considering data from Saudi Arabia. In addition, they performed a validation process using an independent dataset of the aforementioned meteorological variables and information publicly available from NASA. Recently, Nwokolo and Ogulezie (2018) reported a complete set of models and characteristics for the estimation of diffuse radiation (301 empirical models divided into 28 groups); including those based on support vector machine (SVM) combined with wavelet transform (WT) and radial basis function (RBF) NN.

Once the direct radiation, diffuse radiation, and the reflected radiation over a horizontal surface have been estimated, an important topic is selecting the appropriate tilt and azimuth for the installation of the PV generator. Tang and Tong (2004), and Jacobson and Jadhav (2018) have made significant contributions in this field.

Regarding the estimation of wind resources, many researchers have also made significant efforts to create algorithms and formulations to find wind speed profiles in locations where limited information is available.

In this sense, Aksoy et al. (2004) developed an algorithm for wind speed simulation using WT. This approach is an alternative technique to methods such as the random number simulation from a Weibull distribution, autoregressive models, and Markov process. The applicability of synthetic wind data generation was verified using data from Turkey from 1994 to 1997. Shamshad et al. (2005) developed a model based on a Markov process using data from Malaysia. The algorithm uses the transition probability matrices to generate wind speed data, retaining the original series' most important statistical properties.

Suomalainen et al. (2012) discussed the importance of incorporating low-frequency information on the wind speed simulation and its relationship with diurnal and seasonal patterns. As a result, they developed a model that considers wind speed occurrence and the corresponding days. They verified their approach using data from Portugal, observing reasonably good results. Carapellucci and Giordano (2013) developed a model to generate wind data from general-purpose information frequently found in wind atlases, such as average and maximum speed. Using as a reference of comparison the Markov process approach, they assessed their technique's performance using several indicators, observing good agreement.

3.2.4 Measure-correlate-predict method for environmental variables

There are two primary sources of information in the measure-correlate-predict (MCP) method, the reference places and the target place. The reference places have long-term measurements of meteorological variables, while the target place has measures during a limited period, frequently 1 year. Fig. 3.2 shows a general scheme to explain the idea

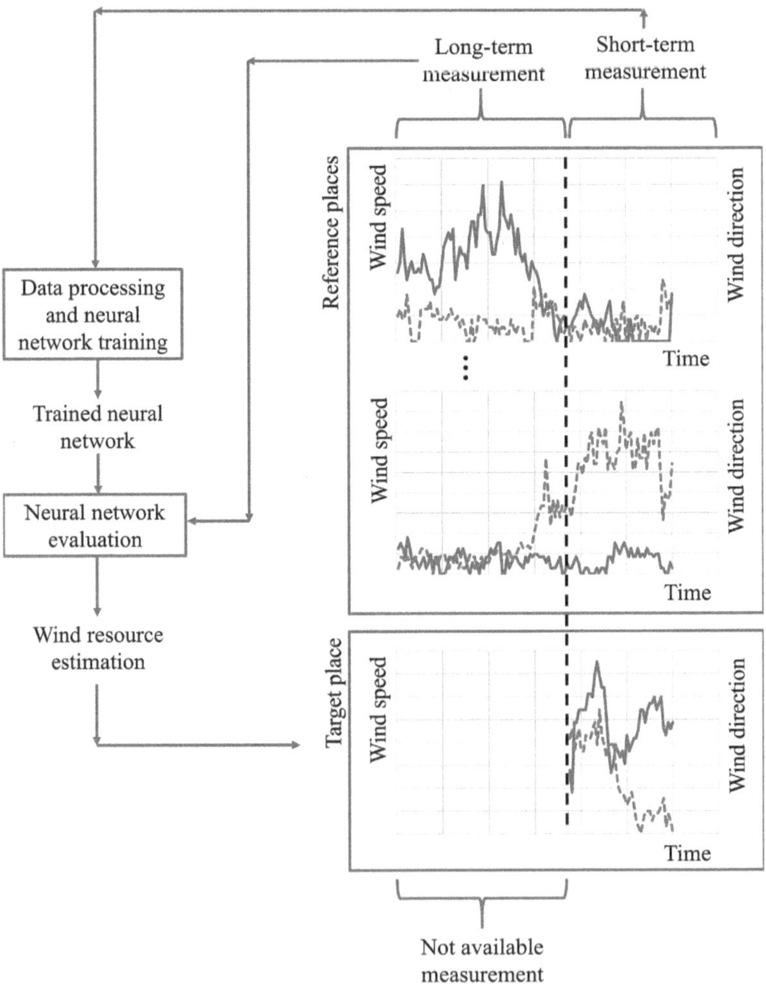

Fig. 3.2 Implementation of MCP methods for wind speed and direction estimation: Scheme of the measure-correlate-predict (MCP) approach.

behind the MCP method. A black dashed line represents the actual moment. The campaign of wind speed and direction measurements started at the time represented by the dashed line. In contrast, the reference places' campaign started many years before, so that complete information is available from these stations. MCP's idea consists of finding out the correlation between the information in the reference places and the target place using the data collected at the beginning of the measuring campaign at the target place (short-term measurement in Fig. 3.2). Such correlation can be determined using a linear or nonlinear function, or an NN, depending on the MCP method adopted. Once the scientist has established the function between the reference and target places, the next step consists of completing the time series at the target place using the long-term data at the reference places. In other words, the information available at the reference sites (long-term measurement in Fig. 3.2) is evaluated in the relationship (function) previously found, using data from both places, to estimate the wind resource at the target place before the beginning of the measuring campaign. Finally, wind engineers can use the complete time series at the target place to enrich the calculation of energy production of the future wind farm.

In this work, we have adopted an MCP approach based on an NN. This method requires a training process using short-term measurement. Then, we evaluate the trained network to estimate the wind resource.

The constant development and promotion of wind energy has motivated many researchers to improve the MCP approach. Thus, many functions and techniques based on data science have been proposed in the literature to minimize the error associated with MCP results. In this regard, Rogers et al. (2005) compared several MCP methods including those models based on the distribution of ratios of wind speed and standard deviations, linear and vector regressions, considering different indicators such as mean wind speed, wind speed, and direction distributions, and annual energy production of a typical wind turbine. They concluded that those models based on the distributions of wind speed ratios and standard deviations are among the best.

Velázquez et al. (2011) assessed the performance of NNs for wind speed estimation using the MCP technique. They used several error metrics, such as the mean absolute percentage error (MAPE) and Pearson's correlation coefficients. According to the results, NN reported better performance than the linear method. Carta et al. (2013) presented a complete review of MCP techniques, including a pertinent description and discussion about the linear, nonlinear, and probabilistic formulations used to relate the reference and target locations and the theoretical basis and hypothesis behind each formulation. In addition, other relevant topics related to MCP results' uncertainty and its effect on the energy yield were discussed.

Díaz et al. (2017) presented a comparison of several MCP techniques. They paid attention to several approaches such as support vector regression (SVR) and multiple linear regression (MLR). The authors trained some models to estimate the wind power density, while other models estimated wind speed and air density. Using meteorological data of wind speed and direction and air density from Spain, they observed promising results by estimating wind power density from the approximation of wind speed and air density separately. Similarly, Díaz et al. (2018) incorporated wind speed and direction measurements and air density data with the MCP approach to estimate wind power generation. Among other techniques, the authors considered NN and SVM during MCP implementation. The analysis of data measured in Spain identified the importance of air density information on MCP performance for wind power approximation and using wind speed and air density information differently to improve MCP performance.

Mifsud et al. (2018) used measurements made using lidar installed on a coastal building in Malta to estimate wind resources with the MCP technique. Among other methods, they considered an NN and SVR as mean approaches for MCP implementation. According to the results, MCP performance depends on the installation height of the lidar. Furthermore, they encourage engineers and practitioners to consider multiple MCP approaches because identifying the best techniques is difficult due to the conditions and variables involved. In the same manner, Sharma et al. (2019) combined information obtained from a meteorological mast and lidar to estimate wind shear using the MCP approach. In specific, wind shear between 10 and 220 m was measured using lidar, while the measuring devices determined wind shear between 10 and 100 m from the meteorological mast data. Then, they used the MCP method to extrapolate the lidar measurements to improve the wind shear estimation.

Ali et al. (2018) discussed their practical experience in South Korea of implementing the MCP approach. They used several evaluation metrics, including the mean wind speed and Weibull distribution on a seasonal and annual basis. Romo et al. (2011) proposed using three different functions to perform MCP, one based on a nonlinear regression and two based on kernel methods, which use conditional probability density functions. They compared their models with the linear regression, and the variance ratio approaches, finding that the kernel methods are among the best.

Carta et al. (2011) proposed an MCP technique using probabilistic Bayesian networks to estimate the wind speed distribution considering a limited amount of reference data. Specifically, the proposed an MCP technique that uses the joint probability function of all variables involved, which allows us to approximate conditional probabilities, enhancing the predictive capabilities of the MCP approach. Equally, Carta and Velázquez (2011) developed a model that employs the reference and target sites' probability density function. Specifically, the authors proposed the contingency-type bivariate function with specified marginal distributions.

Saavedra-Moreno et al. (2013) paid particular attention to reducing the computational burden related to the training of NNs to be employed on MCP techniques. Specifically, the authors illustrated the capabilities of the group method of data handling (GMDH) and extreme learning machines (ELM). Dinler (2013) used multiple principal least squares (MPLS) on the MCP approach. The investigator suggested this approach for places with low correlation. They also performed a comparative analysis, concluding that the proposed MCP method overcomes the variance ratio approach.

Weekes and Tomlin (2014) compared three regression techniques used on the MCP method using 22 locations in the UK. They paid particular attention to wind energy on a small scale. They employed measurements of only 3 months for estimating wind resources and 10 years for MCP performance assessment. MCP evaluation was carried out using mean wind speed, wind power density, and Weibull distribution. Zhang et al.

(2014) proposed a hybrid MCP technique in which its distance and height determine each reference site's weight. The authors compared their approach with several well-known MCP methods based on linear regression, variance ratio, Weibull distribution, NNs, and SVR. The analysis of data measured in the US concluded that their approach's performance depends on the adopted MCP algorithm and the meteorological reference stations considered.

Weekes et al. (2015) compared the forecast model UK4 with an MCP method using the information of 37 meteorological stations in the UK. The monthly and hourly linear correlation analysis concluded that UK4 could estimate wind resources more accurately than the implemented MCP approach. Ishihara and Yamaguchi (2015) implemented the MCP method and Monte Carlo simulation approach to estimate extreme wind speed. They verified the effectiveness of their system using data from Japan. Similarly, Nam and Oh (2020) proposed a mutually complementary MCP technique to estimate wind resources and extreme wind speed. Analyzing data from Jordan, they concluded that the energy density is more sensitive in those locations with interannual variations than the wind speed.

Regarding MCP methods' intrinsic uncertainty, Lackner et al. (2008) identified the sources of uncertainty of MCP implementation and reported a manner to quantize it. Specifically, they developed a methodology to estimate the global uncertainty related to wind resource assessment, the estimation of wind power generation, and the annual energy production evaluation. They concluded that the traditional methods of considering uncertainty are based on oversimplified hypotheses, and provide useful information to carry out a sensitivity factor over the Weibull distribution parameters. Gass et al. (2011) implemented the MCP approach using the variance ratio method to estimate the risk associated with MCP performance. Analyzing data from Austria, they incorporated the internal rate of return and the conditional value at risk, obtaining a techno-economical assessment of uncertainty related to MCP results.

3.3 Training a neural network using a genetic algorithm

In recent years, the NN structure and its training algorithm have been notably improving. From the early-developed feedforward to the more recent long short-term memory (LSTM), NNs have extended their applicability in many areas of science and engineering. Our purpose is to present a methodology to train an NN using a real-coded GA. This approach will then be applied as an MCP technique to estimate wind resources in a nearby location.

In very plain words, researchers can use an NN to represent a determined complex function from a limited amount of data points expressed in the input–output format. Thus, an NN is composed of several matrices whose values have to be determined to represent the function described by the available input–output data. Scientists formulate the optimization problem to minimize the error between the NN evaluation's output and the measured output data.

Historically, some algorithms to train NNs are the backpropagation (BP) algorithm to train FFNNs and the backpropagation through time (BPTT) to train RNNs. During the training of RNNs with BPTT, the error signals can blow up or vanish. This problem was overcome by LSTM (Hochreiter & Schmidhuber, 1997). However, other investigators have tried to deal with it using a GA.

GA is a very flexible optimization technique, so that it is possible to represent many combinatorial problems using convenient coding. Training NNs using a GA is a problem that has been analyzed by many researchers. In this regard, Montana and Lawrence (1989) developed a technique to train an FFNN using a GA. They represented the matrix of the FFNN in a vectorial form and defined the resulting vector as an individual of the GA. Then, crossover and mutation operators over each individual were applied to minimize the sum of the error (SSE) squares. Blanco et al. (2000) also contributed to this field of GA application. First, they developed a methodology to determine the optimal topology of an RNN; and second, they proposed a real-coded GA to train RNNs (Blanco et al., 2001).

In this section, we will explain how to implement a GA to train FFNNs and RNNs. First, we assume that a family of data-points in the input–output format is already available. Scientists can represent the input and output data as two different matrices with T rows. The number of columns of each matrix (M for the input data matrix and Q for the output data matrix) can be understood as the number of features on the input and output data, respectively. Thus, the index $m=1, \ldots, M$ represents a determined feature in the input data, and the index $q=1, \ldots, Q$ represents the feature in the output data.

Under these conditions, we can represent an input data-point t using the vector $\boldsymbol{u}_{(t)}$ as shown in Eq. (3.1), while the corresponding output data-point can be represented using the vector $\boldsymbol{w}_{(t)}$, as shown in Eq. (3.2).

$$\boldsymbol{u}_{(t)} = \begin{bmatrix} \dot{u}_{(t,1)} \\ \vdots \\ \dot{u}_{(t,m)} \\ \vdots \\ \dot{u}_{(t,M)} \end{bmatrix} \tag{3.1}$$

$$\boldsymbol{w}_{(t)} = \begin{bmatrix} \dot{w}_{(t,1)} \\ \vdots \\ \dot{w}_{(t,q)} \\ \vdots \\ \dot{w}_{(t,Q)} \end{bmatrix} \tag{3.2}$$

For the specific problem under study in this work, implementing the MCP method using an NN trained by a GA, we can organize the input and output data points according to the schemes shown in Figs. 3.3 and 3.4 (Carta et al., 2013). Fig. 3.3 considers the wind speed and direction measurements on the reference places, while Fig. 3.4 only considers the wind speed and direction measurements on the target place, so that $Q=2$.

In the next section, we describe the general structure of an FFNN and an RNN, including their mathematical formulation (Rojas, 1996).

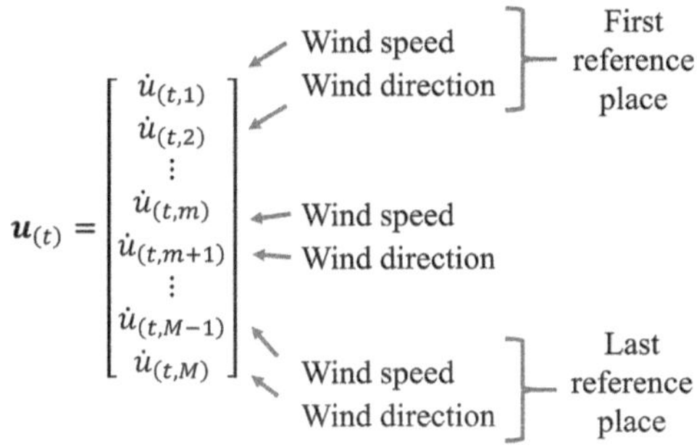

Fig. 3.3 Relationship between the reference data and the NN input: Structure of the NN input data.

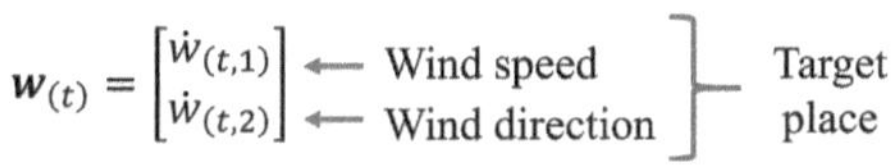

Fig. 3.4 Relationship between the NN output and the target place data: Structure of the NN output data.

3.3.1 Feedforward neural network

In this section, we briefly describe the general structure of a feedforward neural network (FFNN), which is shown in Fig. 3.5. The structure includes three main layers; the first layer has M units, and it receives each input data-point ($t=1, \ldots, T$) in the format previously explained in Eq. (3.1) and Fig. 3.3. The second layer is the hidden one with P units; each unit in the hidden layer is represented by the index $p=1, \ldots, P$. Finally,

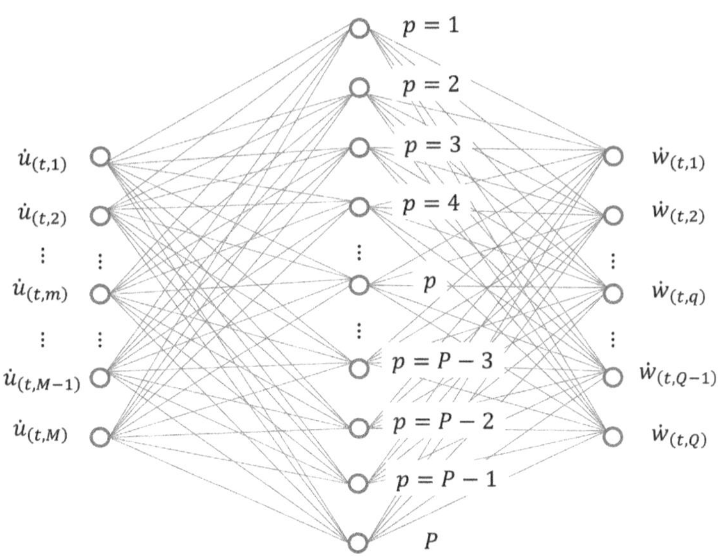

Fig. 3.5 General architecture of an FFNN: Structure of a simple FFNN.

the third layer represents the output values calculated by the FFNN according to Eq. (3.2) and Fig. 3.4. The output layer has Q units. However, as we only consider a single target site, Q equals two.

Considering a determined data-point (t), the relationship between the input and the output in the FFNN shown in Fig. 3.5 is described in Eqs. (3.3) and (3.4). The vectors $\boldsymbol{u}_{(t)}$ and $\boldsymbol{w}_{(t)}$ represent the input and output, respectively. Thus, the values of the matrices $\boldsymbol{W}_{hx}$, $\boldsymbol{b}_h$, $\boldsymbol{W}_{yh}$, $\boldsymbol{b}_y$ have to be found to minimize a determined error index (MAPE, SEE, etc.) calculated between the vectors $\boldsymbol{w}_{(t)}\,\forall t=1, \ldots, T$ and the output data-points of the available training dataset.

Regarding Eqs. (3.3) and (3.4), the input vector ($\boldsymbol{u}_{(t)}$) has M rows and one column. The matrix $\boldsymbol{W}_{\mathbf{hx}}$ has P rows and M columns. The vector $\boldsymbol{b}_{\mathbf{h}}$ has P rows and one column. All these matrices are involved in some matrix operations whose result is evaluated in a sigmoid function, obtaining the value of the vector $\boldsymbol{h}_{(t)}$ that represents the information through the hidden layer. The vector $\boldsymbol{h}_{(t)}$ has P rows and one column. The matrix $\boldsymbol{W}_{\mathbf{yh}}$ has Q rows and P columns. The vector $\boldsymbol{b}_{\mathbf{y}}$ has Q rows and one column. Finally, the vector of output-data ($\boldsymbol{w}_{(t)}$) has Q rows and one column.

$$\boldsymbol{h}_{(t)} = \text{sigmoid}\left(\boldsymbol{W}_{\mathbf{hx}} \times \boldsymbol{u}_{(t)} + \boldsymbol{b}_{\mathbf{h}}\right) \tag{3.3}$$

$$\boldsymbol{w}_{(t)} = \text{sigmoid}\left(\boldsymbol{W}_{\mathbf{yh}} \times \boldsymbol{h}_{(t)} + \boldsymbol{b}_{\mathbf{y}}\right) \tag{3.4}$$

3.3.2 Recurrent neural network

In this section, we briefly describe the general structure of a recurrent neural network (RNN), which is shown in Fig. 3.6. In a general sense, the advantage of an RNN over

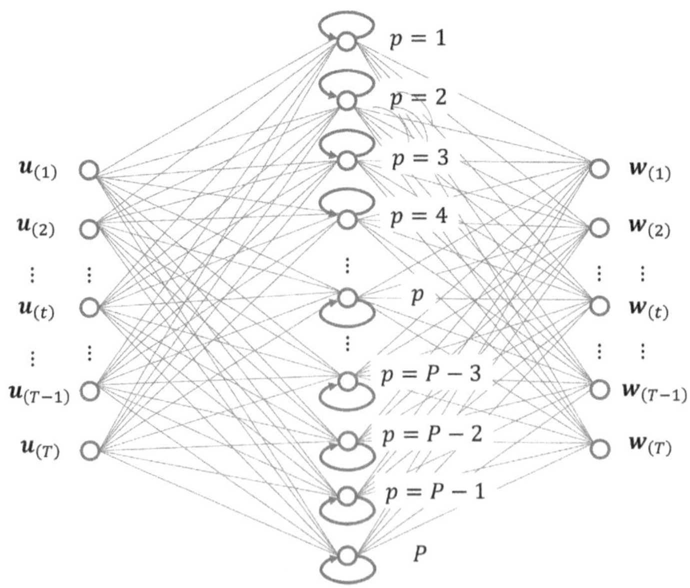

Fig. 3.6 General architecture of an RNN: Structure of a simple RNN.

an FFNN is that RNN can record results previously found using a recycling link (Rojas, 1996).

According to Fig. 3.6, we can observe that all of the input data-points ($t = 1, \ldots, T$) of the training set have been directly provided to the RNN so that the output data-points can be directly calculated. In the case of the FFNN, each data-point in the training set was provided one at a time, sequentially, so that the relationship between two different vectors cannot be considered.

The evaluation of the RNN shown in Fig. 3.6 is carried out using Eqs. (3.5) and (3.6), where the relationship between a determined data-point (t) and its predecessor ($t-1$) is explicitly represented. The matrix $\boldsymbol{W}_{\mathbf{hh}}$ has P rows and P columns. The initial value of the vector $\boldsymbol{h}_{(t)}$ at $t = 0$ can be assigned randomly.

$$\boldsymbol{h}_{(t)} = \text{sigmoid}\left(\boldsymbol{W}_{\mathbf{hh}} \times \boldsymbol{h}_{(t-1)} + \boldsymbol{W}_{\mathbf{hx}} \times \boldsymbol{u}_{(t)} + \boldsymbol{b}_{\mathbf{h}}\right) \tag{3.5}$$

$$\boldsymbol{w}_{(t)} = \text{sigmoid}\left(\boldsymbol{W}_{\mathbf{yh}} \times \boldsymbol{h}_{(t)} + \boldsymbol{b}_{\mathbf{y}}\right) \tag{3.6}$$

3.3.3 *Training feedforward and recurrent neural networks using a genetic algorithm*

The coding, or in other words, how we represent individuals during GA implementation, is a crucial step. As stated before, some authors have proposed to code the NN matrix elements as a vector. In this work, we are going to use a slightly different approach to represent a possible ensemble of matrices ($\boldsymbol{W}_{\mathbf{hh}}$, $\boldsymbol{W}_{\mathbf{hx}}$, $\boldsymbol{b}_{\mathbf{h}}$, $\boldsymbol{W}_{\mathbf{yh}}$, and $\boldsymbol{b}_{\mathbf{y}}$) as an individual.

Our goal is to determine the values of the FFNN matrices or the RNN matrices to minimize a determined error index (MAPE, SSE, etc.) calculated from the training dataset. This means considering the error between those values obtained from the NN evaluation (FFNN or RNN) and the available output data-points' values. This is an optimization problem in which the NN matrices' values have to be calculated to minimize the error. To this end, we have defined the structure of an individual using the three-index matrix ($\boldsymbol{\mathcal{L}}_{\mathbf{IND}}$) of Fig. 3.7, which refers to the specific case of an FFNN. The case of RNN is similar because we only need to add the matrix $\boldsymbol{W}_{\mathbf{hh}}$.

The first two indices refer to the rows and columns, while the third index ($g = 1, 2, 3, 4$) refers to the NN matrix ($\boldsymbol{W}_{\mathbf{hx}}$, $\boldsymbol{b}_{\mathbf{h}}$, $\boldsymbol{W}_{\mathbf{yh}}$, and $\boldsymbol{b}_{\mathbf{y}}$). Using this format, we can represent one GA individual using a three-index matrix of real values. This matrix has been called $\boldsymbol{\mathcal{L}}_{\mathbf{IND}}$. The dimensions of the matrix $\boldsymbol{\mathcal{L}}_{\mathbf{IND}}$ (L, O, and G) have to be defined considering the dimensions of all NN matrices. Thus, the maximum number of elements per index of the matrix $\boldsymbol{\mathcal{L}}_{\mathbf{IND}}$ is determined using Eq. (3.7).

$$\begin{aligned} L &= \max\left\{\boldsymbol{\mathcal{L}}_{\mathbf{NN}}(1,1), \boldsymbol{\mathcal{L}}_{\mathbf{NN}}(2,1), \boldsymbol{\mathcal{L}}_{\mathbf{NN}}(3,1), \boldsymbol{\mathcal{L}}_{\mathbf{NN}}(4,1)\right\} \\ O &= \max\left\{\boldsymbol{\mathcal{L}}_{\mathbf{NN}}(1,2), \boldsymbol{\mathcal{L}}_{\mathbf{NN}}(2,2), \boldsymbol{\mathcal{L}}_{\mathbf{NN}}(3,2), \boldsymbol{\mathcal{L}}_{\mathbf{NN}}(4,2)\right\} \\ G &= 4 \end{aligned} \tag{3.7}$$

Fig. 3.7 Architecture of a GA individual: Structure of an individual.

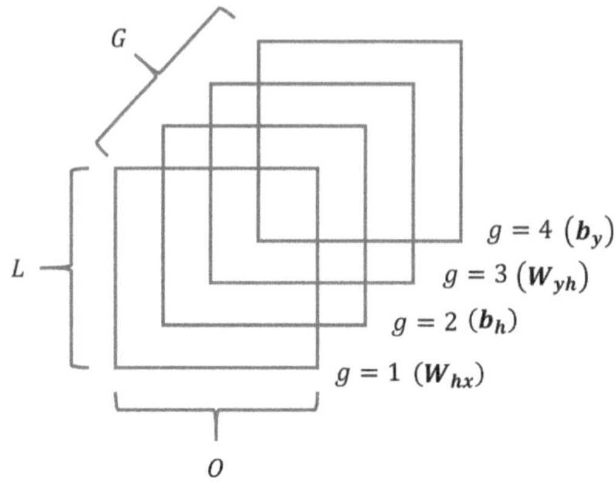

As we are only selecting the maximum rows and columns of the NN matrices, the elements of those matrices with a size lower than that specified in Eq. (3.7) are completed with the symbol of not a number (NaN) or another meaningless character.

Another critical topic is how we retrieve information from an individual to determine its fitness value. To solve this problem, we have to define a matrix $\boldsymbol{\mathcal{L}}_{\mathbf{NN}}$ according to Eq. (3.8):

$$\mathcal{L}_{\mathrm{NN}} = \begin{bmatrix} P & M \\ P & 1 \\ Q & P \\ Q & 1 \end{bmatrix} \tag{3.8}$$

The first row of $\boldsymbol{\mathcal{L}}_{\mathbf{NN}}$ is the number of rows and columns of the matrix $\boldsymbol{W}_{\mathbf{hx}}$. The second row is the number of rows and columns of the vector $\boldsymbol{b}_{\mathbf{h}}$. The third row is the number of rows and columns of the matrix $\boldsymbol{W}_{\mathbf{yh}}$. Finally, the fourth row is the number of rows and columns of vector $\boldsymbol{b}_{\mathbf{y}}$.

Fig. 3.8 shows an example of how to retrieve the matrix $\boldsymbol{W}_{\mathbf{hx}}$ from a determined individual represented with the matrix $\boldsymbol{\mathcal{L}}_{\mathbf{IND}}$. As can be observed, the elements to

Fig. 3.8 Storing and extracting the NN information in a matrix: Retrieving a matrix of NN from an individual.

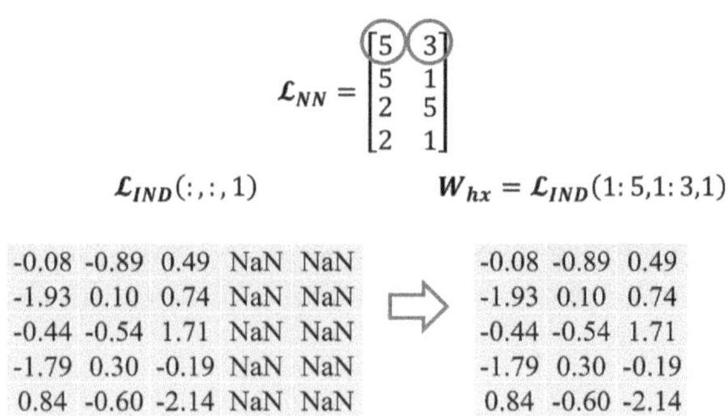

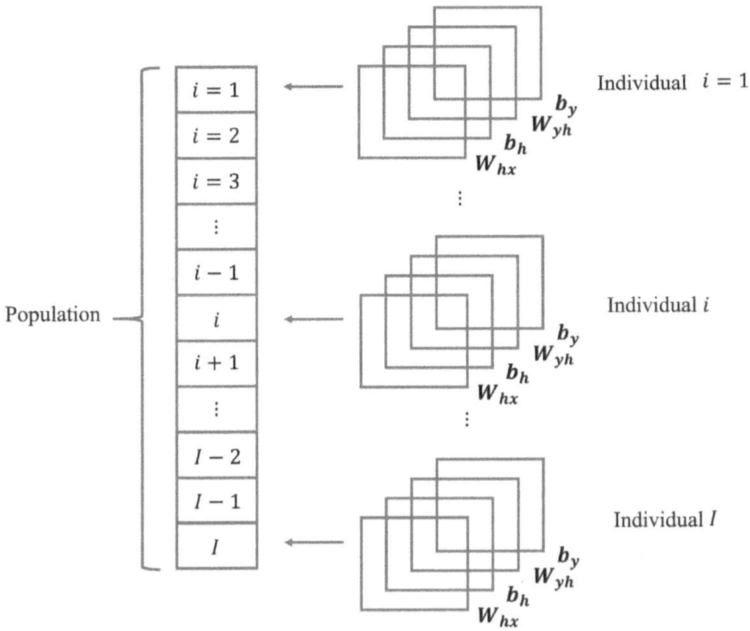

Fig. 3.9 Description of the GA population: Structure of the population.

be retrieved for the matrix $\mathbf{W_{hx}}$ are selected by setting the third index of the matrix $\mathcal{L}_{\mathbf{IND}}$ to one ($g=1$). Then, the elements to be included in the matrix $\mathbf{W_{hx}}$ are chosen considering the size specified in the matrix $\mathcal{L}_{\mathbf{NN}}$. A similar approach is followed to retrieve the remaining FFNN and RNN matrices.

Once we know how to decode an individual to retrieve the NN matrices, we can evaluate the FFNN or the RNN over the entire training dataset. Now, we are going to discuss how to build the population of the GA. The approach followed in this work is illustrated in Fig. 3.9, where the three-index matrix ($\mathcal{L}_{\mathbf{IND}}$) has been stored within a vector, which represents the GA population. This approach can be computationally implemented using a data structure. The index $i=1, \ldots, I$ represents each of the individuals in the GA population.

3.3.4 Training algorithm

As stated before, training an NN can be formulated as an optimization problem in which the NN matrices have to minimize the error between the results obtained from the NN evaluation and the available measured output data-points also available from the training dataset. In this work, we solve the optimization problem mentioned above using the technique proposed by Chuang et al. (2015, 2016).

Regarding the error index, we have used the root mean square deviation (RMSE) index in this work for NN training and, consequently, as the fitness value during the GA implementation.

The real-coded GA technique proposed by Chuang et al. consists of three main processes. The first process is the ranking selection (RS), the second process is the direction-based crossover (DBX), and the third process is the dynamic random mutation (DRM). In addition, the optimization technique requires selecting three parameters: the crossover rate (p_c), the mutation rate (p_m), and the proportional parameter (p_r). (y_{obj} is the objective function obtained from the RMSE, x_{pop} is the population vector of Fig. 3.9 with I elements).

RS allows us to select the best individuals within the population; we implement this procedure by following the method described as follows:

Step 1: Individuals are sorted according to fitness. In this regard, the order in which the individuals should be grouped is saved in a list ($\boldsymbol{\mathcal{L}}_{\mathbf{A}}$).

Step 2: Calculate the number of individuals that have relatively worse fitness. It is made by rounding down the product $I \times p_r$ (floor($I \times p_r$)).

Step 3: Select the first $I -$ floor($I \times p_r$) elements of the list $\boldsymbol{\mathcal{L}}_{\mathbf{A}}$ and save the result in another list ($\boldsymbol{\mathcal{L}}_{\mathbf{B}}$). These are the indices of the relatively best individuals to be kept. The remaining is the index of the individuals to be discarded.

Step 4: Select the first floor($I \times p_r$) elements of the list $\boldsymbol{\mathcal{L}}_{\mathbf{A}}$ and save them in another list ($\boldsymbol{\mathcal{L}}_{\mathbf{C}}$). These are the indices of the individuals used to replace relatively worse individuals.

Step 5: Create a list ($\boldsymbol{\mathcal{L}}_{\mathbf{D}}$) by joining or merging the lists $\boldsymbol{\mathcal{L}}_{\mathbf{B}}$ and $\boldsymbol{\mathcal{L}}_{\mathbf{C}}$ previously built.

Step 6: Sort the individuals in the list $\boldsymbol{\mathcal{L}}_{\mathbf{D}}$ according to fitness and save the results in another list ($\boldsymbol{\mathcal{L}}_{\mathbf{E}}$); this is the result of the RS process.

DBX enhances the traditional crossover operators' capabilities by dividing the group of selected individuals obtained from RS into two subgroups: leading individuals and the group of individuals with a lower ranking. Then, individuals of the two subgroups are crossed sequentially. DBX can be implemented by following the steps explained as follows:

Step 1: Set $u \leftarrow$ floor (I/2).

Step 2: Start the sequential crossing of pairs of individuals. Go to Step 2.1.

Step 2.1: Set $n \leftarrow 1$.

Step 2.2: If $n \leq$ floor(I/2), go to Step 2.3, else end.

Step 2.3: Set $u \leftarrow u + 1$.

Step 2.4: Select the individual n of $\boldsymbol{\mathcal{L}}_{\mathbf{E}}(\boldsymbol{\mathcal{L}}_{\mathbf{E}}(n))$, it will be called x^a_{pop}($x^a_{\text{pop}} \leftarrow x_{\text{pop}}(\boldsymbol{\mathcal{L}}_{\mathbf{E}}(n))$).

Step 2.5: Select the individual u of $\boldsymbol{\mathcal{L}}_{\mathbf{E}}(\boldsymbol{\mathcal{L}}_{\mathbf{E}}(u))$, it will be called x^b_{pop}($x^b_{\text{pop}} \leftarrow x_{\text{pop}}(\boldsymbol{\mathcal{L}}_{\mathbf{E}}(u))$).

Step 2.6: If $RAND > 1 - p_c$ and (max$\{y_{\text{obj}}(i) \forall i = 1, \ldots, I\}$ − min $\{y_{\text{obj}}(i) \forall i = 1, \ldots, I\}$) $> \varepsilon$, calculate the variable S_c using Eq. (3.9). *RAND* is a uniform random number in the interval [0, 1], and ε is a small number (tolerance).

$$S_c = \frac{\left| y_{\text{obj}}\left(x^a_{\text{pop}}\right) - y_{\text{obj}}\left(x^b_{\text{pop}}\right) \right|}{\max\left\{y_{\text{obj}}(i) \forall i = 1, \ldots, I\right\} - \min\left\{y_{\text{obj}}(i) \forall i = 1, \ldots, I\right\}} \tag{3.9}$$

Step 2.7: For each dimension, $g = 1, 2, 3, 4$, calculate the element of the corrective matrix ($\Delta x_{\text{pop}}(g)$) using Eq. (3.10). The corrective matrix ($\Delta x_{\text{pop}}(g)$) is a three-index matrix.

$$\Delta x_{\text{pop}}(g) = \begin{cases} 0; RAND < 0.5 \\ x^a_{\text{pop}(g)} - x^b_{\text{pop}(g)}; \; RAND \geq 0.5 \end{cases} \tag{3.10}$$

Step 2.8: Calculate the offspring of individuals a and b using Eq. (3.11).

$$\begin{aligned} x^a_{\text{pop}} &\leftarrow x^a_{\text{pop}} + S_c \Delta x_{\text{pop}} \\ x^b_{\text{pop}} &\leftarrow x^b_{\text{pop}} + S_c \Delta x_{\text{pop}} \end{aligned} \tag{3.11}$$

Step 2.9: Perform DBX in the next group of individuals by setting $n \leftarrow n+1$ and going to Step 2.2.

DRM is incorporated to improve the diversity of the population. DRM used in this work is slightly different from the original version of Chuang et al. because we are not considering any limit on the NN matrices' values. DRM has been implemented in this work according to the procedure explained as follows:

Step 1: Considering the generation under study (k), calculate the variable S_m using Eq. (3.12), where K is the total number of generations ($k = 1, \ldots, K$).

$$S_m = \left(\frac{1-k}{K}\right)^2 \tag{3.12}$$

Step 2: Apply DRM over the population. Go to Step 2.1.

Step 2.1: Set $i \leftarrow 1$.

Step 2.2: If $i \leq I$, go to Step 2.3, else end.

Step 2.3: If $RAND < p_m$, go to Step 2.4.

Step 2.4: Set $x_{\text{pop}(i)} \leftarrow x_{\text{pop}(i)} + S_m \boldsymbol{RANG}$, where $\boldsymbol{RANG}$ is a random matrix from a Gaussian distribution with mean 0 and standard deviation 1. The rows and columns of $\boldsymbol{RANG}$ are given in Eq. (3.8).

Step 2.5: Set $i \leftarrow i+1$ and go to Step 2.2.

We have implemented an elitist GA. Specifically, we have selected the best individual for each generation. Once RS, DBX, and DRM have been applied, we have inserted the best individual in a position chosen randomly within the population. In other words, we have randomly selected a number in the interval $[1, I]$. Then, we have arbitrarily substituted the existing individual with the best.

3.4 Case study

In this section, we illustrate the methodology previously explained to train an FFNN or an RNN using a GA by applying the MCP method to estimate wind resources. In the next subsections, the specific conditions of the case study, as well as the NN training process using an FFNN and an RNN, will be carefully explained.

3.4.1 Case description

Our case study will be located in the Netherlands. We will use hourly values of wind speed and direction measured by the Dutch meteorological institute (Koninklijk Nederlands Meteorologisch Instituut (KNMI), 2020) to illustrate the MCP method.

Fig. 3.10 The Netherlands: Meteorological stations of the Netherlands (KNMI).

Fig. 3.10 shows the map of Dutch stations. The reference places will be Cadzand, Hoofdplaat, Vlakte van de Raan, and Wilhelminadorp, while the target place will be Vlissingen.

The training dataset has 8760 data-points ($T = 8760$), which corresponds to 1 year. According to the scheme previously discussed in Fig. 3.2, this would be the short-term measurement. The testing dataset has 54,035 hourly values of wind speed and direction. This dataset is the long-term measurement used to complete the target place's limited wind speed and direction series. During the training process, the implemented GA will minimize the RMSE error index using the training dataset. We will then test the performance of the trained NN (FFNN or RNN) by using long-term measurement as an independent dataset.

Regarding the settings of the GA, the crossover rate is set to 0.95 ($p_c = 0.95$), the mutation rate is set to 0.1 ($p_m = 0.1$), while the proportional parameter is set to 0.3 ($p_r = 0.3$). The population size is set to 100 individuals ($I = 100$), and the number of generations is adjusted to 300 ($K = 300$).

An FFNN and an RNN are considered using the same structure, which means that we have chosen the number of hidden units as 25 ($P = 25$). We present the matrices $\mathcal{L}_{\mathrm{NN}}$ in Eq. (3.13) for the FFNN and in Eq. (3.14) for the RNN (sizes of matrices $\boldsymbol{W}_{\mathbf{hh}}$, $\boldsymbol{W}_{\mathbf{hx}}$, $\boldsymbol{b}_{\mathbf{h}}$, $\boldsymbol{W}_{\mathbf{yh}}$, and $\boldsymbol{b}_{\mathbf{y}}$ for each row, 1, 2, 3, 4, and 5, respectively). During the RNN training, we have initialized the vector $\boldsymbol{h}_{(t)}$ to zero at $t = 0$.

$$\mathcal{L}_{\mathrm{NN}} = \begin{bmatrix} 25 & 8 \\ 25 & 1 \\ 2 & 25 \\ 2 & 1 \end{bmatrix} \tag{3.13}$$

$$\mathcal{L}_{\mathrm{NN}} = \begin{bmatrix} 25 & 25 \\ 25 & 8 \\ 25 & 1 \\ 2 & 25 \\ 2 & 1 \end{bmatrix} \tag{3.14}$$

In the next section, we will discuss the main results related to the NN training process. We will consider an FFNN similar to that shown in Fig. 3.5 and an RNN with the structure shown in Fig. 3.6. Before starting the training process, we have scaled the training dataset in the interval [0, 1]; this is a usual NN training procedure. We have implemented the optimization algorithm in a computer provided with i7-3630QM CPU at 2.40GHz, 8GB of RAM, and a 64-bit operating system. The computing language used was MATLAB®. Referring to MATLAB® syntax, the population presented in Fig. 3.9 can be created by using a data structure repetitively. (It can be done using this syntax: *Population = repmat(struct('stratum',NaN.*ones(L,O,G)), I,1).*) Then, a specific individual can be accessed using the corresponding index *i* (the reader can implement it using this syntax: *Individual = Population(i,1).stratum*), which is a three-index matrix.

The next subsections describe the results obtained from implementing the MCP method using an FFNN and an RNN to estimate the wind speed and direction for the target place previously mentioned. MCP results are evaluated in terms of the RMSE error index, wind speed, and direction distributions, as well as the sample and partial autocorrelation functions.

3.4.2 Measure-correlate-predict using a feedforward network

This subsection presents the results related to the MCP method's implementation based on training an FFNN using a real-coded GA. Regarding the wind speed and direction estimation, the RMSE error was 2.495 m/s and 51.467 degree, respectively. Fig. 3.11 shows the probability distribution of estimated wind speed, where we can

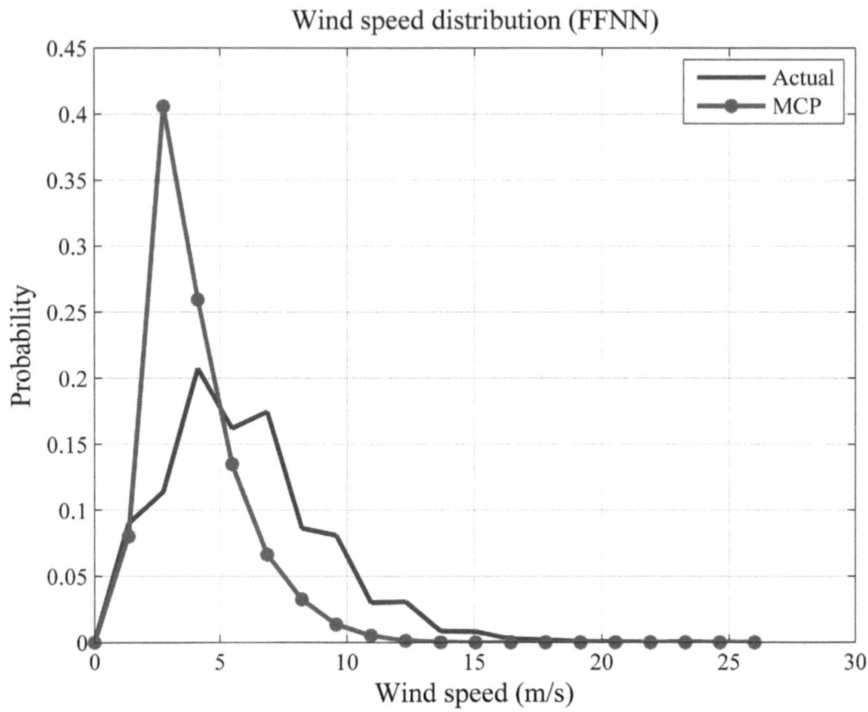

Fig. 3.11 Wind speed probability distribution using an FFNN: Actual and estimated wind speed distribution (FFNN).

Table 3.1 Comparison of Weibull distribution parameters (FFNN).

Data	Scale factor (m/s)	Shape factor	Average (m/s)
Actual	6.652	2.065	5.873
MCP	4.468	2.307	3.946

observe the effects of MPC estimation error. Additionally, Table 3.1 presents the Weibull distribution parameters and mean values. We can identify a difference of 32.8% on the scale factor, 11.7% on the shape factor, and 32.8% on the mean wind speed.

Fig. 3.12 shows the probability distribution of wind direction, where we can observe a reasonable agreement.

Figs. 3.13 and 3.14 present the sample and partial autocorrelation functions for both the actual and estimated values through the MCP method. As can be observed, the MCP method results adequately describe the autoregressive nature of the wind time series.

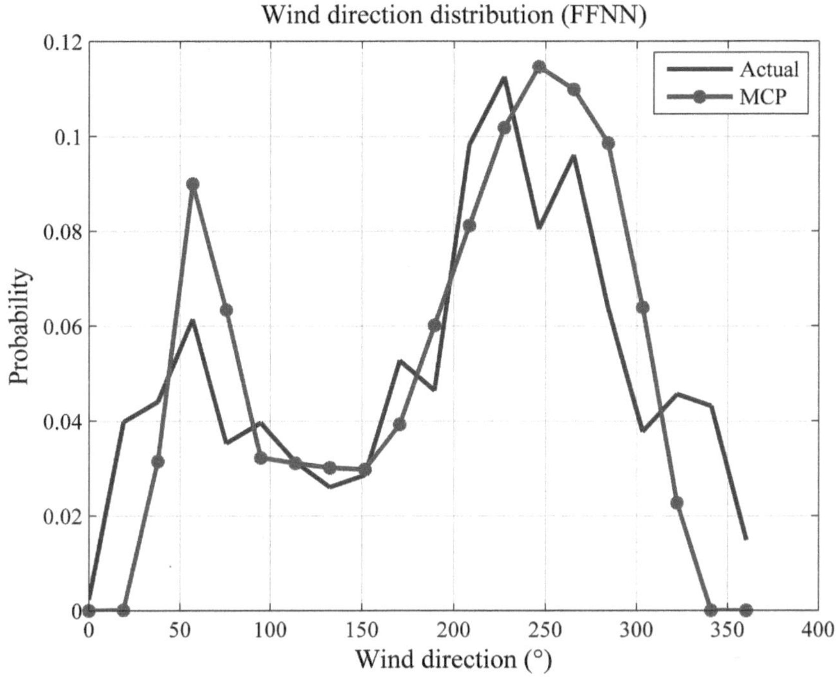

Fig. 3.12 Wind direction probability distribution using an FFNN: Actual and estimated wind direction distribution (FFNN).

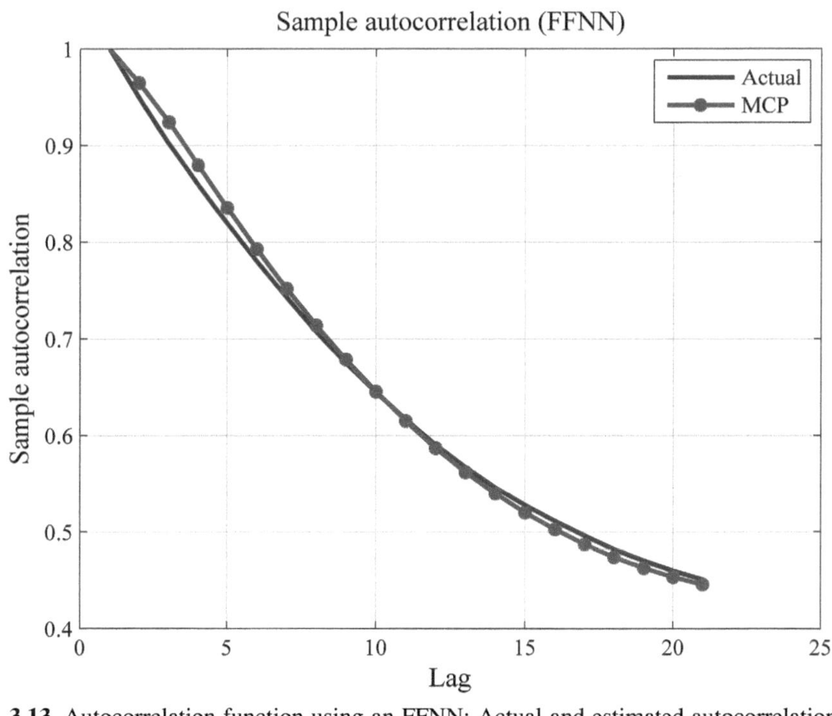

Fig. 3.13 Autocorrelation function using an FFNN: Actual and estimated autocorrelation (FFNN).

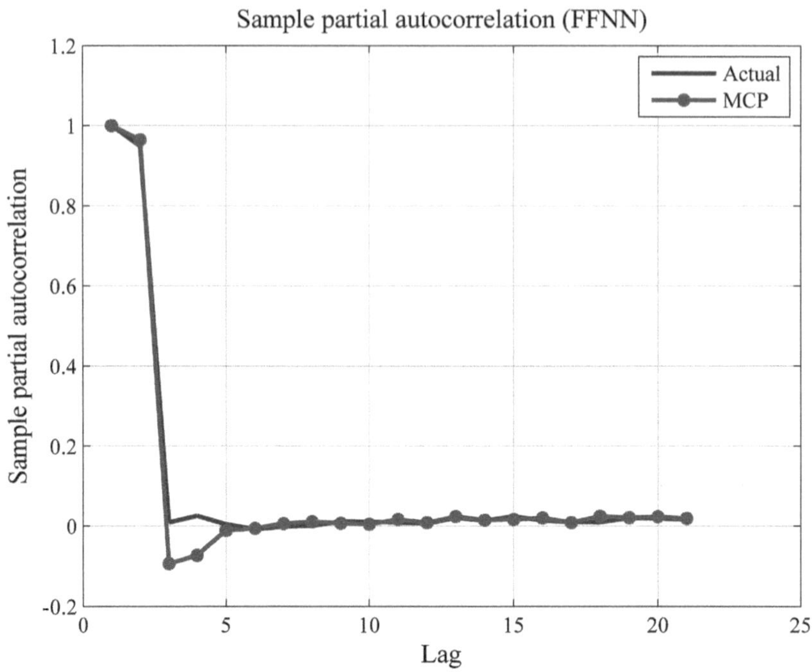

Fig. 3.14 Partial autocorrelation function using an FFNN: Actual and estimated partial autocorrelation (FFNN).

3.4.3 Measure-correlate-predict using a recurrent network

This subsection presents the results obtained when an RNN is trained for wind resource estimation through the MCP method. Concerning the error index, the RMSE obtained was 2.331 m/s for wind speed and 49.417 degree for wind direction estimation. Fig. 3.15 shows the probability distribution of wind speed obtained from the RNN. These results are complemented by the information in Table 3.2, where the Weibull distribution parameters and the mean wind speed are reported. According to these results, the difference in the scale factors is 29.6%, in the shape factors is 5.9%, and in the mean wind speed is 29.7%.

Fig. 3.16 shows the probability distribution of wind direction, where we can observe a reasonable agreement with the measured data.

Figs. 3.17 and 3.18 show the sample and partial autocorrelation functions. The MCP technique can successfully estimate the purely autoregressive behavior of the measured wind speed.

Fig. 3.19 shows the probability distribution of the error related to the wind speed estimation for both NNs. We can observe that both distributions have a small bias. For the FFNN, the bias is 1.927 m/s, while for the RNN, the bias is 1.743 m/s—a little lower. Fig. 3.20 presents the probability distribution of the error related to the wind direction estimation for both NNs. In this case, the bias is lower than in the case of the

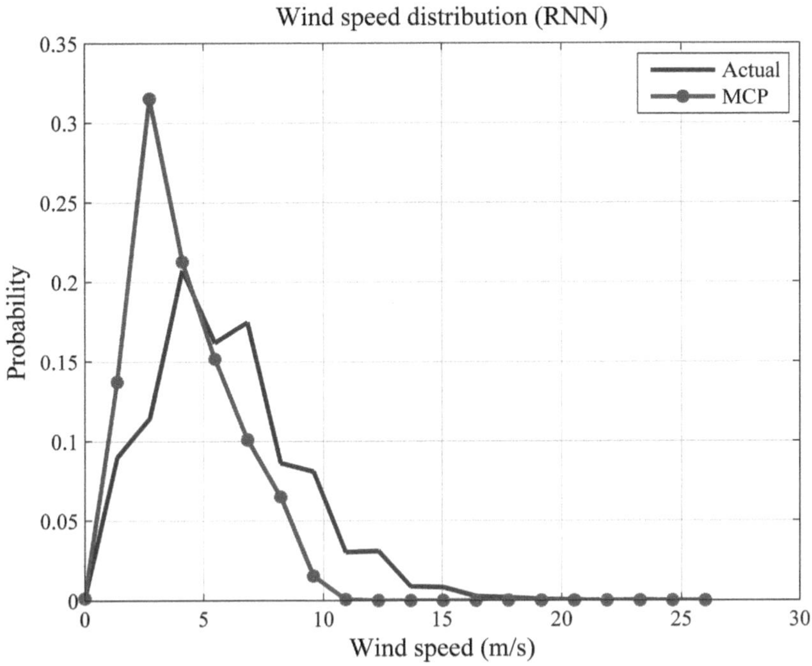

Fig. 3.15 Wind speed probability distribution using an RNN: Actual and estimated wind speed distribution (RNN).

Table 3.2 Comparison of Weibull distribution parameters (RNN).

Data	Scale factor (m/s)	Shape factor	Average (m/s)
Actual	6.652	2.065	5.873
MCP	4.682	2.186	4.129

wind speed estimation. For the FFNN, the bias is 0.6105 degree, and for the case of the RNN, the bias is 3.3206 degree.

3.4.4 Genetic algorithm performance and measure-correlate-predict error estimation

In this section, we will discuss the performance of the real-coded GA implemented to train the FNN and the RNN with the characteristics previously described. Regarding the training process of the FFNN, the computational time spent was 194.473 min, while the time required by the RNN was 230.526 min. Fig. 3.21 presents the evolution of the optimization process, where we can observe that the RNN offers a slightly better result compared to the FFNN. We want to reiterate that the training dataset was scaled

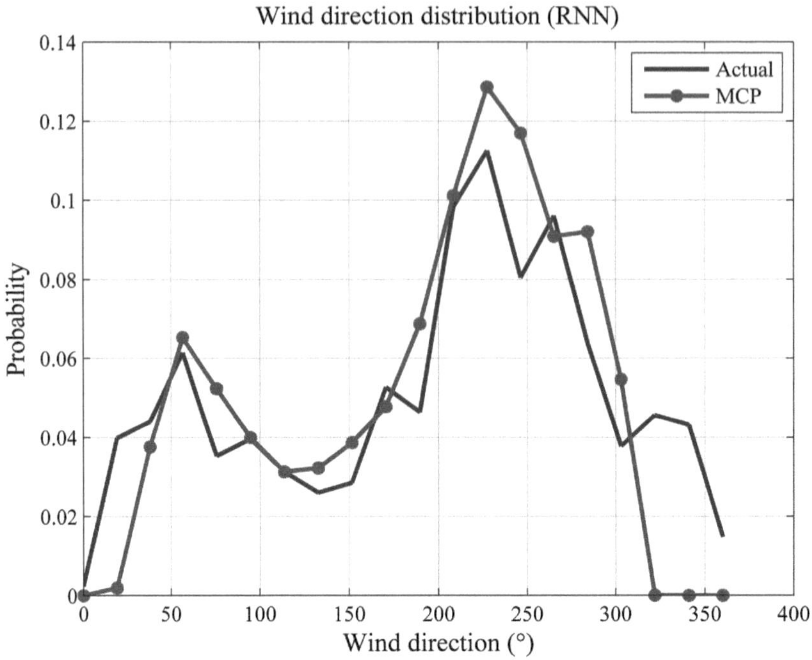

Fig. 3.16 Wind direction probability distribution using an RNN: Actual and estimated wind direction distribution (RNN).

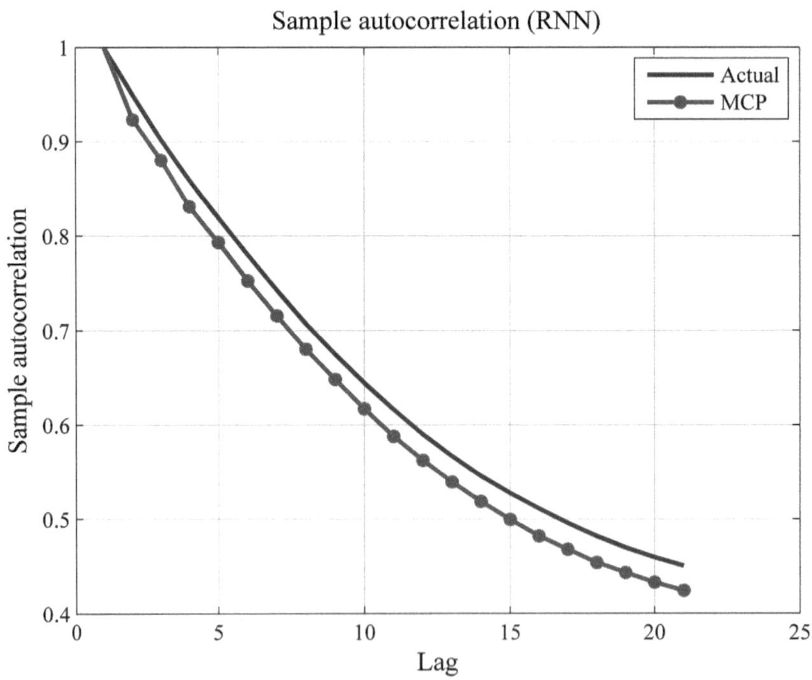

Fig. 3.17 Autocorrelation function using an RNN: Actual and estimated autocorrelation (RNN).

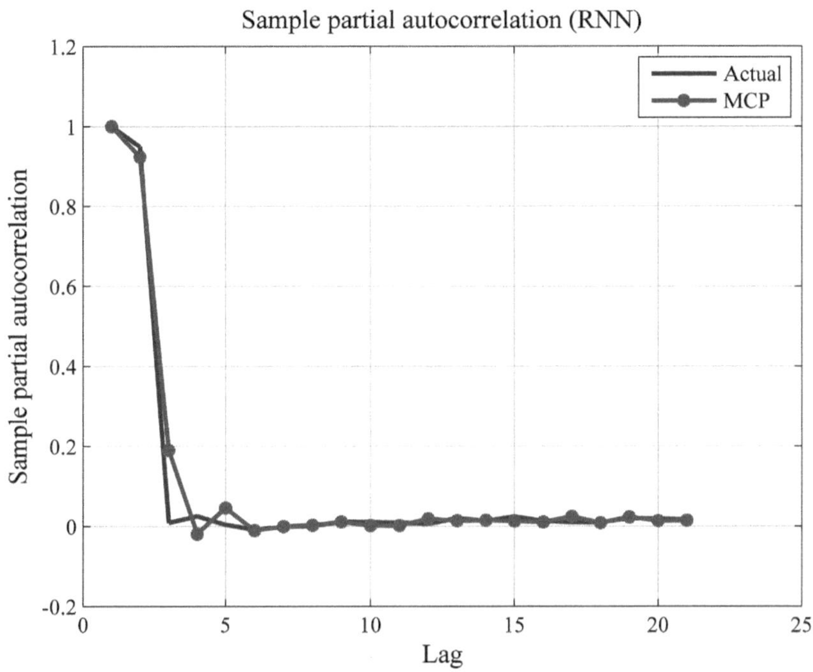

Fig. 3.18 Partial autocorrelation function using an RNN: Actual and estimated partial autocorrelation (RNN).

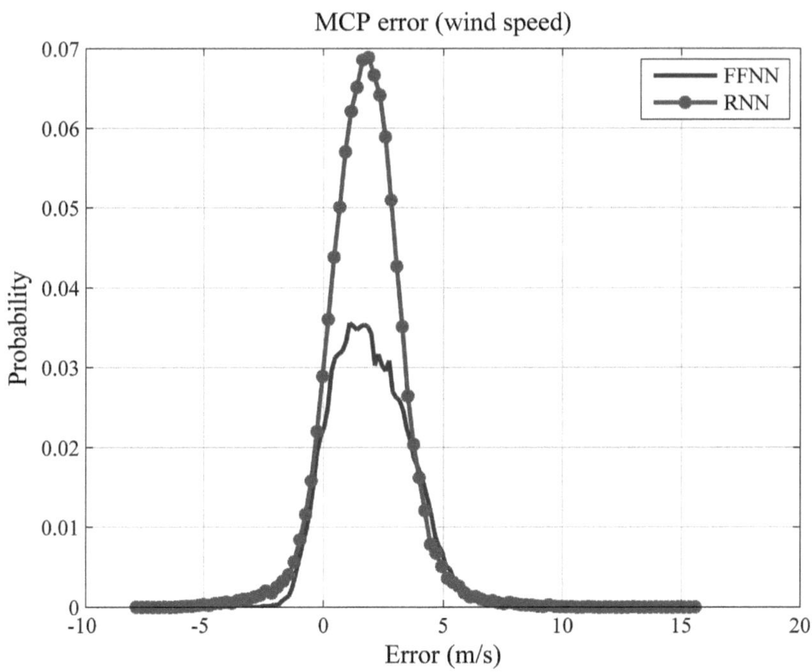

Fig. 3.19 Probability distribution of the error (wind speed): Error distribution of wind speed estimation.

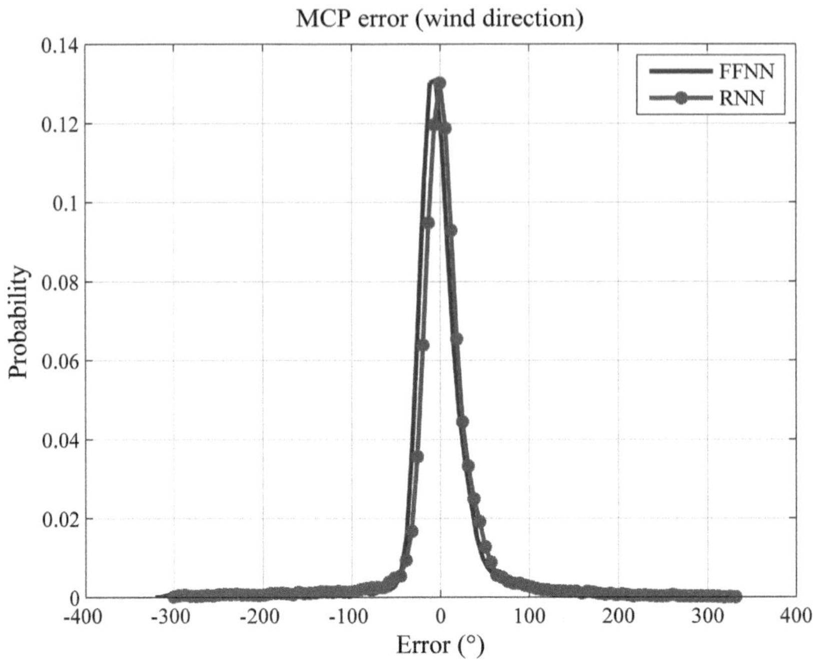

Fig. 3.20 Probability distribution of the error (wind direction): Error distribution of wind direction estimation.

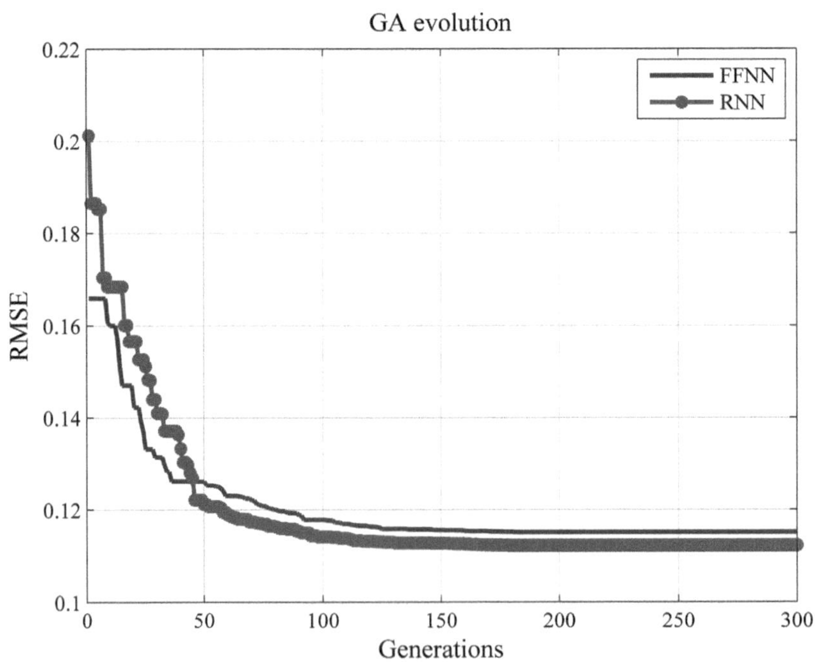

Fig. 3.21 Training an FFNN and an RNN using a GA: Evolution of the GA for training an FFNN and an RNN.

in the range [0, 1]. Thus, the RMSE results shown in Fig. 3.21 were calculated by the GA using the scaled values.

In general, heuristic techniques such as ant colony optimization (ACO) are population-based approaches with high versatility and robustness. However, depending on the problem under analysis, heuristic optimization methods can be outperformed by specialized optimization procedures (Dorigo et al., 1996). NN training is an example of this situation. BP and Levenberg–Marquardt (LM) algorithms are well-known and effective methods that offer better results than some of the most popular heuristic techniques. Under this circumstance, we will employ the LM algorithm as a reference for comparison.

The reference training method is the LM algorithm, considering an FFNN with a hyperbolic tangent sigmoid transfer function and a linear activation function. The training process was performed over 300 epochs, considering a learning rate equal to 0.0001. The RMSE of the wind speed and direction estimations are 1.215 m/s and 56.275 degree, respectively. These results are obtained by spending only 1.357 min, which is negligible when compared to GA implementation. According to these results, we can observe that a GA requires a considerable amount of computational resources to reach a reasonably good solution.

On the other hand, as with the remainder of heuristic techniques, ACO could be initialized with a near-optimal solution rapidly obtained from implementing another approach, thus increasing computational efficiency (Dorigo & Gambardella, 1997). In other words, the capabilities and searching power of a GA can be used to improve the quality of an already known solution obtained by implementing another technique or heuristic. This idea will be discussed in Chapter 4 for a binary-coded GA and Chapter 5 for NN training by a real-coded GA.

3.5 Conclusions

This chapter has explored essential concepts related to natural resources estimation, emphasizing wind and solar resources. We have described the measurement process, the required instruments, the available databases, and the empirical models developed to estimate natural resources in those locations where the information is limited, and the MCP method. We centered our attention on training an NN (FFNN or RNN) using a real-coded GA. Specifically, we describe an individual's structure, starting from the set of matrices that constitute the NN under study; these matrices are all embedded into a three-index matrix, which is later embedded into a vector to construct the entire population. Once we have built the GA population, the optimization problem is solved using a real-coded GA that requires RS, DBX, and DRM.

We have conducted a case study of wind resource estimation using data from the Netherlands. Computational experiments with an FFNN and an RNN revealed similar results for both of them. In wind speed estimation, the error observed in the mean wind speed was 32.8% for the FFNN, while for the RNN, the error was 29.7%. In wind direction estimation, the error observed from the implementation of the FFNN was 51.46 degree. For the case of the RNN, the error was 49.42 degree. For both variables,

RNN offered an error slightly lower than the FFNN. In addition, both techniques presented a low bias on the probability distribution of the estimation error. The comparison with an FFNN trained by LM algorithm revealed that using a GA as a training method requires considerable computational resources. Using the LM algorithm, the RMSE during wind speed and direction estimations are 1.215 m/s and 56.275 degree, respectively, requiring only 1.357 min.

Despite these results, GA searching power can be employed to enhance the performance of a solution already estimated by an alternative method. In other words, GA can be initialized with a high-quality individual obtained by an alternative technique in order to guide the algorithm in a promising direction. This idea will be explored later in Chapters 4 and 5.

References

Akinoğlu, B., & Ecevit, A. (1990). Construction of a quadratic model using modified Ångström coefficients to estimate global solar radiation. *Solar Energy*, *45*(2), 85–92. https://doi.org/10.1016/0038-092X(90)90032-8.

Aksoy, H., Toprak, Z., Aytek, A., & Ünal, N. (2004). Stochastic generation of hourly mean wind speed data. *Renewable Energy*, *29*(14), 2111–2131. https://doi.org/10.1016/j.renene.2004.03.011.

Ali, S., Lee, S.-M., & Jang, C.-M. (2018). Forecasting the long-term wind data via measure-correlate-predict (MCP) methods. *Energies*, *11*(6), 1541. https://doi.org/10.3390/en11061541.

Almorox, J., & Hontoria, C. (2004). Global solar radiation estimation using sunshine duration in Spain. *Energy Conversion and Management*, *45*(9–10), 1529–1535. https://doi.org/10.1016/j.enconman.2003.08.022.

Almorox, J., Hontoria, C., & Benito, M. (2011). Models for obtaining daily global solar radiation with measured air temperature data in Madrid (Spain). *Applied Energy*, *88*(5), 1703–1709. https://doi.org/10.1016/j.apenergy.2010.11.003.

AWS Scientific, Inc. (1997). *Wind resource assessment handbook: Fundamentals for conducting a successful monitoring program*. National Renewable Energy Laboratory. NREL subcontract no. TAT-5-15283-01 https://www.nrel.gov/docs/legosti/fy97/22223.pdf.

Barbier, E. (2002). Geothermal energy technology and current status: An overview. *Renewable and Sustainable Energy Reviews*, *6*(1–2), 3–65. https://doi.org/10.1016/S1364-0321(02)00002-3.

Besharat, F., Dehghan, A., & Faghih, A. (2013). Empirical models for estimating global solar radiation: A review and case study. *Renewable and Sustainable Energy Reviews*, *21*, 798–821. https://doi.org/10.1016/j.rser.2012.12.043.

Blanco, A., Delgado, M., & Pegalajar, M. (2000). A genetic algorithm to obtain the optimal recurrent neural network. *International Journal of Approximate Reasoning*, *23*, 67–83. https://doi.org/10.1016/S0888-613X(99)00032-8.

Blanco, A., Delgado, M., & Pegalajar, M. (2001). A real-coded genetic algorithm for training recurrent neural networks. *Neural Networks*, *14*, 93–105. https://doi.org/10.1016/S0893-6080(00)00081-2.

Borawski, P., Beldycka-Borawska, A., Jankowski, K., Dubis, B., & Dunn, J. (2020). Development of wind energy market in the European Union. *Renewable Energy*, *161*, 691–700. https://doi.org/10.1016/j.renene.2020.07.081.

Boyd, M. (2016). *NIST weather station for photovoltaic and building system research*. National Institute of Standards and Technology. NIST Technical Note 1913 https://nvlpubs.nist.gov/nistpubs/TechnicalNotes/NIST.TN.1913.pdf.
Bulut, H., & Buyukalaca, O. (2007). Simple model for the generation of the daily global solar-radiation data in Turkey. *Applied Energy*, *84*(5), 477–491. https://doi.org/10.1016/j.apenergy.2006.10.003.
Carapellucci, R., & Giordano, L. (2013). A new approach for synthetically generating wind speeds: A comparison with the Markov chains method. *Energy*, *49*, 298–305. https://doi.org/10.1016/j.energy.2012.10.032.
Carta, J., & Velázquez, S. (2011). A new probabilistic method to estimate the long-term wind speed characteristics at a potential wind energy conversion site. *Energy*, *36*(5), 2671–2685. https://doi.org/10.1016/j.energy.2011.02.008.
Carta, J., Velázquez, S., & Cabrera, P. (2013). A review of measure-correlate-predict (MCP) methods used to estimate long-term wind characteristics at a target site. *Renewable and Sustainable Energy Reviews*, *27*, 362–400. https://doi.org/10.1016/j.rser.2013.07.004.
Carta, J., Velázquez, S., & Matías, J. (2011). Use of Bayesian networks classifiers for long-term mean wind turbine energy output estimation at potential wind energy conversion site. *Energy Conversion and Management*, *52*(2), 1137–1149. https://doi.org/10.1016/j.enconman.2010.09.008.
Chen, R., Ersi, K., Yang, J., Lu, S., & Zhao, W. (2004). Validation of five global radiation models with measured daily data in China. *Energy Conversion and Management*, *45* (11 – 12), 1759–1769. https://doi.org/10.1016/j.enconman.2003.09.019.
Chuang, Y.-C., Chen, C.-T., & Hwang, C. (2015). A real-coded genetic algorithm with a direction-based crossover operator. *Information Sciences*, *305*, 320–348. https://doi.org/10.1016/j.ins.2015.01.026.
Chuang, Y.-C., Chen, C.-T., & Hwang, C. (2016). A simple and efficient real-coded genetic algorithm for constrained optimization. *Applied Soft Computing*, *38*, 87–105. https://doi.org/10.1016/j.asoc.2015.09.036.
Compo, G. P., Whitaker, J. S., Sardeshmukh, P. D., Matsui, N., Allan, R. J., Yin, X., Gleason, B. E., Vose, R. S., Rutledge, G., Bessemoulin, P., Brönnimann, S., Brunet, M., Crouthamel, R. I., Grant, A. N., Groisman, P. Y., Jones, P. D., Kruk, M. C., Kruger, A. C., Marshall, G. J., … Worley, S. J. (2011). The twentieth century reanalysis project. *Quarterly Journal of the Royal Meteorological Society*, *137*, 1–28. https://doi.org/10.1002/qj.776.
De Miguel, A., Bilbao, J., Aguiar, R., Kambezidis, H., & Negro, E. (2001). Diffuse solar irradiation model evaluation in the North Mediterranean belt area. *Solar Energy*, *70*(2), 143–153. https://doi.org/10.1016/S0038-092X(00)00135-3.
Dee, D., Uppala, S., Simmons, A., Berrisford, P., Poli, P., Kobayashi, S., Andrae, U., Balmaseda, M., Balsamo, G., Bauer, P., Bechtold, P., Beljaars, A., van de Berg, L., Bidlot, J., Bormann, N., Delsol, C., Dragani, R., Fuentes, M., Geer, A., … Vitart, F. (2011). The ERA-interim reanalysis: Configuration and performance of the data assimilation system. *Quarterly Journal of the Royal Meteorological Society*, *137*, 553–597. https://doi.org/10.1002/qj.828.
Delucchi, M., & Jacobson, M. (2011). Providing all global energy with wind, water, and solar power, part II: Reliability, system and transmission costs, and policies. *Energy Policy*, *39* (3), 1170–1190. https://doi.org/10.1016/j.enpol.2010.11.045.
Despotovic, M., Nedic, V., Despotovic, D., & Cvetanovic, S. (2015). Review and statistical analysis of different global solar radiation sunshine models. *Renewable and Sustainable Energy Reviews*, *52*, 1869–1880. https://doi.org/10.1016/j.rser.2015.08.035.

Díaz, S., Carta, J., & Matías, J. (2017). Comparison of several measure-correlate-predict models using support vector regression techniques to estimate wind power densities. A case study. *Energy Conversion and Management*, *140*, 334–354. https://doi.org/10.1016/j.enconman.2017.02.064.

Díaz, S., Carta, J., & Matías, J. (2018). Performance assessment of five MCP models proposed for the estimation of long-term wind turbine power outputs at a target site using three machine learning techniques. *Applied Energy*, *209*, 455–477. https://doi.org/10.1016/j.apenergy.2017.11.007.

Dinler, A. (2013). A new low-correlation MCP (measure-correlate-predict) method for wind energy forecasting. *Energy*, *63*, 152–160. https://doi.org/10.1016/j.energy.2013.10.007.

Dorigo, M., & Gambardella, L. M. (1997). Ant colonies for the travelling salesman problem. *Biosystems*, *43*(2), 73–81. https://doi.org/10.1016/S0303-2647(97)01708-5.

Dorigo, M., Maniezzo, V., & Colorni, A. (1996). Ant system: Optimization by a colony of cooperating agents. *IEEE Transactions on Systems, Man, and Cybernetics. Part B, Cybernetics*, *26*(1), 29–41. https://doi.org/10.1109/3477.484436.

Duffie, J., & Beckman, W. (1991). *Solar engineering of thermal processes* (4th ed.). Wiley. https://doi.org/10.1002/9781118671603.

Ellabban, O., Abu-Rub, H., & Blaabjerg, F. (2014). Renewable energy resources: Current status, future prospects and their enabling technology. *Renewable and Sustainable Energy Reviews*, *39*, 748–764. https://doi.org/10.1016/j.rser.2014.07.113.

El-Sebaii, A., Al-Hazmi, F., Al-Ghamdi, A., & Yaghmour, S. (2010). Global, direct and diffuse solar radiation on horizontal and tilted surfaces in Jeddah, Saudi Arabia. *Applied Energy*, *87*(2), 568–576. https://doi.org/10.1016/j.apenergy.2009.06.032.

Erbs, D., Klein, S., & Duffie, J. (1982). Estimation of the diffuse radiation fraction for hourly, daily, and monthly-average global radiation. *Solar Energy*, *28*(4), 293–302. https://doi.org/10.1016/0038-092X(82)90302-4.

European Centre for Medium-Range Weather Forecasts (ECMWF). (2020). https://www.ecmwf.int/.

Falnes, J. (2007). A review of wave-energy extraction. *Marine Structures*, *20*(4), 185–201. https://doi.org/10.1016/j.marstruc.2007.09.001.

Fan, J., Wu, L., Zhang, F., Cai, H., Zeng, W., Wang, X., & Zou, H. (2019). Empirical and machine learning models for predicting daily global solar radiation from sunshine duration: A review and case study in China. *Renewable and Sustainable Energy Reviews*, *100*, 186–212. https://doi.org/10.1016/j.rser.2018.10.018.

Feng, S., Hu, Q., & Qian, W. (2004). Quality control of daily meteorological data in China, 1951-2000: A new dataset. *International Journal of Climatology*, *24*(7), 853–870. https://doi.org/10.1002/joc.1047.

Fick, S., & Hijmans, R. (2017). WorldClim 2: New 1-km spatial resolution climate surfaces for global land areas. *International Journal of Climatology*, *37*(12), 4302–4315. https://doi.org/10.1002/joc.5086.

Gass, V., Strauss, F., Schmidt, J., & Schmid, E. (2011). Assessing the effect of wind power uncertainty on profitability. *Renewable and Sustainable Energy Reviews*, *15*(6), 2677–2683. https://doi.org/10.1016/j.rser.2011.01.024.

Graham, V., & Hollands, K. (1990). A method to generate synthetic hourly solar radiation globally. *Solar Energy*, *44*(6), 333–341. https://doi.org/10.1016/0038-092X(90)90137-2.

Gueymard, C. (2008). REST2: High-performance solar radiation model for cloudless-sky irradiance, illuminance, and photosynthetically active radiation—Validation with a benchmark dataset. *Solar Energy*, *82*(3), 272–285. https://doi.org/10.1016/j.solener.2007.04.008.

Hassan, M., Khalil, A., Khalil, S., & Kassem, M. (2018). Independent models for estimation of daily global solar radiation: A review and a case study. *Renewable and Sustainable Energy Reviews*, *82*, 1565–1575. https://doi.org/10.1016/j.rser.2017.07.002.

Herbert, G., Iniyan, S., Sreevalsan, E., & Rajapandian, S. (2007). A review of wind energy technologies. *Renewable and Sustainable Energy Reviews*, *11*(6), 1117–1145. https://doi.org/10.1016/j.rser.2005.08.004.

Hersbach, H., Bell, B., Berrisford, P., Hirahara, S., Horányi, A., Muñoz-Sabater, J., Nicolas, J., Peubey, C., Radu, R., Schepers, D., Simmons, A., Soci, C., Abdalla, S., Abellan, X., Balsamo, G., Bechtold, P., Biavati, G., Bidlot, J., Bonavita, M., … Thépaut, J.-N. (2020). The ERA5 global reanalysis. *Quarterly Journal of the Royal Meteorological Society*, *146*, 1999–2049. https://doi.org/10.1002/qj.3803.

Hochreiter, S., & Schmidhuber, J. (1997). Long short-term memory. *Neural Computation*, *9*(8), 1735–1780. https://doi.org/10.1162/neco.1997.9.8.1735.

Ishihara, T., & Yamaguchi, A. (2015). Prediction of the extreme wind speed in the mixed climate region by using Monte Carlo simulation and measure-correlate-predict method. *Wind Energy*, *18*(1), 171–186. https://doi.org/10.1002/we.1693.

Jacobson, M., & Delucchi, M. (2011). Providing all global energy with wind, water, and solar power, part I: Technologies, energy resources, quantities and areas of infrastructure, and materials. *Energy Policy*, *39*(3), 1154–1169. https://doi.org/10.1016/j.enpol.2010.11.040.

Jacobson, M., & Jadhav, V. (2018). World estimates of PV optimal tilt angles and ratios of sunlight incident upon tilted and tracked PV panels relative to horizontal panels. *Solar Energy*, *169*, 55–66. https://doi.org/10.1016/j.solener.2018.04.030.

Jacovides, C., Tymvios, F., Assimakopoulos, V., & Kaltsounides, N. (2006). Comparative study of various correlations in estimating hourly diffuse fraction of global solar radiation. *Renewable Energy*, *31*(15), 2492–2504. https://doi.org/10.1016/j.renene.2005.11.009.

Kanamitsu, M., Ebisuzaki, W., Woollen, J., Yang, S.-K., Hnilo, J., Fiorino, M., & Potter, G. (2002). NCEP-DOE AMIP-II reanalysis (R-2). *American Meteorological Society*, *83* (11), 1631–1644. https://doi.org/10.1175/BAMS-83-11-1631.

Kobayashi, S., Ota, Y., Harada, Y., Ebita, A., Moriya, M., Onoda, H., Onogi, K., Kamahori, H., Kobayashi, C., Endo, H., Miyaoka, K., & Takahashi, K. (2015). The JRA-55 reanalysis: General specifications and basic characteristics. *Journal of the Meteorological Society of Japan*, *93*(1), 5–48. https://doi.org/10.2151/jmsj.2015-001.

Koninklijk Nederlands Meteorologisch Instituut (KNMI). (2020). https://www.knmi.nl/over-het-knmi/about.

Lackner, M., Rogers, A., & Manwell, J. (2008). Uncertainty analysis in MCP-based wind resource assessment and energy production estimation. *Journal of Solar Energy Engineering-Transactions of the ASME*, *130*(3). https://doi.org/10.1115/1.2931499, 031006.

Li, J., & Huang, J. (2020). The expansion of China's solar energy: Challenges and policy options. *Renewable and Sustainable Energy Reviews*, *132*. https://doi.org/10.1016/j.rser.2020.110002, 110002.

Liu, B., & Jordan, R. (1960). The interrelationship and characteristic distribution of direct, diffuse and total solar radiation. *Solar Energy*, *4*(3), 1–19. https://doi.org/10.1016/0038-092X(60)90062-1.

Lu, X., McElroy, M., Peng, W., Liu, S., Nielsen, C., & Wang, H. (2016). Challenges faced by China compared with the US in developing wind power. *Nature Energy*, *1*, 1–6. https://doi.org/10.1038/nenergy.2016.61.

Lund, H. (2007). Renewable energy strategies for sustainable development. *Energy*, *32*(6), 912–919. https://doi.org/10.1016/j.energy.2006.10.017.

Lundquist, J., Clifton, A., Dana, S., Huskey, A., Moriarty, P., van Dam, J., & Herges, T. (2019). *Wind energy instrumentation atlas*. National Renewable Energy Laboratory. Technical Report NREL/TP-5000-68986 https://www.nrel.gov/docs/fy19osti/68986.pdf.

Maurer, E., Wood, A., Adam, J., Lettenmaier, D., & Nijssen, B. (2002). A long-term hydrologically based dataset of land surface fluxes and states for the conterminous United States. *Journal of Climate*, *15*(22), 3237–3251. https://doi.org/10.1175/1520-0442(2002)015<3237:ALTHBD>2.0.CO;2.

McKendry, P. (2002). Energy production from biomass (part 1): Overview of biomass. *Bioresource Technology*, *83*(1), 37–46. https://doi.org/10.1016/S0960-8524(01)00118-3.

Meza, F., & Varas, E. (2000). Estimation of mean monthly solar global radiation as a function of temperature. *Agricultural and Forest Meteorology*, *100*(2–3), 231–241. https://doi.org/10.1016/S0168-1923(99)00090-8.

Mifsud, M., Sant, T., & Farrugia, R. (2018). A comparison of measure-correlate-predict methodologies using LiDAR as a candidate site measurement device for the Mediterranean Island of Malta. *Renewable Energy*, *127*, 947–959. https://doi.org/10.1016/j.renene.2018.05.023.

Molod, A., Takacs, L., Suarez, M., & Bacmeister, J. (2015). Development of the GEOS-5 atmospheric general circulation model: Evolution from MERRA to MERRA2. *Geoscientific Model Development*, *8*, 1339–1356. https://doi.org/10.5194/gmd-8-1339-2015, 2015.

Montana, D., & Lawrence, D. (1989). *Training feedforward neural networks using genetic algorithms* (pp. 762–767). Morgan Kaufmann Publishers Inc. https://www.ijcai.org/Proceedings/89-1/Papers/122.pdf.

Murtugudde, R., & Busalacchi, A. (1999). Interannual variability of the dynamics and thermodynamics of the tropical Indian Ocean. *Journal of Climate*, *12*(8), 2300–2326. https://doi.org/10.1175/1520-0442(1999)012<2300:IVOTDA>2.0.CO;2.

Murtugudde, R., Seager, R., & Busalacchi, A. (1996). Simulation of the tropical oceans with an ocean GCM coupled to an atmospheric mixed layer model. *Journal of Climate*, *9*(8), 1795–1815. https://doi.org/10.1175/1520-0442(1996)009<1795:SOTTOW>2.0.CO;2.

Nam, W., & Oh, K.-Y. (2020). Mutually complementary measure-correlate-predict method for enhanced long-term wind-resource assessment. *Mathematics*, *8*(10), 1795. https://doi.org/10.3390/math8101795.

National Centers for Environmental Information National Oceanic and Atmospheric Administration (NOAA). (2020). https://www.ncdc.noaa.gov/.

Nwokolo, S. (2017). A comprehensive review of empirical models for estimating global solar radiation in Africa. *Renewable and Sustainable Energy Reviews*, *78*, 955–995. https://doi.org/10.1016/j.rser.2017.04.101.

Nwokolo, S., & Ogulezie, J. (2018). A qualitative review of empirical models for estimating diffuse solar radiation from experimental data in Africa. *Renewable and Sustainable Energy Reviews*, *92*, 353–393. https://doi.org/10.1016/j.rser.2018.04.118.

Onogi, K., Tsutsui, J., Koide, H., Sakamoto, M., Kobayashi, S., Hatsushika, H., Matsumoto, T., Yamazaki, N., Kamahori, H., Takahashi, K., Kadokura, S., Wada, K., Kato, K., Oyama, R., Ose, T., Mannoji, N., & Taira, R. (2007). The JRA-25 reanalysis. *Journal of the Meteorological Society of Japan*, *85*(3), 369–432. https://doi.org/10.2151/jmsj.85.369.

Orgill, J., & Hollands, K. (1977). Correlation equation for hourly diffuse radiation on a horizontal surface. *Solar Energy*, *19*(4), 357–359. https://doi.org/10.1016/0038-092X(77)90006-8.

Panwar, N., Kaushik, S., & Kothari, S. (2011). Role of renewable energy sources in environmental protection: A review. *Renewable and Sustainable Energy Reviews*, *15*(3), 1513–1524. https://doi.org/10.1016/j.rser.2010.11.037.

Pfenninger, S., & Staffell, I. (2016). Long-term patterns of European PV output using 30 years of validated hourly reanalysis and satellite data. *Energy*, *114*, 1251–1265. https://doi.org/10.1016/j.energy.2016.08.060.

Poli, P., Hersbach, H., Dee, D., Berrisford, P., Simmons, A., Vitart, F., Laloyaux, P., Tan, D., Peubey, C., Thépaut, J.-N., Trémolet, Y., Hólm, E., Bonavita, M., Isaksen, L., & Fisher, M. (2016). ERA-20C: An atmospheric reanalysis of the twentieth century. *Journal of Climate*, *29*(11), 4083–4097. https://doi.org/10.1175/JCLI-D-15-0556.1.

Rajagopalan, B., & Lall, U. (1999). A k-nearest-neighbor simulator for daily precipitation and other weather variables. *Water Resources Research*, *35*(10), 3089–3101. https://doi.org/10.1029/1999WR900028.

Renné, D., George, R., Wilcox, S., Stoffel, T., Myers, D., & Heimiller, D. (2008). *Solar resource assessment*. National Renewable Energy Laboratory. NREL/TP-581-42301 https://www.nrel.gov/docs/fy08osti/42301.pdf.

Rienecker, M., Suarez, M., Gelaro, R., Todling, R., Bacmeister, J., Liu, E., Bosilovich, M., Schubert, S., Takacs, L., Kong Kim, G., Bloom, S., Chen, J., Collins, D., Conaty, A., da Silva, A., Gu, W., Joiner, J., Koster, R., Lucchesi, R., … Woollen, J. (2011). MERRA: NASA's modern-era retrospective analysis for research and applications. *Journal of Climate*, *24*(14), 3624–3648. https://doi.org/10.1175/JCLI-D-11-00015.1.

Robertson, A., & Mechoso, C. (2000). Interannual and interdecadal variability of the South Atlantic convergence zone. *Monthly Weather Review*, *128*(8), 2947–2957. https://doi.org/10.1175/1520-0493(2000)128<2947:IAIVOT>2.0.CO;2.

Rogers, A., Rogers, J., & Manwell, J. (2005). Comparison of the performance of four measure-correlate-predict algorithms. *Journal of Wind Engineering and Industrial Aerodynamics*, *93*(3), 243–264. https://doi.org/10.1016/j.jweia.2004.12.002.

Rojas, R. (1996). *Neural networks. A systematic introduction*. Springer-Verlag Berlin Heidelberg. https://page.mi.fu-berlin.de/rojas/neural/.

Romo, A., Amezcua, J., & Probst, O. (2011). Validation of three new measure-correlate-predict models for the long-term prospection of the wind resource. *Journal of Renewable and Sustainable Energy*, *3*(2). https://doi.org/10.1063/1.3574447, 023105.

Ruiz, P., Nijs, W., Tarvydas, D., Sgobbi, A., Zucker, A., Pilli, R., Jonsson, R., Camia, A., Thiel, C., Hoyer-Klick, C., Longa, F., Kober, T., Badger, J., Volker, P., Elbersen, B., Brosowski, A., & Thrän, D. (2016). ENSPRESO—An open, EU-28 wide, transparent and coherent database of wind, solar and biomass energy potentials. *Energy Strategy Reviews*, *26*. https://doi.org/10.1016/j.esr.2019.100379, 100379.

Saavedra-Moreno, B., Salcedo-Sanz, S., Carro-Calvo, L., Gascón-Moreno, J., Jiménez-Fernández, S., & Prieto, L. (2013). Very fast training neural-computation techniques for real measure-correlate-predict wind operations in wind farms. *Journal of Wind Engineering and Industrial Aerodynamics*, *116*, 49–60. https://doi.org/10.1016/j.jweia.2013.03.005.

Saha, S., Moorthi, S., Wu, X., Wang, J., Nadiga, S., Tripp, P., Behringer, D., Hou, Y.-T., Chuang, H.-Y., Iredell, M., Ek, M., Meng, J., Yang, R., Peña Mendez, M., van den Dool, H., Zhang, Q., Wang, W., Chen, M., & Becker, E. (2014). The NCEP climate forecast system version 2. *Journal of Climate*, *27*(6), 2185–2208. https://doi.org/10.1175/JCLI-D-12-00823.1.

Samuel, T. (1991). Estimation of global radiation for Sri-Lanka. *Solar Energy*, *47*(5), 333–337. https://doi.org/10.1016/0038-092X(91)90026-S.

Sengupta, M., Habte, A., Gueymard, C., Wilbert, S., Renné, D., & Stoffel, T. (2017). *Best practices handbook for the collection and use of solar resource data for solar energy applications* (2nd ed.). National Renewable Energy Laboratory. NREL/TP-5D00-68886 https://www.nrel.gov/docs/fy18osti/68886.pdf.

Shamshad, A., Bawadi, M., Wan Hussin, W., Majid, T., & Sanusi, S. (2005). First and second order Markov chain models for synthetic generation of wind speed time series. *Energy*, *30* (5), 693–708. https://doi.org/10.1016/j.energy.2004.05.026.

Sharma, P., Warudkar, V., & Ahmed, S. (2019). Application of lidar and measure correlate predict method in offshore wind resource assessments. *Journal of Cleaner Production*, *215*, 534–543. https://doi.org/10.1016/j.jclepro.2018.12.267.

Sheffield, J., Goteti, G., & Wood, E. (2006). Development of a 50-year high-resolution global dataset of meteorological forcings for land surface modeling. *Journal of Climate*, *19*(13), 3088–3111. https://doi.org/10.1175/JCLI3790.1.

Solangi, K., Islam, M., Saidur, R., Rahim, N., & Fayaz, H. (2011). A review on global solar energy policy. *Renewable and Sustainable Energy Reviews*, *15*(4), 2149–2163. https://doi.org/10.1016/j.rser.2011.01.007.

Staffell, I., & Pfenninger, S. (2016). Using bias-corrected reanalysis to simulate current and future wind power output. *Energy*, *114*, 1224–1239. https://doi.org/10.1016/j.energy.2016.08.068.

Suomalainen, K., Silva, C., Ferrão, P., & Connors, S. (2012). Synthetic wind speed scenarios including diurnal effects: Implications for wind power dimensioning. *Energy*, *37*(1), 41–50. https://doi.org/10.1016/j.energy.2011.08.001.

Supit, I., & Van Kappel, R. (1998). A simple method to estimate global radiation. *Solar Energy*, *63*(3), 147–160. https://doi.org/10.1016/S0038-092X(98)00068-1.

Tang, R., & Tong, W. (2004). Optimal tilt-angles for solar collectors used in China. *Applied Energy*, *79*(3), 239–248. https://doi.org/10.1016/j.apenergy.2004.01.003.

Trnka, M., Zalud, Z., Eitzinger, J., & Dubrovský, M. (2005). Global solar radiation in Central European lowlands estimated by various empirical formulae. *Agricultural and Forest Meteorology*, *131*(1–2), 54–76. https://doi.org/10.1016/j.agrformet.2005.05.002.

Uppala, S., KÅllberg, P., Simmons, A., Andrae, U., Da Costa Bechtold, V., Fiorino, M., Gibson, J., Haseler, J., Hernandez, A., Kelly, G., Li, X., Onogi, K., Saarinen, S., Sokka, N., Allan, R., Andersson, E., Arpe, K., Balmaseda, M., Beljaars, A., … Woollen, J. (2005). The ERA-40 re-analysis. *Quarterly Journal of the Royal Meteorological Society*, *131*, 2961–3012. https://doi.org/10.1256/qj.04.176.

US Energy Information Administration. (2020a). *Annual energy outlook 2020*. https://www.eia.gov/outlooks/aeo/.

US Energy Information Administration. (2020b). *International data. Electricity*. https://www.eia.gov/electricity/data.php.

US Energy Information Administration. (2020c). *International overview*. https://www.eia.gov/international/overview/world.

Velázquez, S., Carta, J., & Matías, J. (2011). Comparison between ANNs and linear MCP algorithms in the long-term estimation of the cost per kWh produced by a wind turbine at a candidate site: A case study in the Canary Islands. *Applied Energy*, *88*(11), 3869–3881. https://doi.org/10.1016/j.apenergy.2011.05.007.

Vidal, J., Martin, E., Franchistéguy, L., Baillon, M., & Soubeyroux, J. (2010). A 50-year high-resolution atmospheric reanalysis over France with the Safran system. *International Journal of Climatology*, *30*(11), 1627–1644. https://doi.org/10.1002/joc.2003.

Weekes, S., & Tomlin, A. (2014). Data efficient measure-correlate-predict approaches to wind resource assessment for small-scale wind energy. *Renewable Energy*, *63*, 162–171. https://doi.org/10.1016/j.renene.2013.08.033.

Weekes, S., Tomlin, A., Vosper, S., Skea, A., Gallani, M., & Standen, J. (2015). Long-term wind resource assessment for small and medium-scale turbines using operational forecast data and measure-correlate-predict. *Renewable Energy*, *81*, 760–769. https://doi.org/10.1016/j.renene.2015.03.066.

Wilcox, S., & Marion, W. (2008). *User's manual for TMY3 data sets*. National Renewable Energy Laboratory. Technical report NREL/TP-581-43156 https://www.nrel.gov/docs/fy08osti/43156.pdf.

Wong, L., & Chow, W. (2001). Solar radiation model. *Applied Energy*, *69*(3), 191–224. https://doi.org/10.1016/S0306-2619(01)00012-5.

Yan, Z., Jones, P., Davies, T., Moberg, A., Bergström, H., Camuffo, D., Cocheo, C., Maugeri, M., Demarée, G., Verhoeve, T., Thoen, E., Barriendos, M., Rodríguez, R., Martín-Vide, J., & Yang, C. (2002). Trends of extreme temperatures in Europe and China based on daily observations. *Climate Change*, *53*(1–3), 355–392. https://doi.org/10.1007/978-94-010-0371-1_13.

Yang, K., Huang, G., & Tamai, N. (2001). A hybrid model for estimating global solar radiation. *Solar Energy*, *70*(1), 13–22. https://doi.org/10.1016/S0038-092X(00)00121-3.

Yang, K., Koike, T., & Ye, B. (2006). Improving estimation of hourly, daily, and monthly solar radiation by importing global data sets. *Agricultural and Forest Meteorology*, *137*(1–2), 43–55. https://doi.org/10.1016/j.agrformet.2006.02.001.

Yorukoglu, M., & Celik, A. (2006). A critical review on the estimation of daily global solar radiation from sunshine duration. *Energy Conversion and Management*, *47*(15–16), 2441–2450. https://doi.org/10.1016/j.enconman.2005.11.002.

Zhang, J., Chowdhury, S., Messac, A., & Hodge, B.-M. (2014). A hybrid measure-correlate-predict method for long-term wind condition assessment. *Energy Conversion and Management*, *87*, 697–710. https://doi.org/10.1016/j.enconman.2014.07.057.

Zhang, J., Zhao, L., Deng, S., Xu, W., & Zhang, Y. (2017). A critical review of the models used to estimate solar radiation. *Renewable and Sustainable Energy Reviews*, *70*, 314–329. https://doi.org/10.1016/j.rser.2016.11.124.

Zhang, D., & Zheng, W. (2004). Diurnal cycles of surface winds and temperatures as simulated by five boundary layer parameterizations. *Journal of Applied Meteorology*, *43*(1), 157–169. https://doi.org/10.1175/1520-0450(2004)043<0157:DCOSWA>2.0.CO;2.

Renewable generation and energy storage systems

4

4.1 Wind generation

Wind energy has undergone considerable development over the last few years, and that is why its careful modeling is a matter of study worldwide. In general, characterizing the relationship between wind speed and power production is essential for analyzing a typical power system with wind generation. The manufacturers frequently provide information related to this matter, considering specific laboratory conditions. However, wind power production is influenced by the orography and terrain conditions, which makes estimating the energy yield a complex problem.

The power curve can generally be modeled using parametric and nonparametric techniques. Regarding parametric methods, approaches include the maximum principle method, the dynamical power curve, and the ideal power curve, among many others. The nonparametric methods are those based on copula theory, neural networks (NNs), fuzzy techniques, and data mining, among other methodologies (Carrillo et al., 2013; Lydia et al., 2014).

To analyze the wind resource variability, Khalfallah and Koliub (2007) measured wind speed and power long-term between 1973 and 2001 in Egypt. Recording hourly values at different heights, the authors analyzed the effects of wind variability on the power curve, observing that a variation of $\pm 10\%$ around the average results in a variation of $\pm 25\%$ on the produced energy.

Based on the limitations of analyzing wind energy production using average values over a 10 min interval, Anahua et al. (2008) developed a methodology to describe the relationship between wind speed and power using Markovian theory. The procedure applies a relaxation model using the Langevin equation and a stochastic analysis based on the Markov process.

Üstüntaş and Şahin (2008) paid particular attention to those cases where the wind speed is lower than the cut-in speed or higher than the cut-out speed, because they are critical to model the power curve accurately. In this regard, the authors proposed performing a clustering process based on fuzzy logic to represent the wind power behavior as a function of the wind speed. The authors observed that the power curve could be modeled reasonably well with four or five clusters.

Trivellato et al. (2012) developed a power curve model under ideal conditions. Ideal conditions refer to a wind profile that is steady, laminar, spatially uniform, and undisturbed by the wind turbine (WT) structure. This model is instrumental in estimating the energy yield by extending the conditions of a laboratory test to those with higher turbulence levels. This reasoning was verified by analyzing a small-capacity WT under real operating conditions over 2 years, concluding that the energy estimation offered by the ideal model has a reasonable experimental error.

Genetic Optimization Techniques for Sizing and Management of Modern Power Systems
https://doi.org/10.1016/B978-0-12-823889-9.00005-9

Analyzing measured data has inherent problems related to operational transitions and other issues, which results in some deviations from the actual values. Wang et al. (2014) presented a technique based on copula theory to deal with this problem. The method studies the raw data obtained from the supervisory control and data acquisition (SCADA) system to detect and exclude the outliers using the joint probability distribution. Thus, the model can capture the nonlinear behavior in the data from a probabilistic perspective. This approach is beneficial to detect physical degradation to reduce maintenance costs.

As aforementioned, wind power strongly depends on terrain characteristics and dynamic phenomena. In practice, turbulence intensity (TI) is widely employed to quantize wind fluctuations. Hedevang (2014) developed a method to incorporate TI in estimating wind generation. The technique only requires the power curve under conditions of no turbulence, the corresponding value of TI, and some statistics in the 10 min interval.

Paiva et al. (2014) developed a procedure using local measurement equipment, instruments installed on the nacelle, and the SCADA system. This technique is complementary to the International Electrotechnical Commission (IEC) standard. The authors have a particular interest in wind power forecasting and the influence of the power curve in the prediction error. To this end, they analyzed data from 1 year of operation, concluding that the contribution of power curve estimation in the prediction error is between 10% and 15% of the root mean square error.

Shokrzadeh et al. (2014) discussed some available methodologies for power curve fitting and introduced relevant concepts to enhance them. Specifically, the authors developed the locally weighted polynomial regression, studied the spline regression, and introduced the penalized spline regression after discussing its pros and cons. The locally weighted polynomial regression was developed to overcome the problems of fitting a function to a specific wind speed interval without losing quality in the others. The spline regression was designed to control the shape of the resulting power curve. Finally, the penalized spline regression was introduced to select the number of intervals in spline regression appropriately.

The growing interest in wind energy has applications at both large-scale wind farm and residential scales. In general, adopting wind generation at a small scale has a low impact on distribution system operation, among other benefits. However, significant differences are observed between the power curve provided by the manufacturers and that measured in practice in urban zones. This fact limits the widespread adoption of small-capacity wind generation. Concerning this problem, Pagnini et al. (2015) presented a comparative analysis between small-scale WTs with horizontal and vertical axes operating under the same geographical conditions at high temporal resolution. According to the reported results, the energy produced by the horizontal-axis WT is higher than that with the vertical axis. However, the turbine with a horizontal axis requires more maintenance because it is susceptible to wind gusts and sudden changes in wind direction. Conversely, the turbine with a vertical axis has fewer problems with wind gusts and drastic wind speed variations. The authors also observed differences when comparing with the manufacturers' data, particularly in the interval between 6 and 7 m per second, where the generation loss was considerable. Such behavior might be related to some difficulties in operating under high-turbulence conditions. Furthermore, the authors do not find any particular change in the energy assessment when the resolution was increased to 1 min. They concluded that both technologies are sensitive to TI and should not be installed in zones with complex terrain.

Villanueva and Feijóo (2016) presented a methodology using experimental information collected over 1 year. Their model is based on a standard distribution and incorporates the wind shear, TI, and other factors. The methodology was designed to solve several specific problems, such as fulfilling void data, detecting abnormal working conditions, and simulating wind power or energy yield. To this end, the model uses a process based on Monte Carlo simulation.

Bai et al. (2016) developed a power curve model based on copula theory. The model analyzes the relationship between wind speed and power using wind azimuth, wind elevation, air density, and solar radiation. These factors are incorporated into a recursive formula with conditional copulas to reduce the variability of the functional relationship between velocity and power in a WT. Pelletier et al. (2016) represented the nonlinear relation between wind velocity and power using an NN. The authors incorporated wind speed, air density, TI, wind shear, wind direction, and yaw error as main parameters, and performed a qualitative analysis considering the power production and the power coefficients, finding results comparable with the IEC standard.

Marčiukaitis et al. (2017) proposed a nonlinear regression model that fits the turbine's physical characteristics with a low number of parameters to be estimated. To avoid the processes related to removing outliers from the raw data, the authors introduced a mode approach. In the mode approach, the power curve is estimated considering the modes of power data for each of the wind speed intervals. Villanueva and Feijóo (2018) evaluated a family of logistic functions to model the power curve. After extensive analysis, the authors suggested using a logistic function with three or five parameters to balance the computational effort with the fitting error.

Yesilbudak (2018) proposed a technique composed of three main procedures, namely, clustering, filtering, and modeling. Clustering can be performed using different techniques such as k-means, k-means++, or similar, and is combined with a distance measure such as the squared Euclidean distance. The procedure is applied to partition the raw data. The filtering process is implemented using a multivariate outlier detection algorithm based on the Mahalanobis distance. This method is applied to the results of the clustering analysis previously explained. Finally, to minimize the fitting error, the modeling procedure builds a hybrid curve from the evaluation of Fourier, Gaussian, and other relevant techniques. Saint-Drenan et al. (2020) developed an accurate model that considers specific features of the installation site and WT. Such characteristics are power density, cut-in and cut-out velocities, power coefficient TI, air density, and wind shear. The authors verified the performance of their model by comparison with a large number of WTs, observing promising results.

Virgolino et al. (2020) proposed a methodology that incorporates a Gaussian process regression, a machine learning model, and a standard logistic function. The technique applies a Bayesian reasoning approach to determine the model parameters. It is essential to highlight that the deviations reduce as the wind speed gets close to the rated value in the specific case of power curve modeling. Based on this fact, the authors suggest employing a noise function dependent on the input value, which conducts us to heteroscedasticity. Other authors, such as Rogers et al. (2020), have also followed this research direction. Xu et al. (2021) introduced a stochastic model for the power curve using the quantiles. The quantile power curve consists of several power curves related to the corresponding confidence level. The stochastic relationship is established using NN. This approach provides valuable information about the inherent uncertainty of the power curve.

Even when the manufacturers provide the power curve, some analytical expressions can be found in the literature for different purposes. Giorsetto and Utsurogi (1983) developed a model to characterize a typical WT for power system reliability analysis. The power curve ($P_{W(v)}$) is shown in Eq. (4.1) (Dialynas & Machias, 1989):

$$P_{W(v)} = \begin{cases} 0; 0 \leq v \leq v_i, v > v_o \\ \alpha + \beta v + \delta(v^2); v_i \leq v \leq v_r \\ P_W^{rtd}; v_r \leq v \leq v_o \end{cases} \tag{4.1}$$

where v is the wind speed, v_i, v_r, and v_o are the cut-in, rated, and cut-out wind speed of the turbine, and $P_W{}^{rtd}$ is the rated power. The parameters α, β, and δ are calculated using Eqs. (4.2)–(4.4) (Dialynas & Machias, 1989).

$$\alpha = \frac{1}{(v_i - v_r)^2}\left[v_i(v_i + v_r) - 4v_i v_r\left(\frac{v_i + v_r}{2v_r}\right)^3\right] \tag{4.2}$$

$$\beta = \frac{1}{(v_i - v_r)^2}\left[4(v_i + v_r)\left(\frac{v_i + v_r}{2v_r}\right)^3 - (3v_i + v_r)\right] \tag{4.3}$$

$$\delta = \frac{1}{(v_i - v_r)^2}\left[2 - 4\left(\frac{v_i + v_r}{2v_r}\right)^3\right] \tag{4.4}$$

Karaki et al. (1999) carried out a probabilistic analysis for renewable energy systems in which they employed the simplified model of Eq. (4.5):

$$P_{W(v)} = \begin{cases} 0; 0 \leq v \leq v_i, v > v_o \\ \alpha + \delta(v^\sigma); v_i \leq v \leq v_r \\ P_W^{rtd}; v_r \leq v \leq v_o \end{cases} \tag{4.5}$$

where the parameter σ is the order of the characteristic curve, and the parameters α and β are calculated using Eqs. (4.6) and (4.7) (Karaki et al., 1999).

$$\alpha = \frac{P_W^{rtd}(v_i)^\sigma}{(v_i)^\sigma - (v_r)^\sigma} \tag{4.6}$$

$$\delta = \frac{P_W^{\text{rtd}}}{(v_i)^{\sigma} - (v_r)^{\sigma}} \tag{4.7}$$

Roy (2012) developed a model for power system management, specifically, on the generation power dispatch in a short period. The model development starts by considering the first-order model of Eqs. (4.8) and (4.9):

$$P_{W(v)} = \mu_{(v,\psi)} P_W^{\text{rtd}} \tag{4.8}$$

$$\mu_{(v,\psi)} = 1 - \exp\left[-\left(\frac{v/v_r}{\rho_{(\psi)}}\right)^{\varphi_{(\psi)}}\right] \tag{4.9}$$

where the factor $\mu_{(v,\psi)}$ is the output coefficient under turbulence, v is interpreted in this model as the short-duration-stable mean wind speed, and ψ is TI. The scaling parameter $\rho_{(\psi)}$ and the index parameter $\varphi_{(\psi)}$ are calculated using Eqs. (4.10) and (4.11) (Roy, 2012):

$$\rho_{(\psi)} = 3.49 - 6.01\psi \tag{4.10}$$

$$\varphi_{(\psi)} = 0.71 + 0.21\psi - 1.26\left(\psi^2\right) \tag{4.11}$$

4.2 Solar photovoltaic generation

Solar energy is a promising source of clean energy that is developing rapidly. For this reason, researchers have been creating computational models to represent its behavior to design a solar generation system that is sustainable from the techno-economic perspective. One of the earliest works was presented by Mbewe et al. (1985), in which a solar photovoltaic (PV) system was modeled. Specifically, the authors developed mathematical expressions for PV and PV-thermal systems with optical concentration. They represented the open-circuit voltage, short-circuit current, fill factor, and efficiency considering silicon solar cells, resulting in a model of the solar cell current, which depends on the optical concentration, temperature, and voltage. The model performance offers a fitting error of approximately 10%.

Baruch (1985) modeled the PV cell using a quantum two-level system. Thus, the authors demonstrated that the structure of a two-level system is similar to the ideal diode equation. Alonso-García and Ruíz (2006) developed a simplified model for the PV cell from a linearized analysis of the current–voltage characteristic, incorporating the short-circuit current and the open-circuit voltage, among other parameters. The authors verified the model's effectiveness using a large set of cells, finding good results. De Bernardez and Buitrago (2006) developed a technique to extract the cell parameters from the dark current–voltage characteristic. This technique is instrumental in assessing the degradation mechanisms to forecast the failure process of a typical PV module.

Karatepe et al. (2007) improve the PV simulation model by considering the effect of the bypass diodes. PV modules have two bypass diodes to avoid harmful operating conditions. Thus, these diodes influence the current–voltage characteristic, creating one or several local maxima in the power curve. Consequently, the maximum power point tracking (MPPT) algorithm has problems operating satisfactorily. The authors paid particular attention to analyzing the cell instead of the module. The analysis of each cell allows us to incorporate the influence of the shadow, including its direction and shape. The computational model proposed by the authors is based on an NN, using the irradiation and temperature as input data, and the photocurrent, the saturation current, the diode ideality factor, and the series and parallel resistances as output.

Vazquez and Rey-Stolle (2008) developed a model for the PV generator specially designed for reliability analysis. The degradation of module interconnection, the effects of moisture intrusion, the degradation of the semiconductor, among many other factors, reduce the performance of the PV generator over time. Thus, the authors developed a model to consider module aging using probability theory. Concretely, the power production was assumed to follow a Gaussian distribution, considering the rated power as the reference parameter and the average generation to decrease linearly over time. Under these assumptions, the authors estimated the reliability function and consequently the mean time to failure and power warranty time, among other parameters.

Boyd et al. (2011) performed a complete evaluation of the standard PV simulation models based on short-circuit current, open-circuit voltage, current and voltage at maximum power, and temperature coefficient under short-circuit and open-circuit conditions. The authors used measured data of beam, global, and diffuse horizontal irradiance, and global irradiance over the tilted surface to assess the accuracy of the standard model. Such data was recorded over 30 days. The authors also considered different technologies, including monocrystalline, polycrystalline, amorphous silicon, and copper indium diselenide (CIS). The evaluation revealed a simulation error of 6% for the case of monocrystalline and polycrystalline silicon and a slightly lower precision for CIS and amorphous silicon. The analysis showed that simulation error decreases to 5% when aged modules are considered.

Bühler and Krenzinger (2013) carefully analyzed the deficiencies of the standard PV model and the algorithms applied to extract its parameters. Consequently, the authors developed a specific procedure to estimate the parameters of their model. Thus, the authors introduced a simulation model with two diodes, which belong to the models of diffusion and recombination phenomena. The fitting mechanism showed good results modeling the current–voltage characteristic of amorphous silicon single junction and CIS modules. The maximum error observed was slightly greater than 1%.

Averbukh et al. (2013) developed a simplified model to estimate the current–voltage characteristic considering a dataset with three points and the temperature coefficient of short-circuit current. The authors proposed a technique for parameter estimation that does not require any fitting method. Instead, the simplified approach needs solving a system of four nonlinear equations with four unknown variables through a numerical process. The model is expressed in normalized form using the per-unit system with the photocurrent and thermal voltage as base values. Finally, the results are adjusted to consider any temperature and irradiation values. Concerning the fitting error, it shows a maximum of 2.5%.

Makrides et al. (2013) evaluated the performance of several models such as the single-point efficiency, the single-point efficiency with temperature correction, the model of photovoltaic for utility-scale applications (PVUSA), and the one-diode model. The authors used measurements of 4 years and different technologies to perform a comparative analysis. Such an analysis revealed that incorporating a temperature loss correction to the temperature coefficient of power improves energy production forecasting, specifically for crystalline silicon modules. On the other hand, the one-diode model has a good performance in analyzing monocrystalline and multicrystalline silicon modules. At the same time, the PVUSA showed positive results studying some modules with thin-film technologies. Finally, the energy production of those modules based on copper-indium-gallium-diselenide was better estimated using the single-point efficiency with temperature correction and one-diode models.

Lun et al. (2014) presented a model based on NN training. The authors proposed a time warp invariant echo state network (TWIESN) to estimate the current–voltage characteristic using a two-diode model. The TWIESN is an improved version of the echo state network (ESN) that deals with complex features often observed in dynamic systems. ESN is a training technique that mitigates some of the general problems of NN performance. The authors compared TWIESN with the backpropagation algorithm, concluding that TWIESN is more effective for this specific application.

Chaibi et al. (2019) presented a comparative analysis between the single-diode and double-diode models under different meteorological conditions. The authors considered the Mediterranean and the Semi-Continental climatic zones during the comparative process. They developed a hybrid method to select the appropriate model using the corresponding results. The double-diode model has better performance under low-irradiance conditions; however, the single-diode model offers good results under low-temperature variations. According to the reported results, the estimation error reduces by 54% for cloudy weather while reducing by 21% for sunny days.

This chapter will comment on a simplified model reported by Lorenzo et al. (1994). The model can be applied using the information usually provided by the PV-module manufacturer. The PV cell temperature (T_{pv}) is described by Eq. (4.12) (Lambert et al., 2006):

$$T_{pv} = \frac{T_a + G\left[\dfrac{NOCT - 20°\mathrm{C}}{800\,\mathrm{W/m^2}}\right]\left[1 - \dfrac{\eta_{pv}\left(1 - \alpha_{pv}[25°\mathrm{C}]\right)}{k_{pv}}\right]}{1 + [NOCT - 20°\mathrm{C}]\left[\dfrac{G}{800\,\mathrm{W/m^2}}\right]\left[\dfrac{\alpha_{pv}\eta_{pv}}{k_{pv}}\right]} \tag{4.12}$$

where T_a is the ambient temperature, $NOCT$ is the nominal operating cell temperature, G is the solar radiation, η_{pv} is the cell efficiency, α_{pv} is the temperature coefficient, and k_{pv} is the parameter of solar transmittance and absorptance. On the other hand, the thermal voltage (V_t) is presented in Eq. (4.13) (Lorenzo et al., 1994):

$$V_t = \frac{w(k_B)T_{pv}}{e} \tag{4.13}$$

where w is the photovoltaic cell ideality factor, e is the electron charge, and k_B is the Boltzman constant. The open-circuit voltage (V_{oc}) is shown in Eq. (4.14) (Lorenzo et al., 1994):

$$V_{oc} = \frac{V_{oc}^{stc}}{N_s} + \alpha_{pv}\left[T_{pv} - 25°C\right] + (V_t) \times \ln\left[\frac{G}{1000\,W/m^2}\right] \tag{4.14}$$

where $V_{oc}{}^{stc}$ is the open-circuit voltage under standard test condition (STC) and N_s is the number of cells connected in serial. The relative open-circuit voltage (v_{oc}) is calculated as shown in Eq. (4.15) (Lorenzo et al., 1994):

$$v_{oc} = \frac{V_{oc}}{V_t} \tag{4.15}$$

The maximum fill factor (FF_o) is calculated using Eq. (4.16) (Lorenzo et al., 1994):

$$FF_o = \frac{v_{oc} - \ln(v_{oc} + 0.72)}{v_{oc} + 1} \tag{4.16}$$

The short-circuit current (I_{sc}) depends on the solar radiation according to Eq. (4.17) (Lorenzo et al., 1994):

$$I_{sc} = \frac{I_{sc}^{stc}}{N_p}\left(\frac{G}{1000\,W/m^2}\right) \tag{4.17}$$

where $I_{sc}{}^{stc}$ is the short-circuit current under standard test conditions, N_p is the number of strings connected in parallel. The fill factor (FF) is estimated using Eq. (4.18) (Lorenzo et al., 1994):

$$FF = \frac{V_{max}^{stc} I_{max}^{stc}}{V_{oc}^{stc} I_{sc}^{stc}} \tag{4.18}$$

where $V_{max}{}^{stc}$ is the voltage at maximum power production and $I_{max}{}^{stc}$ is the current at maximum power production. The PV cell resistance (R_s) is calculated using Eq. (4.19) (Lorenzo et al., 1994):

$$R_s = \left(1 - \frac{FF}{FF_o}\right)\left(\frac{V_{oc}}{I_{sc}}\right) \tag{4.19}$$

The PV cell current (I_{pv}) as a function of cell voltage (V_{pv}) is estimated using Eq. (4.20) (Lorenzo et al., 1994):

$$I_{pv} = I_{sc}\left[1 - \exp\left(\frac{V_{pv} - V_{oc} + I_{pv}R_s}{V_t}\right)\right] \tag{4.20}$$

Finally, the PV output power (P_{pv}) as a function of the cell current (I_{pv}) is calculated using Eq. (4.21) (Lorenzo et al., 1994):

$$P_{pv} = I_{pv}\left[V_{oc} - I_{pv}R_s + V_t \ln\left(1 - \frac{I_{pv}}{I_{sc}}\right)\right] \tag{4.21}$$

This model is briefly illustrated by calculating the current–voltage characteristics of a commercial module. Under STC, the open-circuit voltage is equal to 50.34 V ($V_{oc}^{stc} = 50.34$ V), the short-circuit current is equal to 13.66 A ($I_{sc}^{stc} = 13.66$ A), the voltage at maximum power production is equal to 41.67 V ($V_{\max}^{stc} = 41.67$ V), the current at maximum power production is equal to 12.97 A ($I_{\max}^{stc} = 12.97$ A), and the module efficiency is equal to 20.89% ($\eta_{pv} = 20.89\%$). Concerning the ideality factor, it is equal to 1 ($w = 1$). The nominal operating cell temperature is equal to 42.5 °C ($NOCT = 42.5$ °C), the voltage temperature coefficient is equal to −0.262%/°C. The module has 72 cells in series and two strings.

The model previously described has been implemented considering solar radiation of 800 W/m^2 ($G = 800$ W/m^2) and a temperature of 20 °C ($T_a = 20$ °C). Fig. 4.1 shows the current–voltage characteristic, and Fig. 4.2 presents the power production as a function of the voltage.

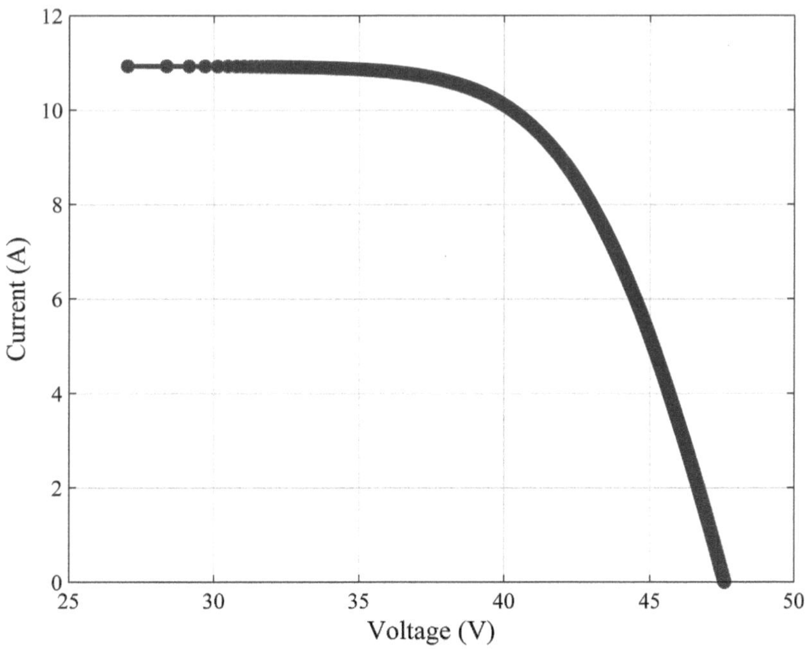

Fig. 4.1 Simulating a commercial PV panel current: Current–voltage characteristic.

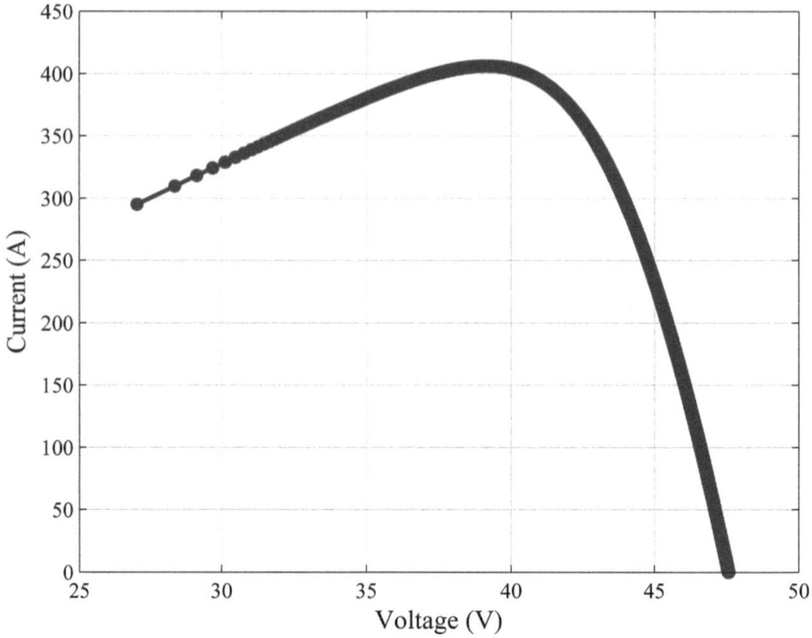

Fig. 4.2 Simulating a commercial PV panel power: Power production versus voltage.

According to this model, the current and voltage at maximum power production equal 10.38 A and 39.14 V, respectively. The manufacturer reports these values as 10.43 A and 38.81 V, respectively. Furthermore, the estimated maximum power is 406.23 W, and the value reported by the manufacturer is 404.78 W. Thus, the results are reasonably suitable for the environmental conditions previously mentioned.

Several simulation models assume that the PV generator is provided with an MPPT algorithm. Thus, it is possible to approximate the power generation using a linear relationship with solar radiation. On the other hand, the effect of MPPT can be incorporated using the golden section search algorithm (GSSA) (Kheldoun et al., 2016).

4.3 Inverter

The PV generator operates using direct current (DC), while distribution systems operate in alternating current (AC). Thus, a conversion device is required to transform the signal from DC to AC; this is the central role of the inverter. The evolution of power electronics has improved conversion efficiency and reliability, reducing acquisition costs. Inverters are provided with an MPPT algorithm, antiislanding operation, automatic synchronization with the distribution system, low harmonic distortion, and a high power factor almost reaching the unity (Rampinelli et al., 2014).

In our discussion, efficiency is a critical variable, defined as the output and input power ratio. However, other aspects also influence efficiencies, such as the DC voltage and temperature. Fig. 4.3 shows the variation of inverter efficiency with the load

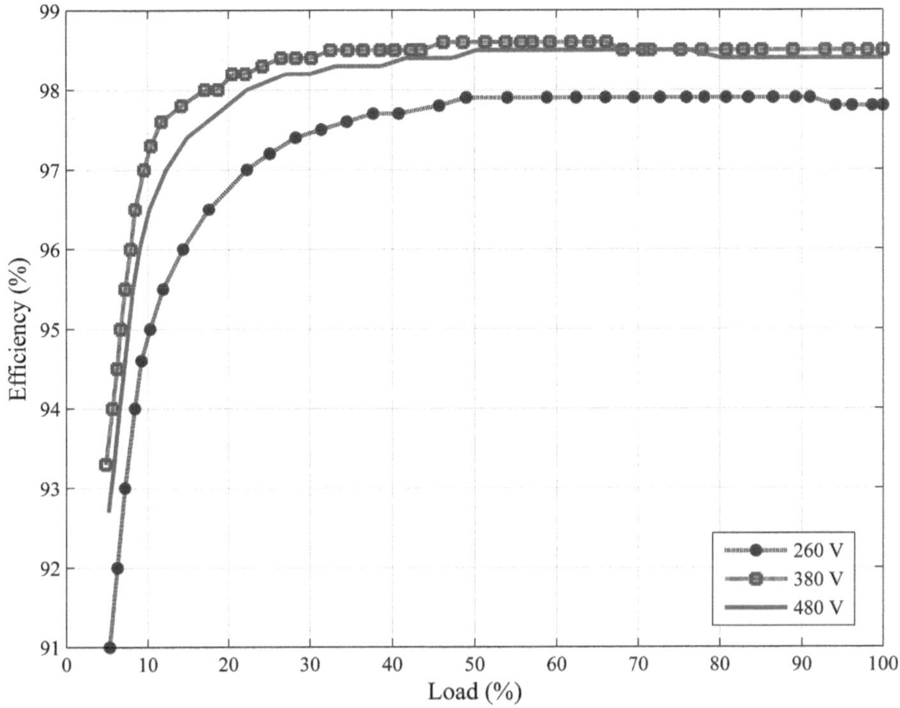

Fig. 4.3 Variable efficiency of a typical inverter: Efficiency curve of a commercial inverter.

and voltage of a commercially available device. Since the efficiency reduces at a low load, approaches frequently applied in the United States and the European Union consider weighted values between 5% and 100% (Rampinelli et al., 2014).

A first approximation consists of using an interpolation process to the measured efficiencies. Assuming the inverter power (P_{inv}) is within the interval [P_a. P_b], the inverter efficiency (η_{inv}) is between η_a and η_b and is approximated using Eq. (4.22) (Rampinelli et al., 2014).

$$\eta_{inv} = \eta_a + \left(\frac{\eta_b - \eta_a}{P_b - P_a}\right)(P_{inv} - P_a) \tag{4.22}$$

Another approach is based on the model of an ideal inverter, including one resistance in serial (R_s) representing the ohmic losses and another resistance in parallel (R_p) modeling the device self-consumption. The technique considers the influence of DC voltage (V_{dc}) and AC voltage (V_{ac}) in the efficiency curve using Eq. (4.23) (Rampinelli et al., 2014).

$$\eta_{inv} = \frac{2R_s P_{inv}}{(V_{dc})^2}\left(\frac{1}{1 - \sqrt{1 - 4\left[R_s/(V_{dc})^2\right]\left[P_{inv} + \left\{(V_{ac})^2/R_p\right\}\right]}}\right) \tag{4.23}$$

Another method incorporates a polynomial function to represent the power losses. Such an approach is shown in Eq. (4.24) (Rampinelli et al., 2014):

$$\eta_{inv} = \frac{P_{inv}/P_{inv}^{rtd}}{P_{inv}/P_{inv}^{rtd} + \left\{\chi + \pi\left(P_{inv}/P_{inv}^{rtd}\right) + \omega\left(P_{inv}/P_{inv}^{rtd}\right)^2\right\}} \tag{4.24}$$

where $P_{inv}{}^{rtd}$ is the inverter-rated power. The parameters χ, π, and ω are coefficients to be adjusted using experimental data.

Some models incorporate the effects of DC voltage and the polynomial characteristic of power losses, as shown in Eq. (4.25) (Rampinelli et al., 2014):

$$P_{inv} = \left[\frac{P_{inv}^{rtd}}{\gamma - \varepsilon} - \zeta(\gamma - \varepsilon)\right][P_{dc} - \varepsilon] + \zeta[P_{dc} - \varepsilon]^2 \tag{4.25}$$

where γ, ε, and ζ are parameters that include the input (P_{dc}) and output power (P_{inv}) of the inverter, and the voltage at the DC side.

Finally, a simplified expression is presented in Eq. (4.26) (Rampinelli et al., 2014), which is particularly useful during the simulation process of batteries and similar devices because the input and output power can be found easily. The parameters λ and ξ are coefficients to be adjusted using experimental data.

$$\eta_{inv} = \frac{P_{inv}/P_{inv}^{rtd}}{P_{inv}/P_{inv}^{rtd} + \lambda + \xi\left(P_{inv}/P_{inv}^{rtd}\right)} \tag{4.26}$$

In general, including the conversion efficiency during low load periods in the simulation and analysis of an energy system is crucial because renewable generation is continuously fluctuating according to the available resources. Under these circumstances, the input power of the inverter is not always at the rated value. Thus, the maximum conversion efficiency is not guaranteed. This is a challenging problem that affects the dimensioning process of the inverter and the associated renewable power sources.

4.4 Conventional generation

Once we have a general idea of the main techniques to simulate wind and solar PV generators, we will briefly describe the traditional model of thermal units, the formulation of economic dispatch (ED), and unit commitment (UC) problems.

4.4.1 The economic dispatch problem

Mathematical representation and management of power systems with conventional generating units is a vital problem, which many authors have studied. Classical and heuristic optimization techniques, such as genetic algorithm (GA), particle swarm optimization (PSO), evolutionary programming (EP), firefly algorithm (FA), consensus algorithm (CA), tabu search (TS), differential evolution (DE), and their combination have been successfully used to solve the economic dispatch (ED) problem.

Solving ED consists of estimating the power production of each committed unit to minimize the operating cost, in the case of mono-objective optimization (MoOO), or the operational cost and the greenhouse gas (GHG) emissions simultaneously, in the case of multiobjective optimization (MuOO).

Concerning the heuristic techniques for MoOO—GA, specifically—Walters and Sheble (1993) presented one of the first works. The authors employed two binary encoding strategies to represent power production. The performance of GA is compared with dynamic programming (DP), which enumerates all possible combinations. The comparative analysis revealed that GA reaches near-optimal solutions. The authors observed that the appropriate selection of scaling parameters for evaluating the fitness function is a crucial procedure in GA implementation. In addition, they realized the relationship between the encoding strategies and the computational burden and suggested additional research efforts in that direction. Bakirtzis et al. (1994) proposed two GA implementations that do not require any convexity constraint to the operating cost. Furthermore, they can be effectively implemented using parallel computing. According to a case study analysis, the approach showed promising results, outperforming the DP method.

Chen and Chang (1995) developed a technique based on a GA using the equal system incremental cost for the encoding. Instead of using the power production with the binary system to encode each individual, the authors proposed using the incremental cost represented by a binary string. The difference between these encoding techniques has an essential impact on computational performance. On the one hand, using power production as the encoding variable makes the GA implementation dependent on the number of units. In other words, the number of elements of each solution increases with the number of generating units, which increases the computational burden. On the other hand, using the incremental cost does not depend on such a relationship, and it means that the number of elements in each solution is the same for small-scale or large-scale systems. Thus, the computational burden remains constant for both optimization problems. A case study analysis revealed that the computational time increases linearly with the system size, making this approach attractive for the analysis of large-scale generating systems.

Orero and Irving (1996) developed an optimization model using a deterministic crowding genetic algorithm (DCGA). Crowding is a nature-inspired concept in which individuals with similar characteristics compete to obtain food and resources. Crowding can be implemented by considering a fraction of the general population to reproduce and die. Thus, the recently born individuals replace those existing with similar features. DCGA is a particular crowding-based GA with promising capabilities to optimize multimodal objective functions. The qualities of DCGA allow us to overcome the problems related to the limited availability of time and memory of the traditional GA implementation. To this end, DCGA increases the selection pressure within a specific search region while the crossover operator performs the search across different areas. Concretely, DCGA randomly selects pairs of individuals until half of the population is completed. Then, crossover and mutation operators are performed over each pair, obtaining a couple of offspring. After that, the offspring compete with a parent chosen according to a specific similarity criterion. The winner of the competition survives to the next generation.

Chiang (2005) developed an improved genetic algorithm with multiplier updating (IGAMU). On the one hand, IGAMU is provided with an improved evolutionary direction operator and a migration operator to increase computational efficiency. This combination is the improved genetic algorithm (IGA). On the other hand, fulfilling the constraints of the optimization problem can be difficult. Penalty factors are frequently added to deal with the problem restrictions, reducing the fitness of those individuals violating the operational limits. Selecting the penalty factors is a difficult task. If they are too large, the objective function could become ill-conditioned, so that good individuals are challenging to find. On the contrary, if the penalty factors are too small, it is challenging to find feasible individuals. To solve this problem, the authors incorporated multiplier updating (MU) to select the appropriate value of the penalty parameters.

PSO is a widely applied technique also used to solve ED. One of the earliest works was presented by Gaing (2003), who carried out several computational experiments, concluding that PSO is computationally stable and efficient, offering high-quality solutions comparable with the real-coded GA with arithmetic crossover. Park et al. (2005) proposed a modified particle swarm optimization (MPSO) to enhance the ability of classical PSO to deal with equality and inequality constraints of the ED problem. MPSO incorporates an algorithm to generate feasible solutions and includes a space-reduction strategy using the position of the best agent and the minimum and maximum output power of each unit. Thus, MPSO has a higher computational efficiency than the classical PSO. The authors analyzed several study cases, concluding that the results obtained from the MPSO implementation are comparable with those offered by GA and EP, among others.

Selvakumar and Thanushkodi (2007) presented a novel improvement to the classical PSO. The authors proposed splitting the cognitive part of PSO into good and bad experiences. In other words, each agent remembers the best and worst positions and uses this information to estimate the velocity for the next iteration. In addition, the authors included a local random search technique to improve the exploration and exploitation features of classical PSO.

EP, FA, CA, and TS, among many other algorithms, have been successfully implemented to solve the ED problem. Concerning EP, Yang et al. (1996) presented one of the first works. The authors adopted EP to overcome the difficulties of the classical Lagrangian approach. From the analysis of several case studies, the authors concluded that the obtained solutions using EP are similar to those found applying DP, GA, and the lambda iteration method. Following this trend, Sinha et al. (2003) carried out extensive analysis on improving EP. The authors proposed improvements on the basic algorithm, implementing the adaptation procedure according to the scaled costs and an empirical learning rate. They included EP implementations with different mutation strategies, such as the classical EP using a Gaussian distribution, the fast EP using a Cauchy distribution, and the improved-fast EP, which combines the two probability distributions aforementioned. According to the computational analysis presented, considering systems of different sizes and characteristics, the authors observed that the fast EP offered better results for small-scale systems. In contrast, the improved-fast EP provided the highest performance for the analysis of large-scale systems.

Yang et al. (2012)) implemented FA to solve ED and compared it with other algorithms such as classical EP, fast EP, improved-fast EP, IGA, IGAMU, and PSO, observing promising results in terms of quality and reliability of the solutions. Yang et al. (2013) proposed a technique using CA in which optimization problems are solved in a distributed manner. In CA, the optimization process considers a multiagent system, and each agent can exchange information locally through a communication infrastructure strongly connected. Thus, the agents exchange information related to power generation to reach the same incremental cost, the consensus variable.

Lin et al. (2002) analyzed and improved TS using a flexible memory system to avoid the local minimum. In their algorithm, the authors incorporate parallelism to reduce the algorithm dependence with the convergence rate. DE is another crucial algorithm frequently employed to solve ED. DE requires a limited number of parameters and provides a good convergence behavior comparable to other popular algorithms such as GA and PSO. DE has been continuously improved to fulfill the demanding characteristics of ED and other optimization problems (Jebaraj et al., 2017).

The environmental impact of fossil fuels has encouraged the adoption of renewable energies such as wind and solar. These power sources continuously change over time, making their daily forecasting a problematic task, introducing uncertainty into power system operation. Hetzer et al. (2008) presented one of the first models regarding the ED problem under conditions of uncertainty. The authors developed an optimization model including the cost of overestimating and underestimating wind generation. The uncertainty of wind generation was incorporated using a Weibull distribution. Then, the researchers studied the behavior of ED for different scale factors and reserve cost.

Liu and Xu (2010a, 2010b) presented an extensive analysis of ED under wind power uncertainty. On the one hand, the authors studied the ED problem and solved the issue of probabilistic infeasibility. On the other hand, they developed an ED model including the oxides of nitrogen emissions, where wind power was represented using a closed form based on the incomplete gamma function. Roy (2012) paid special attention to the wind speed turbulence and gust and their influence on ED. To this end, the author developed a dispatch model and analyzed the impact of TI on the system cost and power losses. Details of the wind generation model have already been described in Eqs. (4.8)–(4.11).

Another critical topic is the combination of optimization techniques to solve ED. Some examples are the combination of sequential quadratic programming (SQP) with PSO (Victoire & Jeyakumar, 2004) and DE (Coelho & Mariani, 2006). Concerning the heuristic techniques for MuOO, they have become popular because of the adverse effects of conventional generation on the environment. MuOO aims to minimize the operating cost and GHG emissions simultaneously. One of the first works was carried out by Abido (2006) using the strength Pareto evolutionary algorithm.

4.4.2 The unit commitment problem

Another fundamental problem of power system management is unit commitment (UC). Solving the UC problem consists of determining when each generating unit should be activated to minimize the total operating cost over a determined period in advance—1 day, for example. Several methods and optimization algorithms have been applied to this problem. Some examples are priority list (PL), branch-and-bound (BB), DP, GA, PSO, EP, DE, gravitational search algorithm (GSA), simulated annealing (SA), Lagrangian relaxation (LR), Benders decomposition (BD), and mixed-integer linear programming (MILP).

PL is a technique able to find a fast near-optimal solution from the cost information of each unit. The priority of committing each unit is determined according to its average generating cost. One of the first works using this method was presented by Tong et al. (1991). The authors implemented PL utilizing an expert system (ES) that finds a suboptimal feasible solution using several heuristic rules. ES commits generating units at those hours with insufficient spinning reserve (SR). After that, the resulting schedule will have an excess SR at other hours, resulting in a cost increment. Thus, the scheduling is analyzed, and generating units will be de-committed to reduce costs without creating an SR shortage. In this context, the obtained result so far does not need a profound modification to get the global optimum. The authors discussed some ideas to refine the estimated schedule. On the one hand, they suggest reducing the commitment of those units with long operating times. On the other hand, they recommend clustering the units with similar characteristics to reduce the decision space and, consequently, the ES's workload.

Following this research direction, Senjyu et al. (2003) incorporated the effects of those units with long operating time committed at peak hours. They also considered the influence of SR requirements and the excess of generation on the operating cost, obtaining an effective solution.

In one of the first works presented in this field using BB, Lauer et al. (1982) developed a technique to solve large-scale problems with 250 or more units. The methodology is attractive from the mathematical perspective because, as the number of units increases, the computational burden increases linearly, and the algorithm trends to enhance the quality of the solution. The reasoning behind this approach consists of finding the upper and lower bounds of the variables under optimization, reducing the searching space, and improving the computational efficiency. On the one hand, the lower bounds are determined using the duality between the UC problem and a related optimization problem based on the Lagrange multipliers. Even when the result from the solution of the dual problem is not feasible, it offers valuable information. On the other hand, the upper bounds are found to obtain viable solutions according to the system PL. The authors suggested employing a dynamic PL, which enhances the results compared with the classical implementation.

Snyder et al. (1987) presented one of the first DP-based approaches. Essentially, the UC is a multistep optimization problem. Thus, DP systematically analyzes a subset of possible combinations related to a specific sequential problem step. From this analysis, a concrete decision is taken. Going from one decision to another through the transition path has an intrinsic cost. Thus, the decision process involves the costs of the decision taken in previous actions. Finally, the strategy aims to minimize the total cost of all decisions considered. To make the optimization problem tractable from a computational perspective, the authors classify the generating units according to their characteristics, reducing the number of possible combinations. The authors introduce the concepts of class, threshold, and window. The class defines the subdivision of the units, the threshold determines the units to be automatically committed, and the window decides the units to be evaluated to determine if they are committed or not. Finally, the ED problem related to the UC problem is solved using a linearized cost model for each generator.

Ouyang and Shahidehpour (1991) proposed variable window truncated dynamic programming (DPVW). In DPVW, the window size is adjusted according to the hourly variations of the load. Specifically, window size changes are defined according to a concrete relationship to the load variations. The authors considered several functions such as the linear, piecewise, and exponential models concerning this topic. After that, DPVW performs some adjustments to avoid feasibility problems in cases where the window size is insufficient. Finally, the authors introduced tuning and refining procedures to obtain a high-quality solution.

Concerning GA, Kazarlis et al. (1996) presented a solution technique that incorporates additional operators to classical GA implementation. The authors applied such operators acting over a block of bits instead of individual ones. The added operators are the swap-window operator and the window-mutation operator. The swap-window operator is an improved version of the classical mutation operator, and the window-mutation operator performs a mutation in multiple points over consecutive bits. The authors also explored including some benefits from the hill-climbing algorithm. Thus, they proposed the swap-mutation operator and the swap-window hill-climb operator. The swap-mutation operator is similar to the stochastic hill-climbing approach, and the swap-window hill-climb operator carries out a local hill-climbing optimization.

Orero and Irving (1997) proposed a hybrid methodology combining PL and GA. On the one hand, GA is an optimization technique that requires intensive computational resources and time. On the other hand, PL provides a reasonable solution very fast. The authors combined these methods to find a new solution considering these features. They proposed applying PL and including the result as an individual in the initial population of GA. This approach allows us to concentrate all the computational resources on finding a better solution.

Swarup and Yamashiro (2002) employed a GA-based technique enriched with specific operators. Specifically, they used the bit change operator, which modifies the bit positions to create a uniform pattern. Thus, the authors mitigated the feasibility problems of some units with large operating times and improved the computational efficiency of the optimization process. Damousis et al. (2004) presented a methodology using an integer-coded genetic algorithm (ICGA). The main difference with other approaches is in the coding adopted, in which the positive or negative integers represent a sequence commitment/de-commitment operating status. This coding allows us to avoid the feasibility problems related to those units with large operating times. Additionally, the authors incorporated specific functions such as the unit swapping-copying operator, the excessive-reserve elimination operator, and the chromosome length augmentation.

Regarding PSO, Ting et al. (2006) developed a solution technique using a real- and a binary-coded PSO. The real-coded implementation solves the ED problem, and the binary-coded PSO solves the UC problem. Later, Yuan, Nie, et al. (2009) combined PL with the binary-coded PSO, enhancing the computational efficiency. Other heuristic approaches such as EP (Juste et al., 1999), DE (Yuan, Su, et al., 2009), GSA (Ji et al., 2014; Yuan et al., 2014), and SA (Mantawy et al., 1998; Simopoulos et al., 2006; Zhuang & Galiana, 1990) have been successfully implemented to solve the UC problem.

LR employs the Lagrange multipliers to incorporate the power system operational constraints into the objective function. Using the duality theory, LR aims to calculate the Lagrange multipliers to maximize the dual objective function. One of the first works using this technique was proposed by Zhuang and Galiana (1988). The method is implemented in three steps. During the first step, the Lagrangian dual is maximized using subgradient methods. In the second step, a feasible solution to fulfill the reserve constraint is estimated. Finally, in the third step, the ED problem is solved.

Cheng et al. (2000) combined GA and LR to update the Lagrangian multipliers, enhancing the intrinsic capabilities of LR. Ongsakul and Petcharaks (2004) developed an adaptive Lagrangian relaxation (ALR) method. ALR incorporates specific procedures to commit or de-commit generating units, the initialization and adaptive tuning of Lagrangian multipliers, and unit classification, among other processes. A relevant characteristic of ALR is the linear behavior of computational time with the system size.

Ma and Shahidehpour (1999) incorporated the optimal power flow into the UC problem using BD. BD divides the optimization problem into a master problem and a subproblem. The master problem solves the UC problem using LR, and the subproblem removes the infeasibility due to transmission security and voltage violations for the scheduling obtained from the master problem. If this is not possible, Bender cuts are created, generating a unit rescheduling. Following this research direction, Li and Shahidehpour (2005) proposed a security-constrained unit commitment (SCUC) model considering the interaction between energy and ancillary services markets. The authors employed the BD to divide the optimization problem into the master problem related to the UC and a subproblem to check the network security.

MILP is also a widely accepted technique for solving the UC problem. Carrión and Arroyo (2006) developed a mathematical formulation that employs a reduced number of binary variables, achieving high computational efficiency. Borghetti et al. (2008) proposed a MILP model to solve the UC problem in power systems provided with pump-storage hydropower stations. In addition, the authors presented an enhanced linearization approach. Improving computational efficiency by enhancing the problem formulation is an important research direction. In this regard, Ostrowski et al. (2012) and Morales-España et al. (2013) proposed different perspectives on developing a tight formulation for the MILP technique.

The environmental problems related to oil consumption and GHG emissions have motivated the adoption of renewable power sources. Incorporating such power sources has a relevant influence on how the UC problem is formulated and solved due to the uncertainty. Takriti et al. (1996) developed a model based on stochastic programming (SP) to incorporate the load forecasting error and generation failures. The UC is formulated considering a finite number of likely scenarios with their corresponding probabilities of occurrence. Thus, the uncertainty is represented through a set of deterministic subproblems, which are embedded into an optimization problem solved by progressive hedging. Ummels et al. (2007) proposed a different technique to deal with uncertain information on the UC problem. The researchers employed a dataset of actual and forecasted measurements in 15 min of onshore and offshore wind installations. Then, they modified the traditional ED and UC formulations to allow the regular update of the solution for the next 36 h.

Wang et al. (2008) created a model to solve the SCUC based on BD. The investigators start by solving the UC problem considering the forecasted profile. After that, they perform a verification process by evaluating the already available scenario set. If the solution has feasibility problems, a re-dispatch procedure is applied. If such an approach does not remove the infeasibility, Benders cuts are defined. The algorithm is an iterative process that searches for the appropriate solution through a sequential re-dispatch.

Ruiz et al. (2009) analyzed SP to solve the UC problem under uncertainty, concluding that the implementation of SP with an additional requirement of SR offers a robust solution that is attractive from the economic and reliability viewpoints. The technique presented by Pappala et al. (2009) uses PSO to reduce the number of wind power scenarios to be considered in SP approach. Tuohy et al. (2009) analyzed the influence of renewable generation on power system operation using SP. The authors concluded that the mid-merit and peaking generators and the interconnection to an external power system are the sensitive parts of the whole generating system when variable power sources such as wind generation are adopted. In addition, they highlight the benefits of using updated information on wind power prediction. Papavasiliou et al. (2011) implemented the SP approach with scenario generation and reduction, solving the UC problem through a dual decomposition algorithm.

Besides the SP methodologies, techniques based on chance-constrained theory (Wang et al., 2012) and robust optimization (Bertsimas et al., 2013) have also been suggested to solve the SCUC problem.

4.4.3 Economic dispatch and unit commitment formulation

Fig. 4.4 shows the structure of a typical energy system provided with conventional units. The index i is used to represent each unit ($i=1, \ldots, I$), and the index t represents each time step ($t=1, \ldots, T$). The variable $p_{L(t)}$ is the load demand at time t. The power production of unit i at time t is represented by the variable $p_{(i,t)}$. The binary variable $U_{(i,t)}$ is equal to one if the unit i is committed at time t and zero if the generator is de-committed.

During the optimization process, some unfeasible conditions could be revealed. The optimization algorithm could commit insufficient or excessive power generation so that the energy balance cannot be guaranteed. Similar situations could be observed for the SR constraint. The unfeasible configurations can be managed by adding some auxiliary devices. In the case of insufficient generation, we add an infinitely expensive

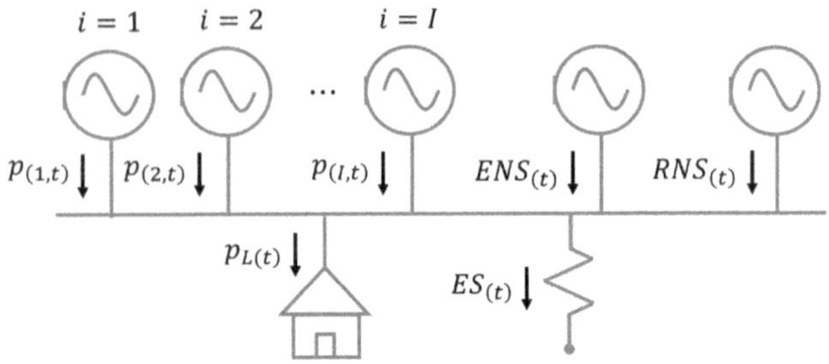

Fig. 4.4 Scheme of a typical energy system: General structure of a typical generating system.

generator to supply the missing power. A similar approach is adopted in the case of inadequate SR. Finally, an additional load is connected to consume the power excess. The generator $ENS_{(t)}$ of Fig. 4.4 supplies the energy not supplied (ENS), the unit $RNS_{(t)}$ provides the reserve not supplied, and the load $ES_{(t)}$ consumes the energy excess at a specific time t.

Conventional generators have to follow specific rules and operational constraints during their life service. When committed ($U_{(i,t)}=1$), the power production of each unit has to be adjusted between a minimum ($p_{(i)}{}^{\min}$) and maximum ($p_{(i)}{}^{\max}$) limit. On the one hand, the typical restrictions on power production are expressed in Eq. (4.27). On the other hand, the output power constraints of units with prohibited operating zones (POZs) are described in Eq. (4.28), where k is the index for each zone (Orero & Irving, 1996).

$$\begin{aligned} & p_{(i)}^{\min} U_{(it)} \leq p_{(it)} \leq p_{(i)}^{\max} U_{(it)} \\ & \forall i = 1,\ldots,I; \\ & t = 1,\ldots,T \end{aligned} \tag{4.27}$$

$$\begin{cases} p_{(i)}^{\min} U_{(i,t)} \leq p_{(i,t)} \leq p_{(i,1)}^{\text{lower}} U_{(i,t)} \\ \vdots \\ p_{(i,k-1)}^{\text{upper}} U_{(i,t)} \leq p_{(i,t)} \leq p_{(i,k)}^{\text{lower}} U_{(i,t)} \forall k = 2,\ldots,K \\ \vdots \\ p_{(i,K+1)}^{\text{upper}} \leq p_{(i,t)} \leq p_{(i)}^{\max} U_{(i,t)} \end{cases} \tag{4.28}$$

The energy balance constraint is presented in Eq. (4.29), while the SR restriction is shown in Eq. (4.30). The variable $SR_{(t)}$ is the SR requirement at time t (Dieu & Ongsakul, 2008).

$$\sum_{i=1}^{I} p_{(i,t)} U_{(i,t)} = p_{L(t)} \forall t = 1,\ldots,T \tag{4.29}$$

$$\sum_{i=1}^{I} p_{(i)}^{\max} U_{(i,t)} - p_{L(t)} \geq SR_{(t)} \forall t = 1,\ldots,T \tag{4.30}$$

The daily cost associated with the fuel consumption of all units (F) is described in Eq. (4.31), where $a_{(i)}$, $b_{(i)}$, $c_{(i)}$, $e_{(i)}$, and $r_{(i)}$ are parameters of the fuel consumption curve of unit i and $SUC_{(i)}$ is its starting-up cost (Dieu & Ongsakul, 2008; Sinha et al., 2003).

$$F=\sum_{t=1}^{T}\left\{\sum_{i=1}^{I}\left\{a_{(i)}+b_{(i)}\left(p_{(i,t)}\right)+c_{(i)}\left(p_{(i,t)}\right)^2+\left|e_{(i)}\times\sin\left(r_{(i)}\left(p_{(i)}^{min}-p_{(i,t)}\right)\right)\right|\right\}\right.$$

$$\left. U_{(i,t)}+\sum_{i=1}^{I}SUC_{(i)}\left(1-U_{(i,t-1)}\right)U_{(i,t)}\right\} \tag{4.31}$$

For each unit ($i=1$, ..., I), starting-up cost ($SUC_{(i)}$) depends on the hot-start cost ($HSC_{(i)}$), the cold start cost ($CSC_{(i)}$), and the minimum down time ($MDT_{(i)}$) according to Eq. (4.32) (Kazarlis et al., 1996).

$$SUC_{(i)}=\begin{cases} HSC_{(i)};MDT_{(i)}\le CST_{(i)} \\ CSC_{(i)};MDT_{(i)}>CST_{(i)} \end{cases}\forall i=1,\ldots,I \tag{4.32}$$

The minimum downtime (MDT) restriction refers to the fact that if a generator is decommitted, it has to be offline for at least $MDT_{(i)}$ hours. The minimum up time (MUT) restriction is similar but considering the online status.

Considering a concrete unit i. if during the optimization process the number of online hours until the time t ($T_{ON(i,t)}$) is lower than the corresponding MUT of that generator ($MUT_{(i)}$), the unit i has to be online. On the contrary, if the number of offline hours until the time t ($T_{OFF(i,t)}$) is lower than the corresponding MDT of that generator ($MDT_{(i)}$), the generator i has to remain disconnected. Finally, if the unit fulfills the MUT or MDT strictly, that generator could be connected or disconnected from the system. This reasoning is described by Eq. (4.33) (Dieu & Ongsakul, 2008).

$$U_{(it)}=\begin{cases} 1;T_{ON(it)}<MUT_{(i)} \\ 0;T_{OFF(it)}<MDT_{(i)} \\ 1\vee 0;T_{ON(it)}=MUT_{(i)}\vee T_{OFF(it)}=MDT_{(i)} \end{cases}$$
$$\forall i=1,\ldots,I;t=1,\ldots,T \tag{4.33}$$

Thermal generators also have to fulfill the ramp restrictions related to the unit's rate to increase or reduce its power production. This constraint can be considered by substituting Eqs. (4.27) or (4.28) with Eq. (4.34) (Damousis et al., 2004).

$$\max\left\{p_{(i)}^{\min}p_{(it-1)}-DR_{(i)}\right\}\le p_{(it)}\le\min\left\{p_{(i)}^{\max}p_{(it-1)}+UR_{(i)}\right\}$$
$$\forall i=1,\ldots,I;t=1,\ldots,T \tag{4.34}$$

4.5 Battery energy storage systems

Battery energy storage systems (BESS) are essential for the mass adoption of renewable energies because they can deal with the problem of renewable power variability. This chapter pays special attention to the simulation of lead-acid batteries (LAB) and vanadium redox flow batteries (VRFB) because they have high applicability in rural and intelligent grids.

4.5.1 Lead-acid batteries

Lead-acid batteries (LAB) are among the most studied technologies for energy storage. Researchers have developed a vast family of mathematical models for LAB simulation, which are critical for the techno-economic analysis of energy systems.

Concerning LAB simulation, Manwell and McGowan (1993) developed a battery model based on chemical kinetics that can represent the variation of battery capacity with the charge and discharge rates. This model has been implemented in many computational programs used for energy system design and optimization. Copetti et al. (1993) and Copetti and Chenlo (1994) proposed a general-purpose model to calculate the battery performance, using normalized expressions of voltage, coulombic efficiency, and state of charge (SOC). Schiffer et al. (2007), Andersson (2006), and Bindner et al. (2005) developed a weighted Ah (amp hours) aging model that can estimate the battery voltage and SOC and its operational lifetime, providing valuable quantitative information for the optimal design of energy systems. This chapter will briefly explain the main steps behind implementing the weighted Ah aging model.

In the weighted Ah aging model, the discharged Ah are weighted to consider the severity of the actual operating situation. Specifically, the weighting factors consider the impact on the positive grid's corrosion and the active mass' degradation. The model was developed from a modified version of the Shepherd model. At a specific hour ($t = 1, \ldots, T$), we need to assess the performance of a battery cell, calculating the voltage, current, and SOC. Then, the positive grid's corrosion and the active mass' degradation are analyzed, and the value of the weighting factors is assigned.

The analysis of the weighted Ah aging model starts with the behavior of a single cell. Eqs. (4.35)–(4.38) calculate the voltage and SOC during charging and discharging conditions of a typical cell of rated capacity C_{10}. At time t. the battery current ($I_{(t)}$) is assumed to be positive if the battery is being charged and negative if it is being discharged. Then, the battery voltage ($U_{(t)}$) is calculated using Eq. (4.35) during the charging and with Eq. (4.36) during discharging. These expressions consider the influence of SOC ($SOC_{(t)}$) and depth of discharge (DOD) ($DOD_{(t)}$) on the voltage (Schiffer et al., 2007).

$$U_{(t)} = U_o - gDOD_{(t)} + R^c_{(t)}\left(\frac{I_{(t)}}{C_{10}}\right) + R^c_{(t)}M_c\left(\frac{I_{(t)}}{C_{10}}\right)\left(\frac{SOC_{(t)}}{C_c - SOC_{(t)}}\right) \quad (4.35)$$

$$\forall I_{(t)} > 0$$

$$U_{(t)} = U_o - gDOD_{(t)} + R^d_{(t)}\left(\frac{I_{(t)}}{C_{10}}\right) + R^d_{(t)}M_d\left(\frac{I_{(t)}}{C_{10}}\right)\left(\frac{DOD_{(t)}}{C^d_{(t)} - DOD_{(t)}}\right) \quad (4.36)$$

$$\forall I_{(t)} \leq 0$$

For each time step (Δt), SOC is calculated using Eq. (4.37), considering the gassing current (${I_{(t)}}^G$) during the charge, which is obtained using Eq. (4.38). We can observe how the effect of ambient temperature ($T_{a(t)}$) is considered in the exponential behavior of the gassing current (Schiffer et al., 2007).

$$SOC_{(t)} = SOC_{(t-\Delta t)} + \int_{t-\Delta t}^{t}\left\{\frac{I_{(\vartheta)} - I^G_{(\vartheta)}}{C_{10}}\right\}d\vartheta \quad (4.37)$$

$$I^G_{(t)} = \left(\frac{C_{10}}{100}\right)\left(I^G_{0(t)}\right)\exp\left(C_U\left[U_{(t)} - U^G_0\right] + C_T\left[T_{a(t)} - T^G_0\right]\right) \quad (4.38)$$

$$\forall I_{(t)} > 0$$

The corrosion of the positive grid depends on the corrosion voltage, the thickness of the corrosion layer, and the resistance increment. The corrosion voltage is estimated using Eq. (4.39) during charging (${U_{(t)}}^c$) and Eq. (4.40) for discharging (${U_{(t)}}^d$) (Schiffer et al., 2007).

$$U^c_{(t)} = U^c_o - \left(\frac{10}{13}\right)gDOD_{(t)} + R^c_{(t)}\left(\frac{I_{(t)}}{2C_{10}}\right) + R^c_{(t)}M_c\left(\frac{I_{(t)}}{2C_{10}}\right)\left(\frac{SOC_{(t)}}{C_c - SOC_{(t)}}\right)$$

$$\forall I_{(t)} > 0 \quad (4.39)$$

$$U^d_{(t)} = U^d_o - \left(\frac{10}{13}\right)gDOD_{(t)} + R^d_{(t)}\left(\frac{I_{(t)}}{2C_{10}}\right) + R^d_{(t)}M_d\left(\frac{I_{(t)}}{2C_{10}}\right)\left(\frac{DOD_{(t)}}{C^d_{(t)} - DOD_{(t)}}\right)$$

$$\forall I_{(t)} \leq 0 \quad (4.40)$$

According to Eq. (4.41), the corrosion layer thickness ($\Delta W_{(t)}$) is calculated using the corrosion voltage. This equation is derived using Arrhenius' law and results published by Lander about the corrosion speed and voltage (Schiffer et al., 2007).

$$\Delta W_{(t)} = \begin{cases} k_{sp}\left(x^{0.6}\right)|x = \left[\Delta W_{(t-\Delta t)}/k_{sp}\right]^{1/0.6} + \Delta t; U^c_{(t)} < 1.74 \\ \Delta W_{(t-\Delta t)} + k_{sp}\Delta t; U^c_{(t)} \geq 1.74 \end{cases} \quad (4.41)$$

Using the corrosion layer thickness previously obtained and the rated battery lifetime (FL), the resistance corrosion layer ($\Delta R_{(t)}$) and the capacity loss due to corrosion (${\Delta C_{(t)}}^c$) are estimated through Eqs. (4.42)–(4.44) (Schiffer et al., 2007). In this case $T = 8760$ h. The corrosion speed parameter (k_{sp}) is calculated following Arrhenius' law. This parameter allows us to estimate the increment of the corrosion layer with time. The parameters $\Delta R_{\max}$ and $\Delta C_{\max}{}^{c}$ are the resistance and capacity at the end of the battery lifetime, respectively, and the parameter $k_{\max}{}^{sp}$ is the corrosion speed for a value of corrosion voltage equal to the battery float voltage (Schiffer et al., 2007).

$$\Delta R_{(t)} = \Delta R_{\max}\left(\frac{\Delta W_{(t)}}{\Delta W_{\max}}\right)\forall t = 1,\ldots,T \tag{4.42}$$

$$\Delta C^{c}_{(t)} = \Delta C^{c}_{\max}\left(\frac{\Delta W_{(t)}}{\Delta W_{\max}}\right)\forall t = 1,\ldots,T \tag{4.43}$$

$$\Delta W_{\max} = T(FL)\left(k^{sp}_{\max}\right) \tag{4.44}$$

Active mass degradation is related to the cycling operation at low SOC or the maximum DOD, among other operational conditions and factors. The SOC impact factor ($f_{(t)}{}^{soc}$) of Eq. (4.45) considers the effect of SOC, the time at low SOC, the discharging current, and the number of cycles that the battery is not completely charged (Schiffer et al., 2007). Fig. 4.5 illustrates a typical partial charging operation between t_A and t_B and between t_B and t, respectively. The partial operation is defined by the parameter $SOC_{\text{lim}}{}^{\max}$ which is equal to 0.9. The effects of operating at low SOC depends on the time under partial cycle operation ($\Delta t_{(t)}{}^{soc}$) defined in Eq. (4.46) and the minimum SOC ($\min\{SOC_{(\vartheta)}|\,\vartheta \in [t_B, t]\}$) (Schiffer et al., 2007).

$$f^{soc}_{(t)} = 1 + \left[c^{soc}_{o} + c^{soc}_{\min}\left(1 - \min\left\{SOC_{(\vartheta)}\,|\,\vartheta \in [t_B, t]\right\}\right) f^{I}_{(t)}\Delta t^{soc}_{(t)}\right] \tag{4.45}$$

$$\Delta t^{soc}_{(t)} = t - t_B \tag{4.46}$$

The impact of discharging current is incorporated with the current factor ($f_{(t)}{}^{I}$) of Eq. (4.47). It is essential to highlight that the discharging current at the beginning of the partial cycle ($I_{(t)}{}^{1th}$) is considered the most important (Schiffer et al., 2007).

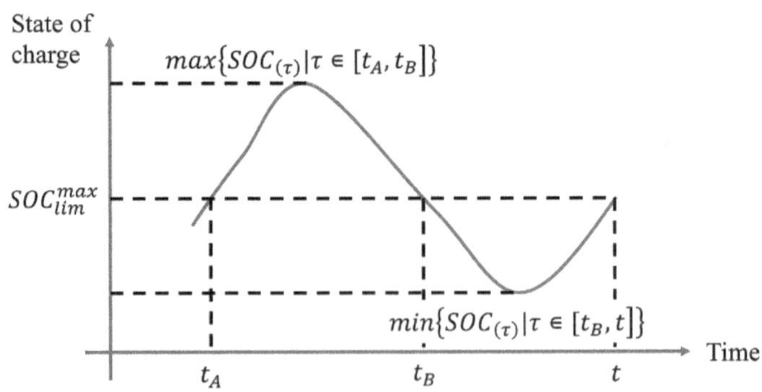

Fig. 4.5 Specifying full and partial charging: General parameters for full and partial charging definition.

The influence of incomplete charges is estimated by weighting the number of bad charges (q) according to the factor (Δq) of Eq. (4.48). The weighting factor depends on the maximum SOC, which should be higher than 0.99 to charge the battery appropriately (Schiffer et al., 2007).

$$f_{(t)}^{I} = \left(\sqrt{\frac{I_r}{I_{(1)}^{1th}}}\right)\left(\sqrt[3]{\exp\left(\frac{q}{3.6}\right)}\right) \tag{4.47}$$

$$\Delta q = \frac{0.0025 - \left(0.95 - \max\left\{SOC_{(\vartheta)} \mid \vartheta \in [t_A, t_B]\right\}\right)^2}{0.0025} \tag{4.48}$$

Acid stratification degree ($f_{(t)}^{ST}$) is estimated using Eq. (4.49), considering the net impact of the gassing and diffusion processes presented in Eqs. (4.50)–(4.53), respectively. According to these expressions, the minimum SOC, the gassing current ($I_{o(t)}^{G}$), and the ambient temperature ($T_{a(t)}$) are relevant factors in the variation of acid stratification. Finally, the total impact of acid stratification ($f_{(t)}^{A}$) is estimated using Eq. (4.54), where the effect of the discharge current with respect to a reference current (I_r) is also considered (Schiffer et al., 2007).

$$f_{(t)}^{ST} = f_{(t-\Delta t)}^{ST} + \int_{t-\Delta t}^{t} \left(\Delta f_{(\vartheta)}^{+} - \Delta f_{(\vartheta)}^{-}\right) d\vartheta \tag{4.49}$$

$$\Delta f_{(t)}^{+} = C_P\left(1 - \min\left\{SOC_{(\vartheta)} \mid \vartheta \in [t_B, t]\right\}\right) \exp\left(-3 f_{(t)}^{ST}\right) \left(\frac{\left|I_{(t)}^{d}\right|}{I_r}\right) \tag{4.50}$$

$$\Delta f_{G(t)}^{-} = C_M \sqrt{\frac{100}{C_{10}}} \left(\frac{I_{o(t)}^{G}}{I_{o}^{G}}\right) \exp\left(C_U\left[U_{(t)} - U_r\right] + C_T\left[T_{a(t)} - T_{o}^{G}\right]\right) \tag{4.51}$$

$$\Delta f_{D(t)}^{-} = \frac{8D}{(Z_D)^2} f_{(t-\Delta t)}^{ST} 2^{\left(T_{a(t)} - 293.15\right)/10} \tag{4.52}$$

$$\Delta f_{(t)}^{-} = \Delta f_{G(t)}^{-} + \Delta f_{D(t)}^{-} \tag{4.53}$$

$$f_{(t)}^{A} = 1 + f_{(t)}^{ST} \sqrt{\frac{I_r}{I_{(t)}^{1th}}} \tag{4.54}$$

Eq. (4.55) reflects the reasoning behind the weighted Ah aging model. The Ah discharged from the battery are weighted ($Z_{(t)}^{W}$), obtaining a more realistic value of the Ah discharged. Eq. (4.56) estimates battery life reduction ($\Delta C_{\max}^{d}$) using the weighted Ah cycles of Eq. (4.55). As can be observed, battery life reduces exponentially with the weighted Ah discharged (Schiffer et al., 2007).

$$Z^W_{(t)} = \frac{1}{C_{10}} \int_0^t \left| I^d_{(\vartheta)} \right| f^{soc}_{(\vartheta)} f^A_{(\vartheta)} d\vartheta \forall t = 1,\dots,T \tag{4.55}$$

$$\Delta C^d_{(t)} = \Delta C^d_{\max} \exp\left(-C_Z \left\{1 - \frac{Z^W_{(t)}}{1.6[Z_I]}\right\}\right) \tag{4.56}$$
$$\forall t = 1,\dots,T$$

Other operational parameters such as the internal resistance (${R_{(t)}}^c$ for charging and ${R_{(t)}}^d$ for discharging), the gassing current, and the battery capacity (${C_{(t)}}^d$) are updated using Eqs. (4.57)–(4.60) (Schiffer et al., 2007).

$$R^c_{(t)} = R^c_{(0)} + \Delta R_{(t)} \forall t = 1,\dots,T \tag{4.57}$$

$$R^d_{(t)} = R^d_{(0)} + \Delta R_{(t)} \forall t = 1,\dots,T \tag{4.58}$$

$$I^G_{o(t)} = I^G_o + \Delta I^G_o \left(\frac{\Delta R_{(t)}}{\Delta R_{\max}}\right) \forall t = 1,\dots,T \tag{4.59}$$

$$C^d_{(t)} = C^d_{(0)} - \Delta C^c_{(t)} - \Delta C^d_{(t)} \forall t = 1,\dots,T \tag{4.60}$$

To summarize, the implementation of the weighted Ah model at each time step consists of simulating the battery performance, evaluating the degree of corrosion of the positive grid, and battery degradation, and then finally, assessing the impact of the aging processes on the internal resistance, gassing current, and battery capacity (Schiffer et al., 2007).

Some parameters of this model are reported in Table 4.1. However, the reader can find more detailed information in the original references (Andersson, 2006; Bindner et al., 2005; Schiffer et al., 2007).

Table 4.1 Parameters of LAB model.

Parameter	Value	Parameter	Value
U_0	2.1 V	U_0^G	2.23 V
g	0.076 V	T_0^G	298 K
$R^c_{(0)}$	0.42 ΩAh	U_0^c	1.75 V
$R^d_{(0)}$	0.699 ΩAh	$k^{sp}_{\max}$	0.0035
M_c	0.888	c_0^{soc}	$6.614 \times 10^{-5}\ h^{-1}$
M_d	0.0464	$c^{soc}_{\min}$	$3.307 \times 10^{-3}\ h^{-1}$
C_c	1.001	C_P	1/30
$C^d_{(0)}$	1.75	C_M	0.1
I_0^G	0.02 A	D	$20 \times 10^{-9}\ m^2\ s^{-1}$
ΔI_0^G	0.06 A	Z_D	20 cm
C_U	$11\ V^{-1}$	U_r	2.5 V
C_T	$0.06\ K^{-1}$	C_Z	5

General-purpose values of LAB coefficients.

Alternatively, analyzing the operating conditions and stress factors is a qualitative manner of assessing battery performance. Svoboda et al. (2007) proposed several factors to carry out this type of study. The authors suggested using the charge factor, the Ah throughput, the highest discharge rate, the time between full charges, the time at low SOC, and the partial cycling.

The charge factor (*CF*) is defined according to Eqs. (4.61) and (4.62). As can be observed, it is the inverse of the Ah efficiency. The charge factor allows us to determine if the battery is under full or partial operation. In addition, it is helpful to find out the aging mechanism. If it is very low, the battery could suffer from sulfation and acid stratification, and a relevant difference between the cells in the bank would be observed. The battery could suffer from active mass shedding, water loss, and corrosion if it is very high. Typically, the charge factor is between 102% and 130% (Svoboda et al., 2007).

$$CF = \frac{Ah_{\text{charged}}}{Ah_{\text{discharged}}} = -\frac{\int_{t=1}^{T} I_{(t)} H\left(I_{(t)}\right) dt}{\int_{t=1}^{T} I_{(t)} H\left(-I_{(t)}\right) dt} \tag{4.61}$$

$$H\left(I_{(t)}\right) = \begin{cases} 0; I_{(t)} < 0 \\ 1; I_{(t)} \geq 0 \end{cases} \forall t = 1, \ldots, T \tag{4.62}$$

The Ah throughput (Q_{thr}) is presented in Eq. (4.63). It is a normalized measure of the Ah discharged from the battery during the entire year. This factor has to be used simultaneously with information about the cycling operation at a partial SOC and its duration to reveal the aging mechanism. Frequently, this factor is between 10 and 100 times the battery capacity in 10h (C_{10}) (Svoboda et al., 2007).

$$Q_{thr} = -\frac{\int_{t=1}^{T} I_{(t)} H\left(-I_{(i)}\right) dt}{C_{10}} \tag{4.63}$$

The highest discharge rate (${I_{1\%}}^{\max}$) is calculated by measuring the highest discharge current so that 1% of the Ah throughput (Q_{thr}) is discharged. To this end, a histogram of frequencies with *S* bins similar to that shown in Fig. 4.6 is built. The intervals ($l_{(s)}$) and their corresponding frequencies ($f_{(s)}$) are used to calculate the number of bins (*L*) required to fulfill 1% of the total Ah throughput discharged. Then, the stress factor is estimated as the average current of all of the bins (*L*) divided by the discharging current in 10h (I_{10}). This reasoning is presented in Eqs. (4.64) and (4.65). Typically, this stress factor is lower than I_{10} (Svoboda et al., 2007).

$$\sum_{s=S-L}^{S} l_{(s)} f_{(s)} = 0.01 (Q_{thr}) C_{10} \tag{4.64}$$

$$I_{1\%}^{\max} = \{0.01 (Q_{thr}) C_{10}\} / \left\{ I_{10} \left(\sum_{s=S-L}^{S} l_{(s)} f_{(s)} \right) \right\} \tag{4.65}$$

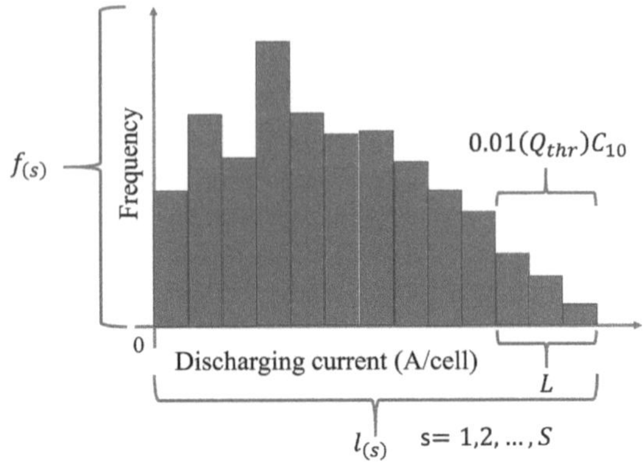

Fig. 4.6 Discrete distribution of discharge current: Histogram of battery discharge current.

The time between full charges (T_F) is calculated by measuring the operating cycles (C_F) and their duration, under which the SOC remains lower or equal to a specific threshold value ($SOC_{\text{lim}}{}^{\text{max}}$) equal to 0.9. This reasoning is mathematically expressed in Eq. (4.66). Frequently, this factor is between 0.7 and 10 days (Svoboda et al., 2007).

$$T_F = \left\{ \int_{t=1}^{T} H\left(SOC_{\text{lim}}^{\text{max}} - SOC_{(t)}\right) dt \right\} / \{24 C_F\} \tag{4.66}$$

The time at low $SOC(T_S)$ analyses the operational situations under which the SOC is lower than a specific threshold ($SOC_{\text{lim}}{}^{\text{low}}$) equal to 0.35. This factor is defined in Eq. (4.67) and is related to the increase in battery sulphation. Typically, it is between 0% and 50% (Svoboda et al., 2007).

$$T_S = 100 \left\{ \int_{t=1}^{T} H\left(SOC_{\text{lim}}^{\text{low}} - SOC_{(t)}\right) dt / T \right\} \tag{4.67}$$

The partial cycling (PC) considers the impact of Ah discharging at a specific SOC value. To estimate the partial cycling, first we need to calculate the Ah discharged at each of the SOC intervals of Table 4.2. This idea is shown in Eq. (4.68). After that,

Table 4.2 Definition of state-of-charge ranges.

x	$pc_{(x)}^{\min}$	$pc_{(x)}^{\max}$
A	0.85	1
B	0.7	0.85
C	0.55	0.7
D	0.4	0.55
E	0	0.4

SOC intervals for partial cycling evaluation.

the stress factor (PC) is determined using the weighted expression of Eq. (4.69), where low SOC intervals are penalized with a high value. Frequently, partial cycling is between 20% and 70% (Svoboda et al., 2007).

$$pc_{(x)} = \frac{-\left\{\int_{t=1}^{T} I_{(t)} H\left(SOC_{(t)} - pc_{(x)}^{min}\right) H\left(pc_{(x)}^{max} - SOC_{(t)}\right) H\left(-I_{(t)}\right) dt\right\}}{\left\{\int_{t=1}^{T} I_{(t)} H\left(-I_{(t)}\right) dt\right\}} \quad (4.68)$$

$$\forall x = A, B, C, D, E$$

$$PC = \left\{pc_{(A)} + 2\left[pc_{(B)}\right] + 3\left[pc_{(C)}\right] + 4\left[pc_{(D)}\right] + 5\left[pc_{(E)}\right]\right\}/5 \quad (4.69)$$

Using the previously described stress factors and their experience, Svoboda et al. (2007) defined an optimal solar home system (SHS) design; its features are summarized in Table 4.3, and this chapter has used them as a reference for the optimal design of energy systems provided with an LAB. In optimal design conditions, the battery is under a low to medium throughput, a medium charge factor, a time between full charge and partial cycling from low to high, and the highest discharge rate between low and medium.

4.5.2 Vanadium redox flow batteries

The vanadium redox flow battery (VRFB) is among the most relevant technologies for energy storage. The model implemented in this chapter was derived by Qiu et al. (2014) and Nguyen et al. (2014, 2015) from the experimental analysis of a commercial product. Specifically, the authors characterized a typical VRFB of 5 kW, 20 kWh, and 50 V. The battery is charging when its power is positive ($P_{(t)} > 0$) and discharging when its power is negative ($P_{(t)} < 0$). Battery disconnection is represented with the power equal to zero ($P_{(t)} = 0$). The model calculates the battery voltage and efficiency at each time step ($\Delta t = 1$ h). The rated capacity (20 kWh) is represented by the variable C_N.

Table 4.3 Reference values of stress factors and risk levels.

Stress factor	Value
CF	115%
Q_{thr}	70
$I_{1\%}^{max}$	1.4
T_F	8 days
T_S	15.50%
PC	70%

Stress factors of an optimally designed SHS.

During the charging process, the battery voltage ($U_{(t)}$) is calculated using Eq. (4.70), and the voltage and power efficiencies ($\eta_{V(t)}{}^{c}$ and $\eta_{P(t)}{}^{c}$) are estimated using Eqs. (4.71) and (4.72). Then, the general efficiency ($\eta_{(t)}$) is calculated through Eq. (4.73) (Nguyen et al., 2014).

$$U_{(t)} = \left(1.895\left[SOC_{(t)}\right] + 1.552\right)P_{(t)} + 6.82\left[SOC_{(t)}\right] + 46.79 \tag{4.70}$$

$$\eta^{c}_{V(t)} = \frac{0.038(T_E)\left(SOC_{(t)} - 1.1755\right) + 61.2674}{\left(1.895\left[SOC_{(t)}\right] + 1.552\right)P_{(t)} + 6.82\left[SOC_{(t)}\right] + 46.79} \tag{4.71}$$

$$\eta^{c}_{P(t)} = \frac{\left(-0.128\left[SOC_{(t)}\right] + 1.05\right)P_{(t)} + 0.19\left[SOC_{(t)}\right] - 0.59}{P_{(t)}} \tag{4.72}$$

$$\eta_{(t)} = \eta^{c}_{V(t)}\eta^{c}_{P(t)} \tag{4.73}$$

Similarly, during the discharging process, the battery voltage ($U_{(t)}$) is estimated using Eq. (4.74), the voltage and power efficiencies ($\eta_{V(t)}{}^{D}$ and $\eta_{P(t)}{}^{D}$) are calculated using Eqs. (4.75) and (4.76), and the general efficiency ($\eta_{(t)}$) is obtained using Eq. (4.77). The variable T_E is the electrolyte temperature (Nguyen et al., 2014).

$$U_{(t)} = -2.72\left|P_{(t)}\right| + 6.3606\left|SOC_{(t)}\right| + 47.335 \tag{4.74}$$

$$\eta^{D}_{V(t)} = \frac{-2.72\left|P_{(t)}\right| + 6.3606\left[SOC_{(t)}\right] + 47.335}{0.038(T_E)\left(SOC_{(t)} - 1.1755\right) + 61.2674} \tag{4.75}$$

$$\eta^{D}_{P(t)} = \frac{\left|P_{(t)}\right|}{1.0334\left|P_{(t)}\right| + 1.727\left(SOC_{(t)}\right)\left(SOC_{(t)} - 1\right) + 0.596} \tag{4.76}$$

$$\eta_{(t)} = \eta^{D}_{V(t)}\eta^{D}_{P(t)} \tag{4.77}$$

Finally, SOC is calculated as a function of the battery power and its efficiency according to Eq. (4.78).

$$\begin{aligned} SOC_{(t)} &= SOC_{(t-\Delta t)} + \int_{t-\Delta t}^{t} \left\{\frac{P_{(\vartheta)}\eta_{(\vartheta)}}{C_N}\right\} d\vartheta \\ \forall t &= 1, \ldots, T \end{aligned} \tag{4.78}$$

The charge controller is a vital component of BESS, and it is in charge of maintaining the battery voltage, current, and power within appropriate limits. Mathematical representation of the charge controller is carried out using the values of Table 4.4 with the corresponding nominal current (LAB) and power (VRFB) (Nguyen et al., 2014; Schiffer et al., 2007).

Table 4.4 Limit values for battery operation.

Control variable	LAB	VRFB
$SOC_{\min}$	0.3	0.15
$SOC_{\max}$	1	0.9
$U_{\min}$	1.9 V	42 V
$U_{\max}$	2.23 V	56.5 V

Operational constraints of BESS.

For LAB, the charging and discharging currents to maintain the battery voltage of Table 4.4 were estimated using the bisection method in the interval between 0 and the discharging current in 10 h (I_{10}). A similar approach was applied to analyze the VRFB but considering the battery power.

Once the mathematical models of renewable power sources, thermal generators, and energy storage devices have been described, the GA implementation to solve the ED and UC problems, the daily battery management, and isolated energy system design are carefully explained in the next sections.

4.6 Economic dispatch by genetic algorithm

As previously explained, the ED problem consists of determining the power production at time (t) of each unit to minimize the total operational costs without any constraint violation. According to Fig. 4.4, additional elements such as the two generators and the load are integrated into the objective function F as described in Eq. (4.79).

$$F = \sum_{i=1}^{I}\left\{a_{(i)} + b_{(i)}\left(p_{(i,t)}\right) + c_{(i)}\left(p_{(i,t)}\right)^2 + \left|e_{(i)} \times \sin\left(r_{(i)}\left(p_{(i)}^{\min} - p_{(i,t)}\right)\right)\right|\right\}$$
$$+ b_{ENS}\left(ENS_{(t)}\right) + c_{ENS}\left(ENS_{(t)}\right)^2 + b_{ES}\left(ES_{(t)}\right) + c_{ES}\left(ES_{(t)}\right)^2 \tag{4.79}$$

The parameters of these devices are set to an enormous value ($b_{ENS} \rightarrow \infty$, $c_{ENS} \rightarrow \infty$, $b_{ES} \rightarrow \infty$, and $c_{ES} \rightarrow \infty$). Thus, these are penalty factors for individuals that do not fulfill the equality constraint of Eq. (4.29).

During the optimization process, if a GA individual does not provide enough power to fulfill the load demand, the generator $ENS_{(t)}$ would produce that energy to guarantee the energy balance and the load $ES_{(t)}$ would be set to zero. On the contrary, if an individual produces more energy than that required by the load, the energy surplus would be consumed by the load $ES_{(t)}$ and the generator $ENS_{(t)}$ would be set to zero. The inequality constraints of Eqs. (4.27) or (4.28) are guaranteed by setting the individual violating these restrictions to its minimum or maximum power.

The real-coded GA developed by Chuang et al. (2015, 2016) will be adopted to solve the ED problem. The algorithm has three operators: the ranking selection (RS), the direction-based crossover (DBX), and the dynamic random mutation (DRM). These operators are applied considering the crossover rate (p_c), the mutation rate (p_m), and the proportional parameter (p_r). n is the index for each individual ($n = 1, \ldots, N$). (y_{obj} is the objective function F, x_{pop} is the population matrix with I rows and N columns.) RS is in charge of extracting the best individuals from the population. This operator is implemented using the procedure explained as follows:

Step 1: Individuals are sorted according to fitness. In this regard, the order in which the individuals should be grouped is saved in a list ($\mathcal{L}_{\mathbf{A}}$).
Step 2: Calculate the number of individuals that have relatively worse fitness. This is done by rounding down the product $N \times p_r$(floor($N \times p_r$)).
Step 3: Select the first $N -$ floor($N \times p_r$) elements of the list $\mathcal{L}_{\mathbf{A}}$ and save the result in another list ($\mathcal{L}_{\mathbf{B}}$). These are the indices of the relatively best individuals to be kept. The remaining is the index of the individuals to be discarded.
Step 4: Select the first floor($N \times p_r$) elements of the list $\mathcal{L}_{\mathbf{A}}$ and save them in another list ($\mathcal{L}_{\mathbf{C}}$). These are the indices of the individuals used to replace relatively worse individuals.
Step 5: Create a list ($\mathcal{L}_{\mathbf{D}}$) by joining or merging the lists $\mathcal{L}_{\mathbf{B}}$ and $\mathcal{L}_{\mathbf{C}}$ previously built.
Step 6: Sort the individuals in the list $\mathcal{L}_{\mathbf{D}}$ according to fitness and save the results in another list ($\mathcal{L}_{\mathbf{E}}$). This is the result of the RS process.

DBX divides the family of individuals obtained from the RS into two sets: the individuals with a high ranking and those with a lower position. These two groups are crossed sequentially, improving the capabilities of the traditional crossover operator. The DBX is implemented by following the procedure:

Step 1: Set $u \leftarrow$ floor($N/2$).
Step 2: Start the sequential crossing of pairs of individuals. Go to Step 2.1.
Step 2.1: Set $m \leftarrow 1$.
Step 2.2: If $m \leq$ floor($N/2$). go to Step 2.3, else end.
Step 2.3: Set $u \leftarrow u + 1$.
Step 2.4: Select the individual m of $\mathcal{L}_{\mathbf{E}}$ $\left(\mathcal{L}_{E(m)}\right)$, it will be called $x_{\text{pop}}^{a}\left(x_{\text{pop}}^{a} \leftarrow x_{\text{pop}}\left(\mathcal{L}_{E(m)}\right)\right)$.
Step 2.5: Select the individual u of $\mathcal{L}_{\mathbf{E}}$ ($\mathcal{L}_{E(u)}$), it will be called $x_{\text{pop}}{}^{b}$($x_{\text{pop}}^{b} \leftarrow x_{\text{pop}}\left(\mathcal{L}_{E(u)}\right)$).
Step 2.6: If $RAND > 1 - p_c$ and $(\max\{y_{\text{obj}(n)} \forall n = 1, \ldots, N\} - \min\{y_{\text{obj}(n)} \forall n = 1, \ldots, N\}) > \varepsilon$ calculate the variable S_c using Eq. (4.80). $RAND$ is a uniform random number in the interval [0, 1], and ε is a small number (tolerance) (Chuang et al., 2015, 2016).

$$S_c = \frac{\left| y_{\text{obj}}\left(x_{\text{pop}}^{a}\right) - y_{\text{obj}}\left(x_{\text{pop}}^{b}\right) \right|}{\max\left\{y_{\text{obj}(n)} \forall n = 1, \ldots, N\right\} - \min\left\{y_{\text{obj}(n)} \forall n = 1, \ldots, N\right\}} \tag{4.80}$$

Step 2.7: Calculate the element of the corrective vector (Δx_{pop}), which has I elements, using Eq. (4.81) (Chuang et al., 2015, 2016). $RAND$ is a random number from a uniform distribution within the interval [0, 1].

$$\Delta x_{\text{pop}(i)} = \begin{cases} 0; RAND < 0.5 \\ x_{\text{pop}(i)}^{a} - x_{\text{pop}(i)}^{b}; RAND \geq 0.5 \end{cases} \tag{4.81}$$
$$\forall i = 1, \ldots, I$$

Step 2.8: Calculate the offspring of individuals a and b using Eq. (4.82) (Chuang et al., 2015, 2016).

$$\begin{aligned} x_{\text{pop}}^{a} &\leftarrow x_{\text{pop}}^{a} + S_c \Delta x_{\text{pop}} \\ x_{\text{pop}}^{b} &\leftarrow x_{\text{pop}}^{b} + S_c \Delta x_{\text{pop}} \end{aligned} \tag{4.82}$$

Step 2.9: Perform DBX in the next group of individuals by setting $m \leftarrow m+1$ and going to Step 2.2.

DRM is incorporated to increase the diversity among the individuals in the population. This operator is implemented by the following steps:

Step 1: Considering the generation under study (j), calculate the variable S_m using Eq. (4.83), where J is the total number of generations ($j=1, \ldots, J$) (Chuang et al., 2015, 2016).

$$S_m = \left(\frac{1-j}{J}\right)^2 \tag{4.83}$$

Step 2: Apply DRM over the population. Go to Step 2.1.
Step 2.1: Set $n \leftarrow 1$.
Step 2.2: If $n \leq N$ go to Step 2.3, else end.
Step 2.3: If $RAND < p_m$ go to Step 2.4.
Step 2.4: Set $x_{\text{pop}(n)} \leftarrow x_{\text{pop}(n)} + S_m \boldsymbol{RAND}(p_{(i)}{}^{\max} - p_{(i)}{}^{\min} \forall i=1,\ldots,I)$. where $\boldsymbol{RAND}$ is a random vector with I elements taken from a uniform distribution within the interval [0, 1].
Step 2.5: Set $n \leftarrow n+1$ and go to Step 2.2.

Some individuals may violate the constraints of Eqs. (4.27) or (4.28). In such a case, the individuals are adjusted to the closest POZ's minimum or maximum values.

Elitism is incorporated into this implementation by inserting the best individual in a position randomly selected from the interval [1, N].

4.7 Unit commitment combining priority list and genetic algorithm

This section briefly describes the PL (priority list) method and its use on GA initialization. It is similar to the approach developed by Orero and Irving (1997).

4.7.1 Priority list method

First, we need to initialize the matrix $U_{(i,t)}$ to zero and create a vector to store the committed capacity at each hour ($p_{(t)}{}^{\text{cap}} \forall t=1, \ldots, T$) and initialize it to zero. Then, the primary unit scheduling can be obtained using the following algorithm (Dieu & Ongsakul, 2008).

Step 1: Build the PL. Calculate the average power ($p_{(i)}{}^{\text{avg}}$) and then the average generation cost ($F_{(i)}$) for each unit ($i=1,\ldots,I$) using Eqs. (4.84) and (4.85) (Dieu & Ongsakul, 2008). The PL ($\mathcal{L}_{\mathbf{F}}$) is obtained by sorting the average cost in ascending order.

$$p_{(i)}^{\text{avg}} = \frac{p_{(i)}^{\min} + p_{(i)}^{\max}}{2} \forall i = 1, \ldots, I \tag{4.84}$$

$$F_{(i)} = \frac{a_{(i)} + b_{(i)}\left(p_{(i)}^{\text{avg}}\right) + c_{(i)}\left(p_{(i)}^{\text{avg}}\right)^2}{p_{(\text{i})}^{\text{avg}}} \forall i = 1, \ldots, I \tag{4.85}$$

Step 2: Set $t \leftarrow 1$.
 Step 2.1: While $t \leq T$ go to Step 2.2, else end.
 Step 2.2: Set $i \leftarrow 1$.
 Step 2.3: While $i \leq I$ go to Step 2.4, else go to Step 2.1.
 Step 2.4: Set $U(\mathcal{L}_{F(i)}, t) \leftarrow 1$ and set $p_{(t)}^{\text{cap}} \leftarrow p_{(t)}^{\text{cap}} + p^{\max}\left(\mathcal{L}_{F(i)}\right)$.
 Step 2.5: If $p_{(t)}{}^{\text{cap}} < p_{L(t)} + SR_{(t)}$ go to Step 2.6, else go to Step 2.7.
 Step 2.6: Set $i \leftarrow i+1$ and go to Step 2.3.
 Step 2.7: Set $i \leftarrow I+1$ and set $t \leftarrow t+1$ and go to Step 2.3.

The solution stored in the matrix $U_{(i,t)}$ is not feasible because the MUT and MDT constraints have not been considered. Thus, Figs. 4.7 and 4.8 show the repairing strategy (Dieu & Ongsakul, 2008).

The minimum up and down time repairing processes are implemented as follows (Dieu & Ongsakul, 2008):

Step 1: Calculate the number of online and offline hours until the time t ($T_{ON(i,t)}$ and $T_{OFF(i,t)}$) on the scheduling obtained ($U_{(i,t)}$) from the primary unit scheduling previously applied. The definitions of Eqs. (4.86) and (4.87) can be applied (Dieu & Ongsakul, 2008).

$$T_{ON(i,t)} \leftarrow \begin{cases} T_{ON(i,t-1)} + 1; U_{(i,t)} = 1 \\ 0; U_{(i,t)} = 0 \end{cases} \tag{4.86}$$

$$T_{OFF(i,t)} \leftarrow \begin{cases} T_{OFF(i,t-1)} + 1; U_{(i,t)} = 0 \\ 0; U_{(i,t)} = 1 \end{cases} \tag{4.87}$$

Step 2: Set $t \leftarrow 1$.
 Step 2.1: If $t \leq T$ go to Step 2.2, else stop.
 Step 2.2: Set $i \leftarrow 1$.

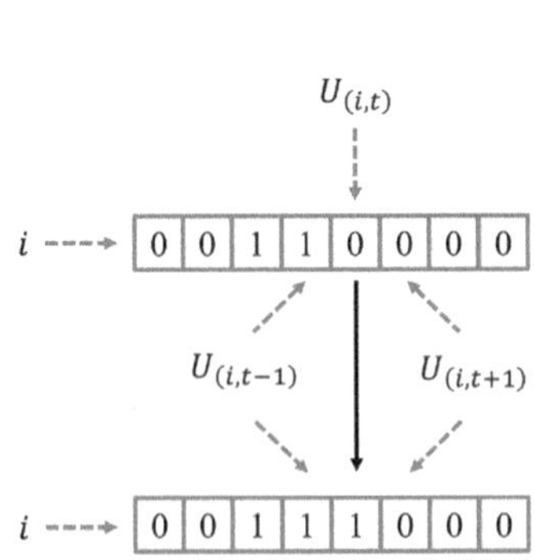

Fig. 4.7 Minimum up time repairing strategy: Implementing the minimum up time repairing process.

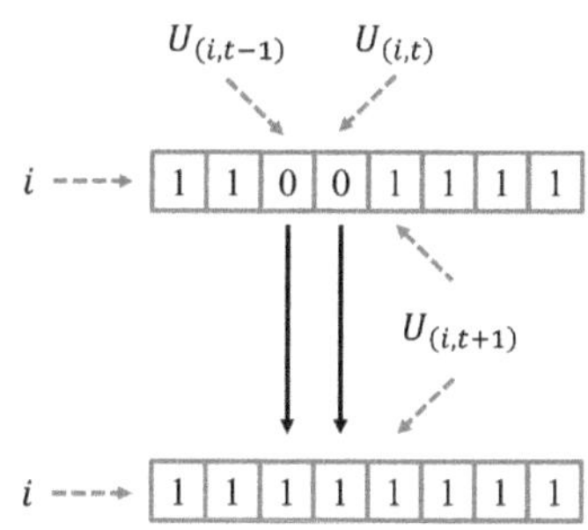

Fig. 4.8 Minimum down time repairing strategy: Implementing the minimum down time repairing process.

Step 2.3: If $i \leq I$ go to Step 2.4, else set $t \leftarrow t+1$ and go to Step 2.1.
Step 2.4: If $U_{(i,t)}=0$ and $U_{(i,t-1)}=1$ and $T_{ON(i,t-1)}<MUT_{(i)}$ set $U_{(i,t)} \leftarrow 1$.
Step 2.5: If $U_{(i,t)}=0$ and $U_{(i,t-1)}=1$ and $t+MDT_{(i)}-1 \leq T$ and T_{ON} $(i,t+MDT_{(i)}-1)<MDT_{(i)}$. set $U_{(i,t)} \leftarrow 1$.
Step 2.6: If $U_{(i,t)}=0$ and $U_{(i,t-1)}=1$ and $t+MD\,T_{(i)}-1>T$ and $\sum_{\vartheta=t}^{T} U_{(i,\vartheta)}>0$ set $U_{(i,t)} \leftarrow 1$.
Step 2.7: Update the matrices $T_{ON(i,t)}$ and $T_{OFF(i,t)}$ using Eqs. (4.86) and (4.87).
Step 2.8: Set $i \leftarrow i+1$ and go to Step 2.3.

At this stage, the UC solution is feasible but expensive. The PL method employs the unit substitution process to improve the quality of the solution (Dieu & Ongsakul, 2008). However, we are going to use a binary-coded GA to this end.

4.7.2 Genetic algorithm

The GA implemented in this chapter for the solution of the UC problem uses binary coding, where 1 means committing and 0 means de-committing. The implementation of the GA operators using binary coding has already been explained in Chapter 2. Thus, we will pay special attention to the objective function evaluation and GA individual structure.

Fig. 4.9 illustrates how to build a GA individual from a specific schedule ($U_{(i,t)}$). As can be observed, we only need to reshape the scheduling matrix ($U_{(i,t)}$) to a row vector with $I \times T$ elements. In this case $T=24$ h.

We are using the GA to improve the quality of an available UC solution obtained by PL so that we only apply the classical selection, crossover, and mutation operators. Once these operators have been applied, an unfeasible solution with respect to the MUT and MDT constraints would be observed. Then, the repairing process used by PL to deal with MUT and MDT constraints is applied again.

During a specific generation we need to determine the objective function of a specific individual expressed as a vector with $I \times T$ elements. Thus, we reshape the individual in a matrix ($U_{(i,t)}$) with I rows and T. For each column ($t=1, \ldots, T$) of the matrix

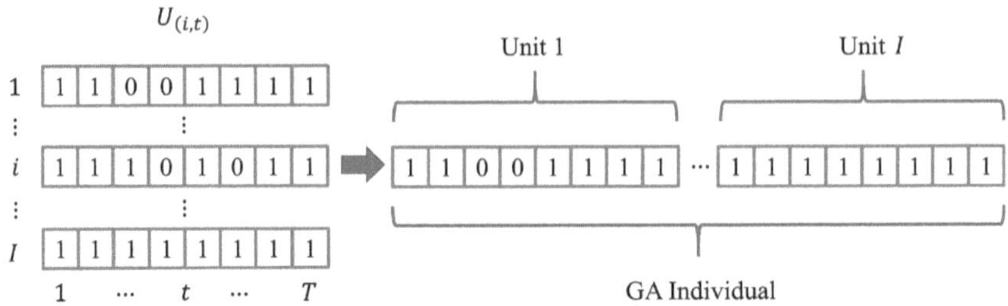

Fig. 4.9 Creating a GA individual to solve the UC: Structure of a GA individual to solve the UC problem.

$U_{(i,t)}$ we solve the ED problem (Section 4.6) only considering the committed units ($U_{(i,t)}$ =1) to minimize the objective function of Eq. (4.79).

4.8 Day-ahead battery management by genetic algorithm

In the modern electricity grid, it is assumed that electricity prices resulting from market transactions can be directly applied to consumers. Thus, the prices at peak-load hours would be higher than the valley-load hours. In this context, energy arbitrage using BESS can be formulated as an optimization problem. We buy or sell electricity from or to the grid to maximize the economic benefit. To this end, we need to apply a forecasting technique on a day-ahead basis and a battery model to estimate such benefits or their associated cost.

The solution to the energy arbitrage problem consists of finding the optimal battery operation to minimize the daily net cost (DNC) (*DNC*), which is calculated according to Eq. (4.88):

$$DNC = \sum_{t=1}^{T} E_{(t)} P_{(t)} \tag{4.88}$$

where $E_{(t)}$ is the electricity price at time t. The electricity price signal is obtained from a forecasting process. Information about this topic is available in Chapter 5. Three states define the battery operation. The charging period is represented as +1, discharging period is characterized as −1, and battery disconnection is expressed with a 0. Battery power ($P_{(t)}$) is obtained via simulation according to its technology (LAB or VRFB), and the electricity prices are estimated via forecasting 24 h ahead.

When the battery is discharged, the power is negative, and we are selling energy to the grid. Thus, we need to find the combination of states (+1, −1, and 0) to minimize DNC. DNC has to be as negative as possible. If DNC is negative, the economic benefit obtained from the sold energy is higher than the economic loss related to the energy bought throughout the day. On the contrary, if DNC is positive, financial losses are higher than the benefits, and the transaction with the grid is not profitable.

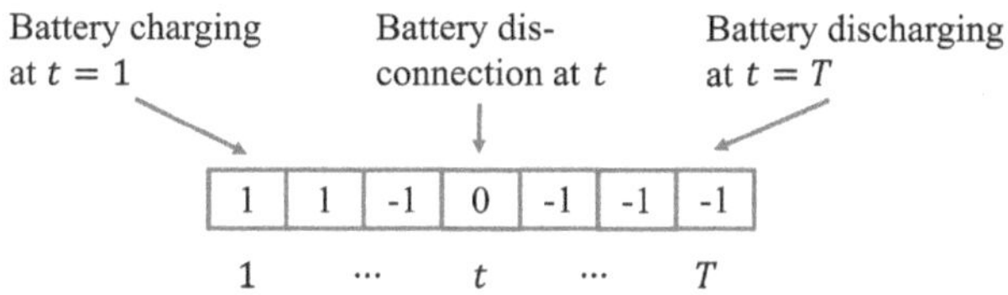

Fig. 4.10 GA individual for energy arbitrage: Structure of a GA individual for battery management.

To solve this optimization problem, we employed an ICGA. ICGA implementation is similar to the binary-coded GA explained in Chapter 2. For this reason, we will center our attention on describing the characteristics of the GA individual. The general characteristics of a typical individual are shown in Fig. 4.10. The elements of each individual correspond to hourly decisions, which can be +1, −1, or 0.

According to Fig. 4.10, the battery has to be charged at the first hour, it has to be disconnected at hour t and discharged at time T. In this case, $T = 24$ h. Then, the battery is simulated, as previously explained in Section 4.5, obtaining the power. Finally, the battery power is used to evaluate Eq. (4.88) to calculate DNC. DNC defines the fitness for this GA individual.

4.9 Design of hybrid systems by genetic algorithm

The optimal design of small-scale energy systems with renewable energies is vital for rural electrification. GA offers promising results in selecting the configuration of power sources to minimize the lifetime costs. Load demand can be fulfilled by combining conventional units such as diesel generators, renewable power sources such as solar PV and wind power, and a BESS. Renewable sources and BESS reduce the number of operating hours of the conventional unit, reducing the fuel consumption and its operating and maintenance costs.

The design problem involves finding out the size of the renewable sources, conventional generation, and BESS to minimize the net present cost (NPC) over the project lifetime. Information about the environmental conditions and load demand are available as hourly time series. A detailed discussion is available in Chapter 3. It is assumed that many manufacturers of WTs, PV panels, and batteries are available, including their acquisition costs. Depending on the number of elements to be selected, the number of possible combinations could be high, so that adopting an ICGA could be a good option.

ICGA implementation is similar to the binary-coded one explained in Chapter 2. We will center our attention on explaining the calculation of the objective function and GA individual structure. Fig. 4.11 shows the general features of a GA individual. Each element is an integer number with a different meaning, depending on its position. As shown in Fig. 4.11, the first element indicates the WT manufacturer, the second shows the number of turbines, the third indicates the battery manufacturer, and the fourth indicates the number of batteries connected in parallel (strings). This structure could be extended to incorporate a solar PV generation and a diesel unit.

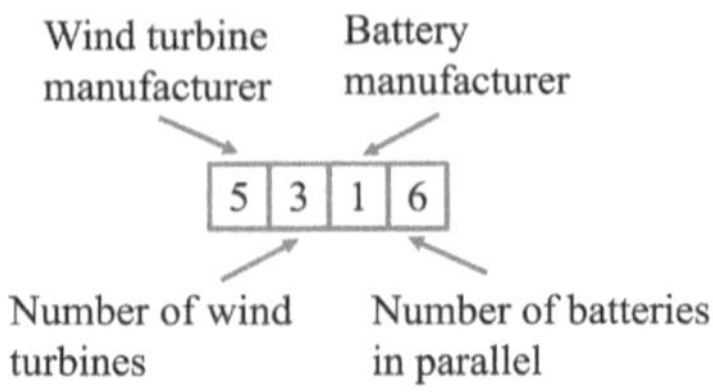

Fig. 4.11 GA individual for small-scale system design: Structure of a GA individual to design a hybrid power system.

Continuing with the individual shown in Fig. 4.11, the configuration is simulated through the year ($T=8760$h) to calculate the battery lifetime (Section 4.6), the fuel consumption, the number of operating hours of the conventional unit, and the energy index of unreliability (EIU). EIU has to be lower than a prespecified value indicated by the decision-maker (DM). EIU is defined as the total load not supplied during the year divided by the total energy demand. It is a normalized measure of the ENS.

The battery life and the operating hours of diesel units define their replacement period, which has a relevant weight on NPC. Providing energy with a specific EIU is also a mandatory constraint of the optimization problem. In rural electrification projects, the BESS is typically based on an LAB. Therefore, we will consider the quantitative approach suggested by Schiffer et al. (2007), Andersson (2006), and Bindner et al. (2005), and the qualitative method proposed by Svoboda et al. (2007).

To assess the fitness of a specific GA individual, we simulate the energy system only considering a simplified LAB model of Eqs. (4.35)–(4.38). Then, we evaluate the stress factors of Eqs. (4.61)–(4.69) and compare the results with the reference values of Table 4.3 and the target EIU established by the DM. Suppose they are lower than those reported in Table 4.3, and the estimated EIU is lower than required. In that case, the configuration under consideration offers a feasible and healthy operation of the battery bank, similar to a battery operating in an optimally designed SHS. Then, we evaluate the complete battery model to obtain a refined value of the battery life, which has a high impact on the NPC. On the contrary, if the results from the simulation with the simplified model are higher than the reference values of Table 4.3 or the calculated EIU is more elevated than required, we assign an artificially high value of the NPC, reducing the GA-individual's fitness.

Now we have discussed the models of renewable power sources, thermal generators, and energy storage devices and their management and design using a GA, we illustrate the GA capabilities by analyzing some examples typically examined in the literature.

4.10 Case studies

The optimization problems and their solution algorithm previously explained are illustrated in this section. Examples frequently used in the literature related to the ED, the UC, the BESS management, and off-grid energy system design are discussed here to illustrate the applicability of GAs. We have implemented the optimization algorithms previously described in a computer with i7-3630QM CPU at 2.40 GHz, 8GB of RAM, and a 64-bit operating system. The computing language used was MATLAB®.

4.10.1 Economic dispatch problem

To illustrate the solution technique of the ED problem using a real-coded GA, we have selected different examples from the literature. The first example was taken from the reference (Orero & Irving, 1996). It is a system with 15 units, with a load demand of 2650 MW at a determined hour ($p_{L(t)}=2650$ MW) and a reserve of 265 MW ($SR_{(t)}=265$ MW). This example does not consider the effects of valve-point loading; however, it includes the POZ. The second example was taken from the reference (Sinha et al., 2003). It is a system with 3 units, in which the influence of valve-point loading has been considered, but not the POZ. The load demand is 850 MW ($p_{L(t)}=850$ MW) with a reserve of 85 MW ($SR_{(t)}=85$ MW). The data of the third example also appears in the reference (Sinha et al., 2003). The system has 13 units. The valve-point loading effect is considered in this case, but the POZs are not included. The load demand is 1800 MW ($p_{L(t)}=1800$ MW) with a reserve of 180 MW ($SR_{(t)}=180$ MW). The fourth example was also taken from reference (Sinha et al., 2003). It is a system with 40 units only considering the influence of valve-point loading. The load demand is 10,500 MW ($p_{L(t)}=10{,}500$ MW) with a reserve of 1050 MW ($SR_{(t)}=1050$ MW). The complete data of all of these systems have been reported in Chapter 1. They have been analyzed using PSO.

The real-coded GA explained in Section 4.6 was implemented considering a proportional parameter of 0.3 ($p_r=0.3$), crossover rate of 0.95 ($p_c=0.95$), and a mutation rate of 0.1 ($p_m=0.1$). The population size was set to 100 individuals ($N=100$), and the generations considered were 300 ($J=300$). The results obtained from implementing the real-coded GA were compared with those offered by PSO. Furthermore, the results reported in references (Orero & Irving, 1996; Sinha et al., 2003) were used as reference values. Comparative analysis is presented in Table 4.5, where a difference between 0.98% and 9.1% can be observed. It is crucial to highlight that the optimization process could reach a lower operating cost by using a better setting for each algorithm (GA and PSO). However, we would like to emphasize how the difficulties of the ED problem increase with the problem size and characteristics. In general, the computational time spent was around 2.6 s.

Table 4.5 Operating cost for each case study.

Case	GA ($)	PSO ($)	Reference ($)
1	32,834.05	32,808.82	32,514
2	8359.13	8411.01	8234.07
3	18,859.72	18,879.62	18,127.06
4	13,4751.58	13,7036.34	12,3489.74

Comparative analysis of GA and PSO.

4.10.2 Unit commitment problem

This subsection analyzes the classical UC problem using the combined PL-GA approach explained in Section 4.7. The system under study was analyzed by Kazarlis et al. (1996) and has been widely used in the literature.

The system has 10 units, and the completed information is presented in Tables 4.6–4.8. The variable $IS_{(i)}$ is the initial state of unit i; positive or negative means that the unit has been online or offline for the specified number of hours, respectively. This information is required to evaluate the MUT and MDT constraint.

Solving the UC problem by GA requires solving the ED problem $T=24$ h times for each individual. According to Table 4.6, the fuel consumption cost of all units is quadratic so that we can use the SQP instead of GA, reducing the computational time. In

Table 4.6 Generators information.

i	$P_{(i)}^{min}$ (MW)	$P_{(i)}^{max}$ (MW)	$a_{(i)}$ $(\$/h)$	$b_{(i)}$ $\left(\frac{\$}{MWh}\right)$	$c_{(i)}$ $\left(\frac{\$}{MW^2h}\right)$	$IS_{(i)}$ (h)	$MUT_{(i)}$ (h)	$MDT_{(i)}$ (h)
1	150	455	1000	16.19	0.00048	8	8	8
2	150	455	970	17.26	0.00031	8	8	8
3	20	130	700	16.6	0.002	−5	5	5
4	20	130	680	16.5	0.00211	−5	5	5
5	25	162	450	19.7	0.00398	−6	6	6
6	20	80	370	22.26	0.00712	−3	3	3
7	25	85	480	27.74	0.00079	−3	3	3
8	10	55	660	25.92	0.00413	−1	1	1
9	10	55	665	27.27	0.00222	−1	1	1
10	10	55	670	27.79	0.00173	−1	1	1

Fuel consumption data.

Table 4.7 Starting-up cost.

i	$CSC_{(i)}(\$/h)$	$HSC_{(i)}\left(\frac{\$}{h}\right)$	$CST_{(i)}(h)$
1	9000	4500	5
2	10,000	5000	5
3	1100	550	4
4	1120	560	4
5	1800	900	4
6	340	170	2
7	520	260	2
8	60	30	0
9	60	30	0
10	60	30	0

Starting-up cost of each unit.

Table 4.8 Hourly load and spinning reserve.

$t(h)$	$p_{L(t)}(MW)$	$SR_{(t)}(MW)$	$t(h)$	$p_{L(t)}(MW)$	$SR_{(t)}(MW)$
1	700	70	13	1400	140
2	750	75	14	1300	130
3	850	85	15	1200	120
4	950	95	16	1050	105
5	1000	100	17	1000	100
6	1100	110	18	1100	110
7	1150	115	19	1200	120
8	1200	120	20	1400	140
9	1300	130	21	1300	130
10	1400	140	22	1100	110
11	1450	145	23	900	90
12	1500	150	24	800	80

Load demand curve.

other words, the ED problem was solved using SQP, and the UC problem was solved using a binary-coded GA.

Concerning the GA parameters, the population size was set to 10 individuals, and the generations were adjusted to 250. The crossover and mutation rates were adjusted to 0.95 and 0.1, respectively. As previously mentioned, the PL method is first applied to obtain a good solution, which is later improved using the GA. The solution obtained from the PL method (563,734.46$) was not improved by the GA with the actual settings. The solution is presented in Table 4.9, which required a computational time of 4.21 min.

Table 4.9 Solution of the UC for the system with 10 units.

t/i	**1**	**2**	**3**	**4**	**5**	**6**	**7**	**8**	**9**	**10**
1	1	1	0	0	0	0	0	0	0	0
2	1	1	0	0	0	0	0	0	0	0
3	1	1	0	0	1	0	0	0	0	0
4	1	1	0	0	1	0	0	0	0	0
5	1	1	0	1	1	0	0	0	0	0
6	1	1	1	1	1	0	0	0	0	0
7	1	1	1	1	1	0	0	0	0	0
8	1	1	1	1	1	0	0	0	0	0
9	1	1	1	1	1	1	1	0	0	0
10	1	1	1	1	1	1	1	1	0	0
11	1	1	1	1	1	1	1	1	1	0
12	1	1	1	1	1	1	1	1	1	1
13	1	1	1	1	1	1	1	1	0	0
14	1	1	1	1	1	1	1	0	0	0

Continued

Table 4.9 Continued

t/i	**1**	**2**	**3**	**4**	**5**	**6**	**7**	**8**	**9**	**10**
15	1	1	1	1	1	0	0	0	0	0
16	1	1	1	1	1	0	0	0	0	0
17	1	1	1	1	1	0	0	0	0	0
18	1	1	1	1	1	0	0	0	0	0
19	1	1	1	1	1	0	0	0	0	0
20	1	1	1	1	1	1	1	1	0	0
21	1	1	1	1	1	1	1	0	0	0
22	1	1	1	1	1	1	1	0	0	0
23	1	1	0	0	1	0	0	0	0	0
24	1	1	0	0	0	0	0	0	0	0

Unit scheduling.

The computational burden and problem size are vital issues in solving the UC problem. Fig. 4.12 shows the computational time spent analyzing problems of different sizes up to 100 units with the same GA setting as aforementioned. It is possible observing how the time increases polynomially with the problem size.

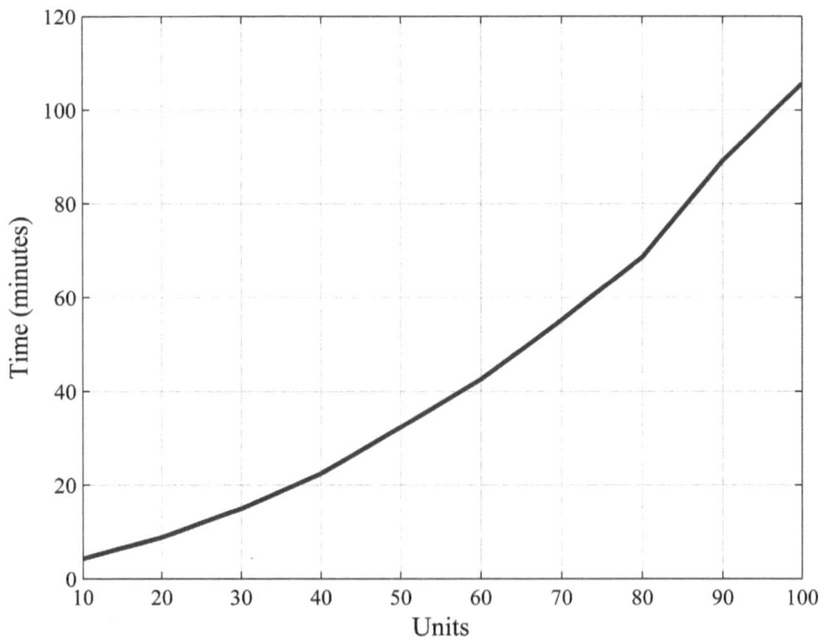

Fig. 4.12 Computational burden: Computational time with the size problem.

4.10.3 Battery energy storage system management

This case illustrates the BESS management procedure of Section 4.8. The storage system is a VRFB with the same characteristics as Section 4.5.2 (5 kW, 20 kWh, and 50 V). We need to manage the battery to perform the energy arbitrage on November 01, 2021. The battery is connected to an intelligent grid in Spain, and thus, it receives the electricity prices from this market. The assumed day-ahead price forecasting is shown in Fig. 4.13.

Under these conditions, we need to decide when to charge or discharge the VRFB to minimize the DNC, considering the price signal of Fig. 4.13. To this end, the integer-coded GA of Section 4.8 was employed using a population size of 100 individuals, 250 generations, a crossover rate of 0.9, and a mutation rate of 0.05. The initial SOC was assumed as 0.15 ($SOC_{(0)} = 0.15$). The evolution of the integer-coded GA is shown in Fig. 4.14, which took 9.98 min. The solution obtained has a DNC equal to −1.765€.

Fig. 4.15 suggests we charge the battery during periods of low price, disconnect the battery during the medium-price period, and discharge the storage system during periods of high electricity price to obtain financial benefits. Figs. 4.16–4.18 show the battery power, SOC, and voltage, respectively. We can observe that the power is reduced as the SOC increases to maintain the voltage at the maximum value (56.5 V). Later, we can observe a similar behavior when the battery is discharged. This is achieved by maintaining the battery voltage at the minimum value (42 V). In

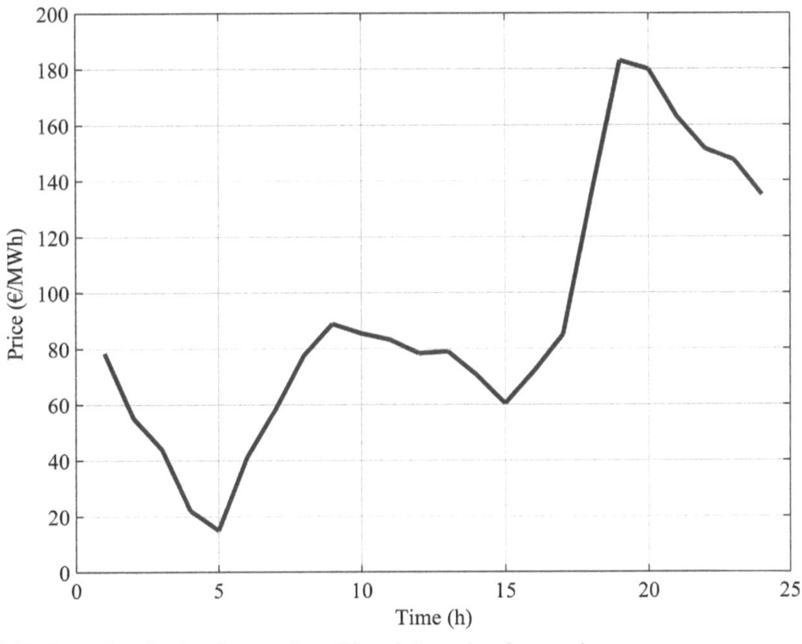

Fig. 4.13 Day-ahead price forecasting: Electricity price forecasting.

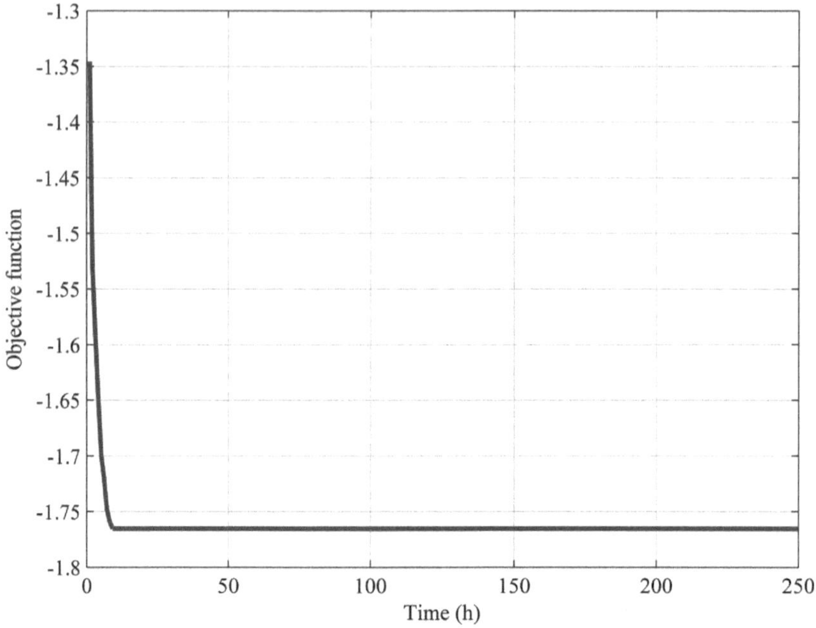

Fig. 4.14 GA optimization process: Evolution of GA for battery management.

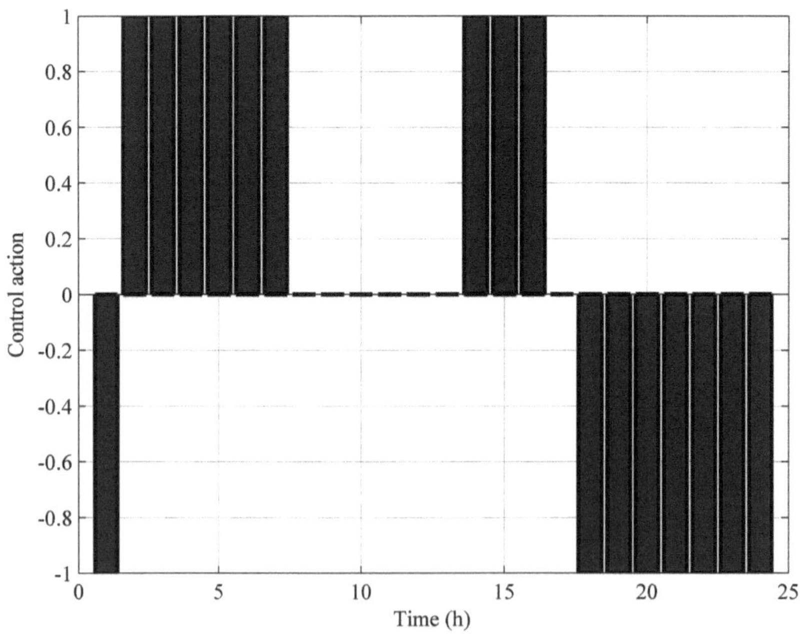

Fig. 4.15 Battery control actions: Day-ahead battery management.

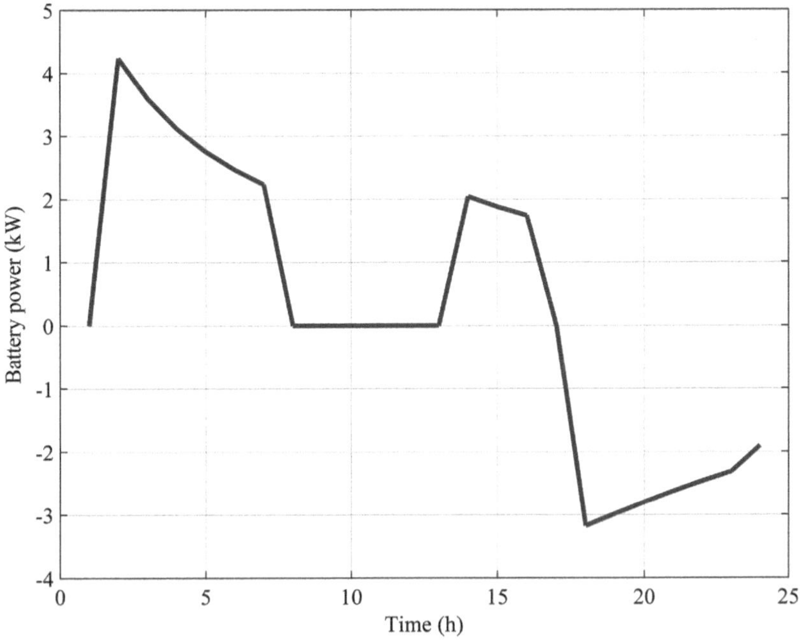

Fig. 4.16 Charging and discharging power: Day-ahead battery power.

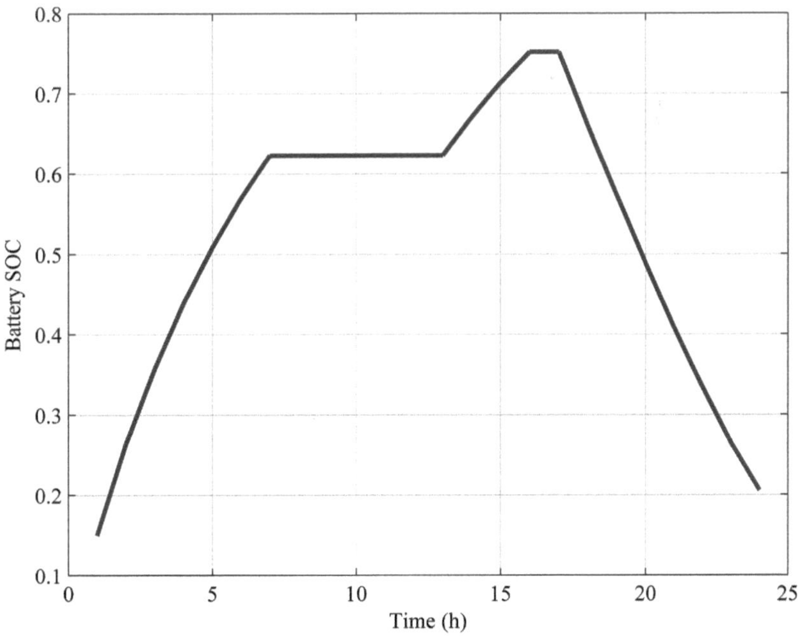

Fig. 4.17 Battery state of charge: Day-ahead battery state of charge.

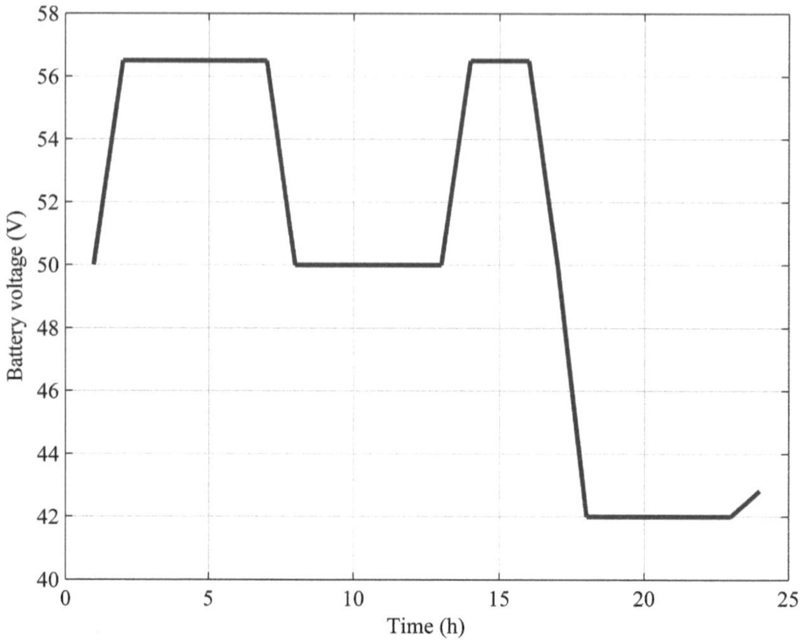

Fig. 4.18 Battery voltage: Day-ahead battery voltage.

general, the feasibility of the solution is guaranteed since it was derived from a simulation process involving the VRFB characteristics.

4.10.4 Hybrid power system design

This case describes how the optimization model of Section 4.9 is used to design an off-grid system for rural electrification. The system will use a WT and a battery bank, and it is supposed to be installed in Tangier (Morocco). It has the load profile shown in Fig. 4.19. The energy system will have an LAB; the list of LAB manufacturers considered and their characteristics are presented in Table 4.10. In addition, the list of WTs considered is shown in Table 4.11, and the techno-economic and optimization parameters are reported in Table 4.12.

The GA evolution is shown in Fig. 4.20, which required 302.815 min. Table 4.13 presents the features of the suggested energy system. Table 4.14 shows the stress factors and their intensity level; we can observe that all of them are either very low or medium.

Some results related to the LAB simulation are shown in Figs. 4.21–4.23. Fig. 4.21 presents the weighted Ah cycles' evolution, which grows fast, finishing the operational battery life in 3.172 years (Fig. 4.22). On the other hand, the nonweighted cycles are also shown, which grow very slowly, indicating that according to this method (nonweighted cycles), the battery life would be its float lifetime. These results create a significant difference in the NPC estimation and system configuration.

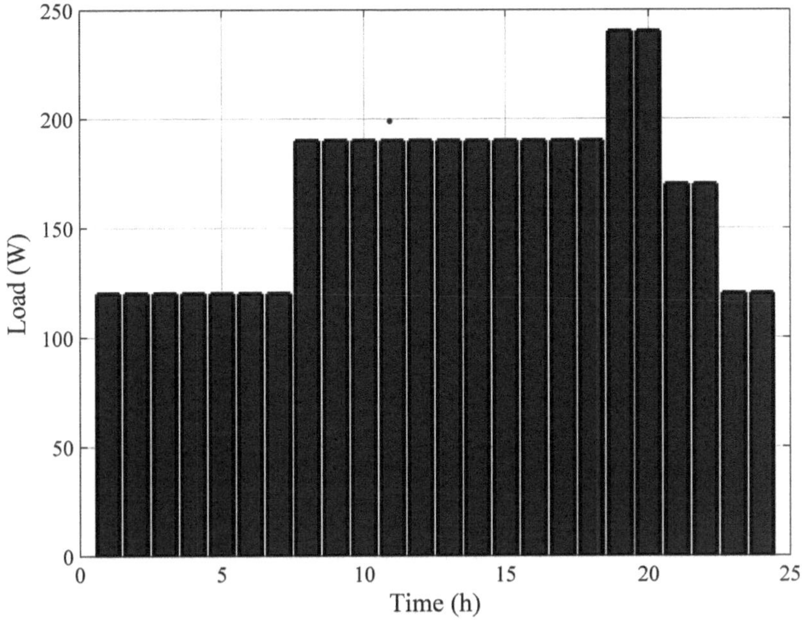

Fig. 4.19 Isolated system load: Daily load profile.

Table 4.10 Lead-acid battery (LAB) characteristics.

$C_{10}(Ah)$	Z_l(cycles)	SOC_{min}	FL(years)
200	1415	0.3	10
300	1415	0.3	10
600	1415	0.3	10
800	1415	0.3	10
1000	1415	0.3	10
1500	1415	0.3	10
2000	1415	0.3	10
2500	1415	0.3	10
3000	1415	0.3	10
460	1479	0.4	12
1156	1479	0.4	12
1900	1479	0.4	12
225	513	0.3	10
360	513	0.3	10
250	511	0.4	10
275	511	0.4	10
305	511	0.4	10
10	363	0.4	10
24	363	0.4	10
55	363	0.4	10
200	363	0.4	10

Batteries considered for hybrid system design.

Table 4.11 Wind turbine (WT) characteristics.

Power (kW)	Lifetime (years)
0.5	15
1	15
1.5	15
2	15
2.5	15

Wind turbines considered for hybrid system design.

Table 4.12 Optimization parameters.

Parameter	Value
Inflation rate (%)	3
Nominal rate (%)	5
Project lifetime (years)	40
Max. turbines	1
Max. strings	10
System voltage (V)	24
Population	25
Generations	100
Crossover (%)	90
Mutation (%)	3
EIU (%)	10

Requirements for optimal hybrid system design.

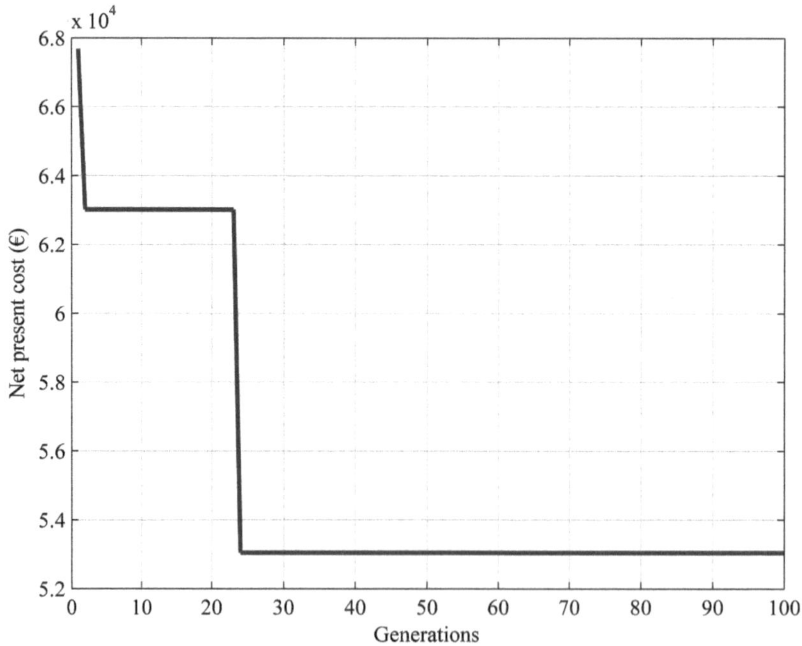

Fig. 4.20 Optimization process evolution: Evolution of GA for hybrid system sizing.

Table 4.13 Suggested hybrid system configuration.

Feature	Value
Type of wind turbine (kW)	1.5
Number of wind turbines	1
Battery type (Ah)	200
Z_I(cycles)	1415
FL(years)	10
Number of strings	2
Net present cost (€)	53,045.503
EIU (%)	11.461
Battery lifetime (year)	3.172

GA-designed hybrid system.

Table 4.14 Qualitative battery lifetime assessment.

Parameter	Value	Intensity level
Charge factor (%)	100.572	Very low
Ah throughput ($\times C_{10}$)	57.57	Medium
Highest discharge rate ($\times I_{10}$)	0.0693	Very low
Time between full charge (days)	1.586	Medium
Partial cycling (%)	49.689	Medium
Time at low SOC (%)	13.687	Medium

Expected LAB performance.

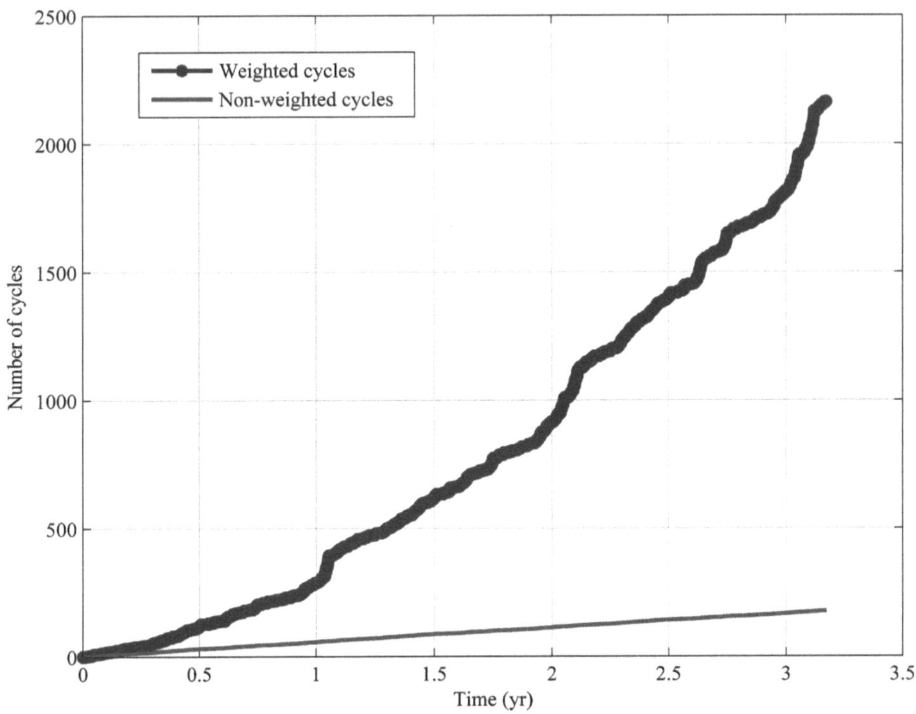

Fig. 4.21 Battery cycles: LAB operational cycles.

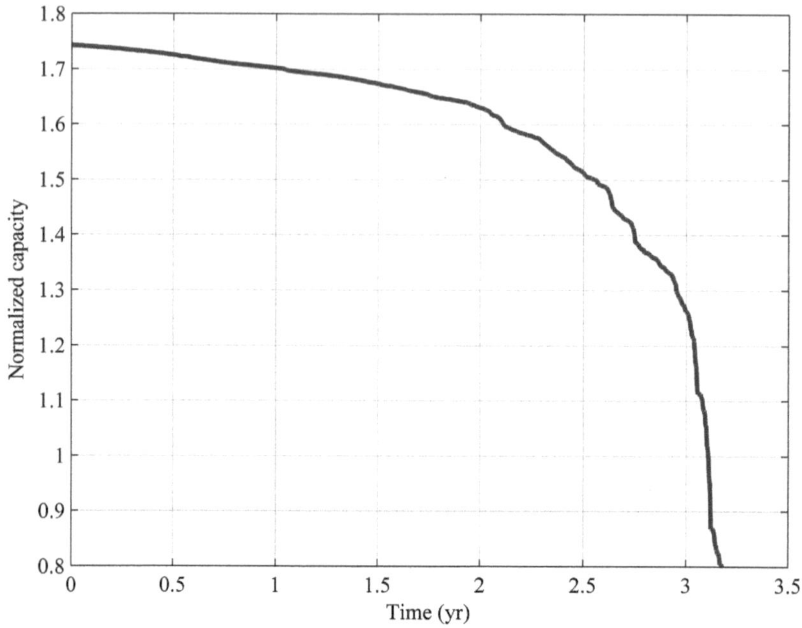

Fig. 4.22 LAB capacity: Evolution of battery capacity.

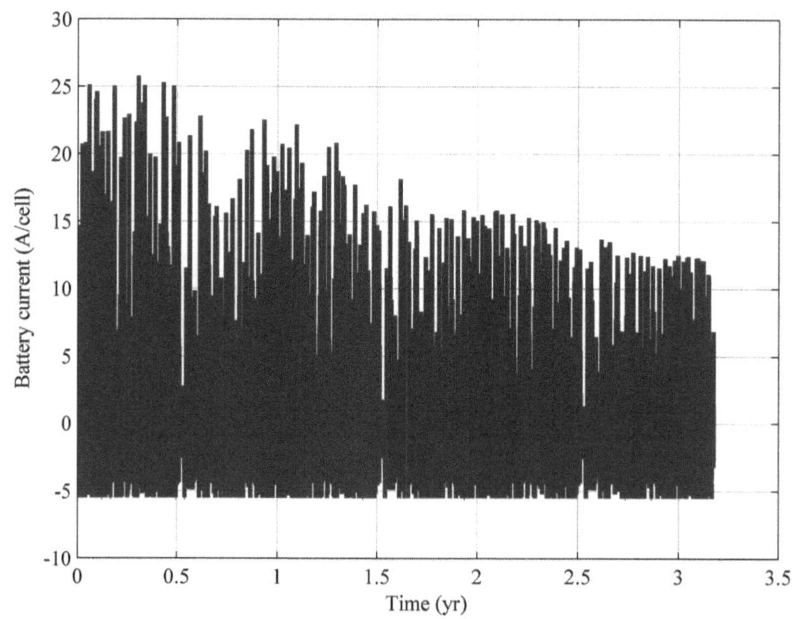

Fig. 4.23 Battery current simulation: LAB battery current.

Finally, the battery current is shown in Fig. 4.23. We can observe how the battery's ability to accept current reduces with the years due to the aging mechanism, affecting the estimation of ENS and EIU.

The EIU index reported in Table 4.13 is slightly higher than the required value of Table 4.12 (10%). This is because we are using a preliminary LAB simulation only using Eqs. (4.35)–(4.38) without considering the variation of internal resistance. Consequently, this approach suggests an optimistic operational condition. However, a more realistic analysis is obtained when the complete aging model is incorporated, resulting in a slightly higher EIU. For the configuration suggested, the EIU estimation using the simplified model of Eqs. (4.35)–(4.38) is 9.68%. On the other hand, if the complete model is used, the EIU estimation is 11.461%. The difference is only 1.781%.

4.11 Conclusions

This chapter explained relevant concepts related to the management and design of energy systems. Modeling wind generation through a wind curve and an NN and the impact of the TI, among other factors, were briefly discussed.

Simulation of a solar PV generation was also explained, considering the information frequently provided by the manufacturers. Estimation of the current–voltage curve and the power-voltage characteristic was briefly described. In addition, the inverter and its modeling using its efficiency curve were briefly explained.

BESS was also studied. Simulation of LABs, including their aging mechanism quantitatively and qualitatively, was intensively discussed. Modeling of a typical VRFB was also described. Conventional generation was also thoroughly discussed. Essential features such as the POZs and the valve-point loading were clearly addressed.

Concerning management, significant optimization problems were described and illustrated. The solution of the ED problem using a real-coded GA was explained and presented using four different systems reported in the technical literature. Factors such as the system size and parameter settings were discussed.

Another significant problem under analysis was the UC. The PL method was used to initialize the GA to enhance computational efficiency. Furthermore, other topics such as the relationship between the system size and the computational burden were discussed.

Managing BESS operating in a real-time pricing environment was another problem analyzed in this chapter. To this end, an integer-coded GA was implemented to minimize the DNC. The effectiveness of this technique was verified by analyzing a storage system operating within the Spanish electricity market.

This chapter also included rural electrification by using an integer-coded GA to design a small-scale energy system. LAB operation and its lifetime were analyzed using the example of a wind-battery system in Tangier (Morocco).

References

Abido, M. A. (2006). Multiobjective evolutionary algorithms for electric power dispatch problem. *IEEE Transactions on Evolutionary Computation*, *10*(3), 315–329. https://doi.org/10.1109/TEVC.2005.857073.

Alonso-García, M. C., & Ruíz, J. A. (2006). A model for the series-parallel association of photovoltaic devices. *Progress in Photovoltaics*, *14*(3), 237–247. https://doi.org/10.1002/pip.656.

Anahua, E., Barth, S., & Peinke, J. (2008). Markovian power curves for wind turbines. *Wind Energy*, *11*(3), 219–232. https://doi.org/10.1002/we.243.

Andersson, A. (2006). *Battery lifetime modelling*. Denmark National Laboratory Risø. https://docplayer.net/7879566-Battery-lifetime-modelling.html.

Averbukh, M., Lineykin, S., & Kuperman, A. (2013). Obtaining small photovoltaic array operational curves for arbitrary cell temperatures and solar irradiation densities from standard conditions data. *Progress in Photovoltaics*, *21*(5), 1016–1024. https://doi.org/10.1002/pip.2199.

Bai, G. H., Fleck, B., & Zuo, M. J. (2016). A stochastic power curve for wind turbines with reduced variability using conditional copula. *Wind Energy*, *19*(8), 1519–1534. https://doi.org/10.1002/we.1934.

Bakirtzis, A., Petridis, V., & Kazarlis, S. (1994). Genetic algorithm solution to the economic dispatch problem. *IEE Proceedings—Generation Transmission, and Distribution*, *141*(4), 377–382. https://doi.org/10.1049/ip-gtd:19941211.

Baruch, P. (1985). A two-level system as a model for a photovoltaic solar-cell. *Journal of Applied Physics*, *57*(4), 1347–1355. https://doi.org/10.1063/1.334486.

Bertsimas, D., Litvinov, E., Sun, X. A., Zhao, J., & Zheng, T. (2013). Adaptive robust optimization for the security constrained unit commitment problem. *IEEE Transactions on Power Systems*, *28*(1), 52–63. https://doi.org/10.1109/TPWRS.2012.2205021.

Bindner, H., Cronin, T., Lundsager, P., Manwell, J. F., Abdulwahid, U., & Baring-Gould, I. (2005). *Lifetime modelling of lead acid batteries*. Denmark National Laboratory Risø. https://www.osti.gov/etdeweb/servlets/purl/20607163.

Borghetti, A., D'Ambrosio, C., Lodi, A., & Martello, S. (2008). An MILP approach for short-term hydro scheduling and unit commitment with head-dependent reservoir. *IEEE Transactions on Power Systems*, *23*(3), 1115–1124. https://doi.org/10.1109/TPWRS.2008.926704.

Boyd, M. T., Klein, S. A., Reindl, D. T., & Dougherty, B. P. (2011). Evaluation and validation of equivalent circuit photovoltaic solar cell performance models. *Journal of Solar Energy Engineering*, *133*(2). https://doi.org/10.1115/1.4003584.

Bühler, A. J., & Krenzinger, A. (2013). Method for photovoltaic parameter extraction according to a modified double-diode model. *Progress in Photovoltaics*, *21*(5), 884–893. https://doi.org/10.1002/pip.2170.

Carrillo, C., Montaño, A. F. O., Cidrás, J., & Díaz-Dorado, E. (2013). Review of power curve modelling for wind turbines. *Renewable and Sustainable Energy Reviews*, *21*, 572–581. https://doi.org/10.1016/j.rser.2013.01.012.

Carrión, M., & Arroyo, J. M. (2006). A computationally efficient mixed-integer linear formulation for the thermal unit commitment problem. *IEEE Transactions on Power Systems*, *21*(3), 1371–1378. https://doi.org/10.1109/TPWRS.2006.876672.

Chaibi, Y., Allouhi, A., Malvoni, M., Salhi, M., & Saadani, R. (2019). Solar irradiance and temperature influence on the photovoltaic cell equivalent-circuit models. *Solar Energy*, *188*, 1102–1110. https://doi.org/10.1016/j.solener.2019.07.005.

Chen, P. H., & Chang, H. C. (1995). Large-scale economic-dispatch by genetic algorithm. *IEEE Transactions on Power Systems*, *10*(4), 1919–1926. https://doi.org/10.1109/59.476058.

Cheng, C. P., Liu, C. W., & Liu, G. C. (2000). Unit commitment by Lagrangian relaxation and genetic algorithms. *IEEE Transactions on Power Systems*, *15*(2), 707–714. https://doi.org/10.1109/59.867163.

Chiang, C. L. (2005). Improved genetic algorithm for power economic dispatch of units with valve-point effects and multiple fuels. *IEEE Transactions on Power Systems*, *20*(4), 1690–1699. https://doi.org/10.1109/TPWRS.2005.857924.

Chuang, Y. C., Chen, C. T., & Hwang, C. (2015). A real-coded genetic algorithm with a direction-based crossover operator. *Information Sciences*, *305*, 320–348. https://doi.org/10.1016/j.ins.2015.01.026.

Chuang, Y. C., Chen, C. T., & Hwang, C. (2016). A simple and efficient real-coded genetic algorithm for constrained optimization. *Applied Soft Computing*, *38*, 87–105. https://doi.org/10.1016/j.asoc.2015.09.036.

Coelho, L. S., & Mariani, V. C. (2006). Combining of chaotic differential evolution and quadratic programming for economic dispatch optimization with valve-point effect. *IEEE Transactions on Power Systems*, *21*(2), 989–996. https://doi.org/10.1109/TPWRS.2006.873410.

Copetti, J. B., & Chenlo, F. (1994). Lead/acid batteries for photovoltaic applications. Test results and modeling. *Journal of Power Sources*, *47*(1–2), 109–118. https://doi.org/10.1016/0378-7753(94)80054-5.

Copetti, J. B., Lorenzo, E., & Chenlo, F. (1993). A general battery model for PV system simulation. *Progress in Photovoltaics*, *1*(4), 283–292. https://doi.org/10.1002/pip.4670010405.

Damousis, I. G., Bakirtzis, A. G., & Dokopoulos, P. S. (2004). A solution to the unit-commitment problem using integer-coded genetic algorithm. *IEEE Transactions on Power Systems*, *19*(2), 1165–1172. https://doi.org/10.1109/TPWRS.2003.821625.

De Bernardez, L., & Buitrago, R. H. (2006). Dark I-V curve measurement of single cells in a photovoltaic module. *Progress in Photovoltaics*, *14*(4), 321–327. https://doi.org/10.1002/pip.665.

Dialynas, E. N., & Machias, A. V. (1989). Reliability modelling interactive techniques of power systems including wind generating units. *Archiv Für Elektrotechnik*, *72*, 33–41. https://doi.org/10.1007/BF01573567.

Dieu, V. N., & Ongsakul, W. (2008). Ramp rate constrained unit commitment by improved priority list and augmented Lagrange Hopfield network. *Electric Power Systems Research*, *78*, 291–301. https://doi.org/10.1016/j.epsr.2007.02.011.

Gaing, Z. L. (2003). Particle swarm optimization to solving the economic dispatch considering the generator constraints. *IEEE Transactions on Power Systems*, *18*(3), 1187–1195. https://doi.org/10.1109/TPWRS.2003.814889.

Giorsetto, P., & Utsurogi, K. F. (1983). Development of a new procedure for reliability modelling of wind turbine generators. *IEEE Transactions on Power Apparatus and Systems*, *102*(1), 134–143. https://doi.org/10.1109/TPAS.1983.318006.

Hedevang, E. (2014). Wind turbine power curves incorporating turbulence intensity. *Wind Energy*, *17*(2), 173–195. https://doi.org/10.1002/we.1566.

Hetzer, J., Yu, D. C., & Bhattarai, K. (2008). An economic dispatch model incorporating wind power. *IEEE Transactions on Energy Conversion*, *23*(2), 603–611. https://doi.org/10.1109/TEC.2007.914171.

Jebaraj, L., Venkatesan, C., Soubache, I., & Rajan, C. C. A. (2017). Application of differential evolution algorithm in static and dynamic economic or emission dispatch problem: A review. *Renewable and Sustainable Energy Reviews*, *77*, 1206–1220. https://doi.org/10.1016/j.rser.2017.03.097.

Ji, B., Yuan, X. H., Chen, Z. H., & Tian, H. (2014). Improved gravitational search algorithm for unit commitment considering uncertainty of wind power. *Energy*, *67*, 52–62. https://doi.org/10.1016/j.energy.2014.02.014.

Juste, K. A., Kita, H., Tanaka, E., & Hasegawa, J. (1999). An evolutionary programming solution to the unit commitment problem. *IEEE Transactions on Power Systems*, *14*(4), 1452–1459. https://doi.org/10.1109/59.801925.

Karaki, S. H., Chedid, R. B., & Ramadan, R. (1999). Probabilistic performance assessment of wind energy conversion system. *IEEE Transactions on Energy Conversion*, *14*(2), 217–224. https://doi.org/10.1109/60.766986.

Karatepe, E., Boztepe, M., & Çolak, M. (2007). Development of a suitable model for characterizing photovoltaic arrays with shaded solar cells. *Solar Energy*, *81*(8), 977–992. https://doi.org/10.1016/j.solener.2006.12.001.

Kazarlis, S. A., Bakirtzis, A. G., & Petridis, V. (1996). A genetic algorithm solution to the unit commitment problem. *IEEE Transactions on Power Systems*, *11*(1), 83–90. https://doi.org/10.1109/59.485989.

Khalfallah, M. G., & Koliub, A. M. (2007). Wind turbines power curve variability. *Desalination*, *209*(1–3), 230–237. https://doi.org/10.1016/j.desal.2007.04.032.

Kheldoun, A., Bradai, R., Boukenoui, R., & Mellit, A. (2016). A new golden section method-based maximum power point tracking algorithm for photovoltaic systems. *Energy Conversion and Management*, *111*, 125–136. https://doi.org/10.1016/j.enconman.2015.12.039.

Lambert, T., Gilman, P., & Lilienthal, P. (2006). Micropower system modeling with HOMER. In F. A. Farret, & M. G. Simões (Eds.), *Integration of alternative sources of energy* (pp. 379–418). Wiley–IEEE Press. https://doi.org/10.1002/0471755621.ch15.

Lauer, G. S., Sandell, N. R., Bertsekas, D. P., & Posbergh, T. A. (1982). Solution of large-scale optimal unit commitment problems. *IEEE Transactions on Power Apparatus and Systems*, *PAS-101*(1), 79–86. https://doi.org/10.1109/TPAS.1982.317243.

Li, Z. Y., & Shahidehpour, M. (2005). Security-constrained unit commitment for simultaneous clearing of energy and ancillary services markets. *IEEE Transactions on Power Systems*, *20*(2), 1079–1088. https://doi.org/10.1109/TPWRS.2005.846063.

Lin, W. M., Cheng, F. S., & Tsay, M. T. (2002). An improved tabu search for economic dispatch with multiple minima. *IEEE Transactions on Power Systems*, *17*(1), 108–112. https://doi.org/10.1109/59.982200.

Liu, X., & Xu, W. (2010a). Economic load dispatch constrained by wind power variability: A here-and-now approach. *IEEE Transactions on Sustainable Energy*, *1*(1), 2–9. https://doi.org/10.1109/TSTE.2010.2044817.

Liu, X., & Xu, W. (2010b). Minimum emission dispatch constrained by stochastic wind power availability and cost. *IEEE Transactions on Power Systems*, *25*(3), 1705–1713. https://doi.org/10.1109/TPWRS.2010.2042085.

Lorenzo, E., Araujo, G., Cuevas, A., Egido, M., Minano, J., & Zilles, R. (1994). *Solar electricity: Engineering of photovoltaic systems*. Progensa.

Lun, S. X., Wang, S., Guo, T. T., & Du, C. J. (2014). An I-V model based on time warp invariant echo state network for photovoltaic array with shaded solar cells. *Solar Energy*, *105*, 529–541. https://doi.org/10.1016/j.solener.2014.04.023.

Lydia, M., Kumar, S. S., Selvakumar, A. I., & Kumar, G. E. P. (2014). A comprehensive review on wind turbine power curve modeling techniques. *Renewable and Sustainable Energy Reviews*, *30*, 452–460. https://doi.org/10.1016/j.rser.2013.10.030.

Ma, H. L., & Shahidehpour, S. M. (1999). Unit commitment with transmission security and voltage constraints. *IEEE Transactions on Power Systems*, *14*(2), 757–764. https://doi.org/10.1109/59.761909.

Makrides, G., Zinsser, B., Schubert, M., & Georghiou, G. E. (2013). Energy yield prediction errors and uncertainties of different photovoltaic models. *Progress in Photovoltaics*, *21*, 500–516. https://doi.org/10.1002/pip.1218.

Mantawy, A. H., Abdel-Magid, Y. L., & Selim, S. Z. (1998). A simulated annealing algorithm for unit commitment. *IEEE Transactions on Power Systems*, *13*(1), 197–204. https://doi.org/10.1109/59.651636.

Manwell, J. F., & McGowan, J. G. (1993). Lead acid battery storage model for hybrid energy systems. *Solar Energy*, *50*, 399–405. https://doi.org/10.1016/0038-092X(93)90060-2.

Marčiukaitis, M., Žutautaitė, I., Martišauskas, L., Jokšas, B., Gecevičius, G., & Sfetsos, A. (2017). Non-linear regression model for wind turbine power curve. *Renewable Energy*, *113*, 732–741. https://doi.org/10.1016/j.renene.2017.06.039.

Mbewe, D. J., Card, H. C., & Card, D. C. (1985). A model of silicon solar-cells for concentrator photovoltaic and photovoltaic thermal system-design. *Solar Energy*, *35*(3), 247–258. https://doi.org/10.1016/0038-092X(85)90104-5.

Morales-España, G., Latorre, J. M., & Ramos, A. (2013). Tight and compact MILP formulation for the thermal unit commitment problem. *IEEE Transactions on Power Systems*, *28*(4), 4897–4908. https://doi.org/10.1109/TPWRS.2013.2251373.

Nguyen, T. A., Crow, M. L., & Elmore, A. C. (2015). Optimal sizing of a vanadium redox battery system for microgrid systems. *IEEE Transactions on Sustainable Energy*, *6*(3), 729–737. https://doi.org/10.1109/TSTE.2015.2404780.

Nguyen, T. A., Qiu, X., Guggenberger, J. D., Crow, M. L., & Elmore, A. C. (2014). Performance characterization for photovoltaic-vanadium redox battery microgrid systems. *IEEE Transactions on Sustainable Energy*, *5*(4), 1379–1388. https://doi.org/10.1109/TSTE.2014.2305132.

Ongsakul, W., & Petcharaks, N. (2004). Unit commitment by enhanced adaptive Lagrangian relaxation. *IEEE Transactions on Power Systems*, *19*(1), 620–628. https://doi.org/10.1109/TPWRS.2003.820707.

Orero, S. O., & Irving, M. R. (1996). Economic dispatch of generators with prohibited operating zones: A genetic algorithm approach. *IEE Proceedings—Generation Transmission, and Distribution*, *143*(6), 529–534. https://doi.org/10.1049/ip-gtd:19960626.

Orero, S. O., & Irving, M. R. (1997). Large scale unit commitment using a hybrid genetic algorithm. *Electrical Power and Energy Systems*, *19*(1), 45–55. https://doi.org/10.1016/S0142-0615(96)00028-2.

Ostrowski, J., Anjos, M. F., & Vannelli, A. (2012). Tight mixed linear programming formulations for the unit commitment problem. *IEEE Transactions on Power Systems*, *27*(1), 39–46. https://doi.org/10.1109/TPWRS.2011.2162008.

Ouyang, Z., & Shahidehpour, S. M. (1991). An intelligent dynamic-programming for unit commitment application. *IEEE Transactions on Power Systems*, *6*(3), 1203–1209. https://doi.org/10.1109/59.119267.

Pagnini, L. C., Burlando, M., & Repetto, M. P. (2015). Experimental power curve of small-size wind turbines in turbulent urban environment. *Applied Energy*, *154*, 112–121. https://doi.org/10.1016/j.apenergy.2015.04.117.

Paiva, L. T., Rodrigues, C. V., & Palma, J. M. L. M. (2014). Determining wind turbine power curves based on operating conditions. *Wind Energy*, *17*(10), 1563–1575. https://doi.org/10.1002/we.1651.

Papavasiliou, A., Oren, S. S., & O'Neill, R. P. (2011). Reserve requirements for wind power integration: A scenario-based stochastic programming framework. *IEEE Transactions on Power Systems*, *26*(4), 2197–2206. https://doi.org/10.1109/TPWRS.2011.2121095.

Pappala, V. S., Erlich, I., Rohrig, K., & Dobschinski, J. (2009). A stochastic model for the optimal operation of a wind-thermal power system. *IEEE Transactions on Power Systems*, *24*(2), 940–950. https://doi.org/10.1109/TPWRS.2009.2016504.

Park, J. B., Lee, K. S., Shin, J. R., & Lee, K. Y. (2005). A particle swarm optimization for economic dispatch with nonsmooth cost functions. *IEEE Transactions on Power Systems*, *20*(1), 34–42. https://doi.org/10.1109/TPWRS.2004.831275.

Pelletier, F., Masson, C., & Tahan, A. (2016). Wind turbine power curve modelling using artificial neural network. *Renewable Energy*, *89*, 207–214. https://doi.org/10.1016/j.renene.2015.11.065.

Qiu, X., Nguyen, T. A., Guggenberger, J. D., Crow, M. L., & Elmore, A. C. (2014). A field validated model of a vanadium redox flow battery for microgrids. *IEEE Transactions on Smart Grid*, *5*(4), 1592–1601. https://doi.org/10.1109/TSG.2014.2310212.

Rampinelli, G. A., Krenzinger, A., & Romero, F. C. (2014). Mathematical models for efficiency of inverters used in grid connected photovoltaic systems. *Renewable and Sustainable Energy Reviews*, *34*, 578–587. https://doi.org/10.1016/j.rser.2014.03.047.

Rogers, T. J., Gardner, P., Dervilis, N., Worden, K., Maguire, A. E., Papatheou, E., & Cross, E. J. (2020). Probabilistic modelling of wind turbine power curves with application of heteroscedastic Gaussian process regression. *Renewable Energy*, *148*, 1124–1136. https://doi.org/10.1016/j.renene.2019.09.145.

Roy, S. (2012). Inclusion of short duration wind variations in economic load dispatch. *IEEE Transactions on Sustainable Energy*, *3*(2), 265–273. https://doi.org/10.1109/TSTE.2011.2179954.

Ruiz, P. A., Philbrick, C. R., Zak, E., Cheung, K. W., & Sauer, P. W. (2009). Uncertainty management in the unit commitment problem. *IEEE Transactions on Power Systems*, *24*(2), 642–651. https://doi.org/10.1109/TPWRS.2008.2012180.

Saint-Drenan, Y. M., Besseau, R., Jansen, M., Staffell, I., Troccoli, A., Dubus, L., Schmidt, J., Gruber, K., Simões, S. G., & Heier, S. (2020). A parametric model for wind turbine power curves incorporating environmental conditions. *Renewable Energy*, *157*, 754–768. https://doi.org/10.1016/j.renene.2020.04.123.

Schiffer, J., Sauer, D. U., Bindner, H., Cronin, T., Lundsager, P., & Kaiser, R. (2007). Model prediction for ranking lead-acid batteries according to expected lifetime in renewable energy systems and autonomous power-supply systems. *Journal of Power Sources*, *168*(1), 66–78. https://doi.org/10.1016/j.jpowsour.2006.11.092.

Selvakumar, A. I., & Thanushkodi, K. (2007). A new particle swarm optimization solution to nonconvex economic dispatch problems. *IEEE Transactions on Power Systems*, *22*(1), 42–51. https://doi.org/10.1109/TPWRS.2006.889132.

Senjyu, T., Shimabukuro, K., Uezato, K., & Funabashi, T. (2003). A fast technique for unit commitment problem by extended priority list. *IEEE Transactions on Power Systems*, *18*(2), 882–888. https://doi.org/10.1109/TPWRS.2003.811000.

Shokrzadeh, S., Jozani, M. J., & Bibeau, E. (2014). Wind turbine power curve modeling using advanced parametric and nonparametric methods. *IEEE Transactions on Sustainable Energy*, *5*(4), 1262–1269. https://doi.org/10.1109/TSTE.2014.2345059.

Simopoulos, D. N., Kavatza, S. D., & Vournas, C. D. (2006). Unit commitment by an enhanced simulated annealing algorithm. *IEEE Transactions on Power Systems*, *21*(1), 68–76. https://doi.org/10.1109/TPWRS.2005.860922.

Sinha, N., Chakrabarti, R., & Chattopadhyay, R. K. (2003). Evolutionary programming techniques for economic load dispatch. *IEEE Transactions on Evolutionary Computation*, *7*(1), 83–94. https://doi.org/10.1109/TEVC.2002.806788.

Snyder, W. L., Powell, H. D., & Rayburn, J. C. (1987). Dynamic-programming approach to unit commitment. *IEEE Transactions on Power Systems*, *2*(2), 339–350. https://doi.org/10.1109/TPWRS.1987.4335130.

Svoboda, V., Wenzl, H., Kaiser, R., Jossen, A., Baring-Gould, I., Manwell, J., et al. (2007). Operating conditions of batteries in off-grid renewable energy systems. *Solar Energy*, *81*(11), 1409–1425. https://doi.org/10.1016/j.solener.2006.12.009.

Swarup, K. S., & Yamashiro, S. (2002). Unit commitment solution methodology using genetic algorithm. *IEEE Transactions on Power Systems*, *17*(1), 87–91. https://doi.org/10.1109/59.982197.

Takriti, S., Birge, J. R., & Long, E. (1996). A stochastic model for the unit commitment problem. *IEEE Transactions on Power Systems*, *11*(3), 1497–1506. https://doi.org/10.1109/59.535691.

Ting, T. O., Rao, M. V. C., & Loo, C. K. (2006). A novel approach for unit commitment problem via an effective hybrid particle swarm optimization. *IEEE Transactions on Power Systems*, *21*(1), 411–418. https://doi.org/10.1109/TPWRS.2005.860907.

Tong, S. K., Shahidehpour, S. M., & Ouyang, Z. (1991). A heuristic short-term unit commitment. *IEEE Transactions on Power Systems*, *6*(3), 1210–1216. https://doi.org/10.1109/59.119268.

Trivellato, F., Battisti, L., & Miori, G. (2012). The ideal power curve of small wind turbines from field data. *Journal of Wind Engineering and Industrial Aerodynamics*, *107*, 263–273. https://doi.org/10.1016/j.jweia.2012.04.026.

Tuohy, A., Meibom, P., Denny, E., & O'Malley, M. (2009). Unit commitment for systems with significant wind penetration. *IEEE Transactions on Power Systems*, *24*(2), 592–601. https://doi.org/10.1109/TPWRS.2009.2016470.

Ummels, B. C., Gibescu, M., Pelgrum, E., Kling, W. L., & Brand, A. J. (2007). Impacts of wind power on thermal generation unit commitment and dispatch. *IEEE Transactions on Energy Conversion*, *22*(1), 44–51. https://doi.org/10.1109/TEC.2006.889616.

Üstüntaş, T., & Şahin, A. D. (2008). Wind turbine power curve estimation based on cluster center fuzzy logic modeling. *Journal of Wind Engineering and Industrial Aerodynamics*, *96*(5), 611–620. https://doi.org/10.1049/iet-rpg.2019.0530.

Vazquez, M., & Rey-Stolle, I. (2008). Photovoltaic module reliability model based on field degradation studies. *Progress in Photovoltaics*, *16*(5), 419–433. https://doi.org/10.1002/pip.825.

Victoire, T. A. A., & Jeyakumar, A. E. (2004). Hybrid PSO-SQP for economic dispatch with valve-point effect. *Electric Power Systems Research*, *71*(1), 51–59. https://doi.org/10.1016/j.epsr.2003.12.017.

Villanueva, D., & Feijóo, A. (2016). Normal-based model for true power curves of wind turbines. *IEEE Transactions on Sustainable Energy*, *7*(3), 1005–1011. https://doi.org/10.1109/TSTE.2016.2515264.

Villanueva, D., & Feijóo, A. (2018). Comparison of logistic functions for modeling wind turbine power curves. *Electric Power Systems Research*, *155*, 281–288. https://doi.org/10.1016/j.epsr.2017.10.028.

Virgolino, G. C. M., Mattos, C. L. C., Magalhães, J. A. F., & Barreto, G. A. (2020). Gaussian processes with logistic mean function for modeling wind turbine power curves. *Renewable Energy*, *162*, 458–465. https://doi.org/10.1016/j.renene.2020.06.021.

Walters, D. C., & Sheble, G. B. (1993). Genetic algorithm solution of economic dispatch with valve point loading. *IEEE Transactions on Power Systems*, *8*(3), 1325–1332. https://doi.org/10.1109/59.260861.

Wang, J. H., Shahidehpour, M., & Li, Z. Y. (2008). Security-constrained unit commitment with volatile wind power generation. *IEEE Transactions on Power Systems*, *23*(3), 1319–1327. https://doi.org/10.1109/TPWRS.2008.926719.

Wang, Q. F., Guan, Y. P., & Wang, J. H. (2012). A chance-constrained two-stage stochastic program for unit commitment with uncertain wind power output. *IEEE Transactions on Power Systems*, *27*(1), 206–215. https://doi.org/10.1109/TPWRS.2011.2159522.

Wang, Y., Infield, D. G., Stephen, B., & Galloway, S. J. (2014). Copula-based model for wind turbine power curve outlier rejection. *Wind Energy*, *17*(11), 1677–1688. https://doi.org/10.1002/we.1661.

Xu, K. Y., Yan, J., Zhang, H., Zhang, H. R., Han, S., & Liu, Y. Q. (2021). Quantile based probabilistic wind turbine power curve model. *Applied Energy*, *296*. https://doi.org/10.1016/j.apenergy.2021.116913.

Yang, H. T., Yang, P. C., & Huang, C. L. (1996). Evolutionary programming based economic dispatch for units with non-smooth fuel cost functions. *IEEE Transactions on Power Systems*, *11*(1), 112–117. https://doi.org/10.1109/59.485992.

Yang, S. P., Tan, S. C., & Xu, J. X. (2013). Consensus based approach for economic dispatch problem in a smart grid. *IEEE Transactions on Power Systems*, *28*(4), 4416–4426. https://doi.org/10.1109/TPWRS.2013.2271640.

Yang, X. S., Hosseini, S. S. S., & Gandomi, A. H. (2012). Firefly algorithm for solving non-convex economic dispatch problems with valve loading effect. *Applied Soft Computing*, *12*(3), 1180–1186. https://doi.org/10.1016/j.asoc.2011.09.017.

Yesilbudak, M. (2018). Implementation of novel hybrid approaches for power curve modeling of wind turbines. *Energy Conversion and Management*, *171*, 156–169. https://doi.org/10.1016/j.enconman.2018.05.092.

Yuan, X. H., Jia, B., Zhang, S. Q., Tian, H., & Hou, Y. H. (2014). A new approach for unit commitment problem via binary gravitational search algorithm. *Applied Soft Computing*, *22*, 249–260. https://doi.org/10.1016/j.asoc.2014.05.029.

Yuan, X. H., Nie, H., So, A. J., Wang, L., & Yuan, Y. B. (2009). An improved binary particle swarm optimization for unit commitment problem. *Expert Systems with Applications*, *36*(4), 8049–8055. https://doi.org/10.1016/j.eswa.2008.10.047.

Yuan, X. H., Su, A. J., Nie, H., Yuan, Y. B., & Wang, L. (2009). Application of enhanced discrete differential evolution approach to unit commitment problem. *Energy Conversion and Management*, *50*(9), 2449–2456. https://doi.org/10.1016/j.enconman.2009.05.033.

Zhuang, F., & Galiana, F. D. (1988). Towards a more rigorous and practical unit commitment by Lagrangian-relaxation. *IEEE Transactions on Power Systems*, *3*(2), 763–773. https://doi.org/10.1109/59.192933.

Zhuang, F., & Galiana, F. D. (1990). Unit commitment by simulated annealing. *IEEE Transactions on Power Systems*, *5*(1), 311–318. https://doi.org/10.1109/59.49122.

Forecasting of electricity prices, demand, and renewable resources 5

5.1 Benefits of an accurate forecasting system

An accurate prediction provides us with valuable knowledge about how to act from now to a determined moment in the future to get the best performance. Decision-makers (DMs) apply this idea in many fields, from social sciences such as sociology and economy to physics and engineering. DMs have presented their experiences and quantified the economic benefit of prediction in various industries such as agriculture and electrical engineering.

In agriculture, the intrinsic variability of the environmental variables has an essential role in food production. An accurate prediction between 3 and 6 months ahead would help DMs make appropriate decisions to mitigate climate variability's adverse effects. However, environmental prediction faces barriers related to many sociopolitical factors and prediction error due to agricultural systems' complexity. For example, the economic profit on a farm in southeast USA or Argentina, where the same crops are grown at a different time of the year, can be up to 28% and 70%, respectively, under perfect forecasting conditions. However, the environmental forecast's uncertainty can reduce these estimations of the expected benefits (Jones et al., 2000).

The incorporation of ensemble forecasting offers a significant improvement on the prediction benefits. An ensemble forecast allows us to use a set of decision levels, while a single forecast only provides a deterministic result. The DM can combine an ensemble forecast with probabilistic decision criteria to increase the decision process's reliability, which generally consists of a cost–benefit relationship (Zhu et al., 2002).

An illustrative example is when two farmers think about whether or not to protect their corresponding crops against the effects of freezing temperature. Both farms are in the same region, but the yield of each farm is different. Furthermore, the economic losses are distinct for each farmer, but the yield protection cost is the same. Under these circumstances, the farmer with the lowest financial losses would invest in yield protection if the cultivator is confident that the freezing temperature will occur. On the contrary, the farmer with the highest financial losses would invest in the yield protection even if the probability of freezing temperature is very low. Thus, the farmers use probabilistic information during the decision-making process using specific criteria depending on the predicted financial losses. Under more realistic weather conditions (specified by a 500-hPa geopotential height over the Northern Hemisphere), most DMs can get an economic benefit using an ensemble forecast when their cost-loss ratio is between 20% and 50% (Zhu et al., 2002).

Another aspect is that the reliability of any forecasting tool depends on a determined circumstance. In the specific field of agriculture, the vulnerability increases

Genetic Optimization Techniques for Sizing and Management of Modern Power Systems
https://doi.org/10.1016/B978-0-12-823889-9.00003-5

due to the yearly climate variability and the difficulty of predicting it. In most cases, the DM has to implement actions several months before the climate conditions appear. Thus, minimizing the prediction error has a crucial role (Hansen, 2002). This is not an easy task, since the yearly climate variation depends on the interactions between the atmosphere, the ocean, and land surfaces. As a consequence of this situation, the decisions taken tend to be conservative because the DM avoids highly unfavorable conditions. However, this could result in poor productivity, efficiency, and profitability (Hansen, 2002).

Agriculture researchers have recognized several requisites for the effective use of prediction tools. The first requirement is that scientists must design prediction tools to address real and perceived variables. The second requirement is that DMs must understand the existing management options and their sensitivity to the forecasting information. The third requirement is that the forecasting product has to have the appropriate scale and accuracy. The fourth requirement is that the forecasting information has to be received and understood by the right audience. Finally, the fifth requirement is that institutional engagement is crucial to maintain a forecasting system beyond the project's operating lifetime (Hansen, 2002). Investigators and DMs have recognized the calculation and use of meteorological predictions as a multidisciplinary problem. Researchers have proposed incorporating several fields such as climate and agricultural sciences, sociology, and DMs to measure climate forecasting benefits reliably (Meinke & Stone, 2005).

Wind energy is among the renewable energy sources with the highest growth. Thus, its impact on power system operation has been a topic of interest for many researchers worldwide. Accommodation of wind power generation results is difficult due to its inherent variability. On the one hand, this variability has a component related to the spatial and geographical characteristics of the place where the windfarm operates. This component also includes the climatic region of the location. On the other hand, variability has a temporal part, related to the long-term changes in wind resources. This consists of the interannual variations of wind power, which can vary as much as 10% within 20 years. Thus, wind power variations are not significant during the lifetime of a typical wind turbine (Albadi & El-Saadany, 2010).

The temporal component also includes the inter-seasonal variations of wind resources which technicians can predict with more accuracy than annual variations. Scientists assess the variations within a daily cycle, or diurnal variations, with more precision than synoptic variations. Researchers analyze synoptic variations only a few days in advance because these are associated with the dynamics of the weather system. Another topic related to the temporal feature of wind variability is turbulence, representing variations in minutes and seconds, influencing the power quality more or less, depending on the network impedance and wind turbine type (Albadi & El-Saadany, 2010).

An aspect related to the spatial component of wind generation is the impact of wind turbine aggregation on the output power variability. This means the interaction of a group of wind turbines can mitigate the intrinsic variability of each turbine in the cluster. The windfarm designer achieves this condition by placing a considerable number of turbines in a delimited geographical area. The resulting windfarm configuration mitigates the diurnal and synoptic components of wind speed variations (Albadi & El-Saadany, 2010).

At the core of the energy management system, we have a forecasting routine that reduces the operating costs related to wind generation uncertainty. The aggregation of wind turbines within a determined area has a significant impact on forecasting error reduction. Indeed, wind engineers have found a directly proportional relationship between the forecasting error enhancement and the diameter of the windfarm installation site (Albadi & El-Saadany, 2010).

Consequently, DMs must consider all of these factors during the wind energy integration assessment. The appropriate amount of wind power integration depends on the flexibility of the power system under analysis. On the one hand, the power system's difficulties with a high wind power injection increase with the penetration level. On the other hand, the capabilities of the power system to accommodate this power injection depend on its flexibility. In this context, engineers associate the flexibility with the ability of the power system to adapt to changes in the operating conditions led by wind generation (Albadi & El-Saadany, 2010).

The adaptation process consists of managing any demand response (DR) program or energy storage system (ESS) to integrate the available wind generation as effectively as possible. DR is incorporated and managed to reduce the problems related to wind uncertainty and contingencies. An example of such situations is the operational violations on the transmission system. In a general sense, the operating cost reduces as wind power penetration increases. However, a portion of the operating cost increases with the forecasting error. In these circumstances, a DR program located in the appropriated region of the system and operating at the right moment would significantly reduce the global operating cost (Sahin et al., 2013).

Some DR programs are implemented by sending a price signal to the consumers; the intention being that they will respond by reducing their power consumption at those times of high price. The capability of the consumers enrolled in these programs is mathematically represented using demand elasticity, which means the change of power consumption as a function of the increment or reduction in the electricity price signal. Inducing the consumer to increase or reduce its power consumption with the electricity price strongly depends on the people's awareness and economic possibilities, which are socioeconomic factors. DR programs not only mitigate the effects of wind uncertainty, but also serve as a mechanism to reduce the peak demand, minimizing the number of hours that expensive generating units are committed to guaranteeing the power balance at peak-load hours. Furthermore, DR programs incentivize power consumption during those hours of low electricity demand and increase the ramping capabilities of conventional units. A high penetration level of wind power requires a high degree of flexibility to accommodate the corresponding power injection. The limitations of ramping capabilities are reflected in the generating costs (Jonghe et al., 2014; Wu et al., 2015).

ESS is a promising technology that allows us to store energy surplus during those periods when renewable resources are abundant, typically during low load hours; following which, the stored energy can be injected into the power system when electricity demand is the highest. The optimal ESS management improves conventional generators' operation, improves their capacity factor and their ramping capabilities, and reduces the negative impact of forecasting errors (Baker et al., 2017).

Deciding on the proper amount of wind power capacity requires a significant effort in quantifying the impact of uncertainty and variability on power system operation. In addition, DMs frequently consider the conditions imposed by the wholesale electricity market. Experts perform this analysis in short and long terms to consider operating and planning constraints, obtaining a whole estimation of the costs related to the fluctuations of accommodating wind generation. Therefore, wind generation variability and uncertainty are reflected in several aspects of power system operation, such as its generation efficiency, reserves, wind power curtailment, energy losses, and voltage magnitudes (Albadi & El-Saadany, 2010).

Intrinsic variability of renewable resources forces conventional units to operate at points of low efficiency, resulting in an increment of the associated greenhouse gas (GHG) emissions. However, system operators can mitigate this issue by employing high-quality forecasting techniques. Reserves represent the operating tool that the power system operator uses to manage the variations of wind generation. Regulation, load following, and operating reserves compensate power fluctuations from seconds to minutes. Power-system designers can mitigate the power fluctuation over a minute (regulation reserve) using wind turbine aggregation. Automatic generation control of some specific units adjusts power production to compensate for load fluctuations from seconds to minutes. The system operator can deal with variations from minutes to an hour (load following condition) using the ramp-up and ramp-down capabilities of conventional generators or the available operating reserve. It also depends on whether the penetration level is high or not. Finally, wind power uncertainty also affects the day-ahead scheduling of the power system. The operator decides to start or stop each generating unit and estimate the expected power production costs, taking into account the electricity market rules (Albadi & El-Saadany, 2010).

Depending on the forecasting error, the system operator could reduce the power injected into the system to guarantee the power balance and system stability and avoid transmission system overloading. This situation could occur during low-load periods with abundant wind generation. The occurrence of this phenomenon depends on the penetration level and system flexibility. Wind power curtailment reduces the amount of green energy consumed by the users, representing a loss of natural resources. On the contrary, if at a determined moment the wind generation suddenly drops, the system operator commits the available reserve units to balance the lack of production, preserving system stability. Thus, the forecasting tools and power system structure play a crucial role in system reliability. This issue directly depends on the penetration level and the availability of DR, ESS, or both (Albadi & El-Saadany, 2010).

Concerning energy losses, integration of wind generation can increase or decrease whole losses in transmission and distribution systems according to the wind power penetration level, the correlation between load and wind resource, and the interconnection points of the wind farms. Furthermore, the adoption of high levels of wind generation can require an expansion of the transmission system and consequently an increment in the energy losses (Albadi & El-Saadany, 2010).

Another relevant issue related to wind power integration is the voltage and reactive power distribution. In the early days, wind generation technology used fixed-speed turbines, which demand a high amount of reactive power, distorting the voltage at the connection point and neighboring regions. With the evolution of power electronics, manufacturers adopted double-fed induction generators to supply reactive power, improving the voltage profile (Albadi & El-Saadany, 2010).

In addition to the aspects previously discussed, implementing a forecasting tool for natural resources can be beneficial for scheduling the maintenance and construction processes. Significant changes to the wind generation plant or incorporation of new equipment could require the momentary interruption of the power supply. This could take hours or weeks, depending on the weather conditions. Thus, employing environmental forecasting would be pertinent. In addition, using forecasting information of precipitation, wind gusts, and freezing temperatures allows us to utilize technicians and maintenance equipment effectively. Incorporating forecasting information of weather conditions in construction, operating, and maintenance results in a cost reduction because engineers can guarantee the deadlines and the wind farm commissioning date. These circumstances are required to achieve a successful economic project and ensure access to government financial incentives. Considering the climate change phenomenon, enhancing the prediction of freezing temperatures and enriching the probabilistic prediction tools' information are among the most challenging aspects (Lerner et al., 2009).

Wind power penetration level also affects the behavior of the electricity market, and consequently, the forecasting error is directly reflected in electricity prices. Scientists have observed relevant nonlinear features in wind power and electricity price predictions, and they have found a non-Gaussian characteristic in the distribution of electricity prices. Thus, deep knowledge of energy and price prediction has improved the ability of market agents, improving its operation, which directly benefits end-users as a whole (Jonsson et al., 2010).

The challenge of reducing forecasting error of power sources based on renewable resources has motivated the private and public sectors worldwide to invest in this field, due to green energies' potential benefits. A representative example is the US Department of Energy and the US National Oceanic and Atmospheric Administration (NOAA). These organizations develop regional measurements to create high-resolution rapid refresh models to evaluate the advantages of an accurate forecast process using economic and reliability indices. The financial assessment includes the analysis of supplied and scheduled energy, the imbalance costs and their associated penalizations, the behavior of electricity market agents, ancillary service costs, and GHG emission reduction (Orwig et al., 2012).

Investigators have reported their estimations of the benefits of forecasting error reduction in many different countries. In the case of the US, researchers of the US National Renewable Energy Laboratory (NREL), in collaboration with a private company, analyzed the benefits of reducing forecasting error using information from the Western Electricity Coordinating Council. Scientists considered penetration levels up to 24%. They used a 3-dimensional model with a spatial resolution of 2 km and a temporal resolution of 10 min; this allowed them to investigate the effects of meteorological patterns. Regarding the forecasting error of renewable power sources, their improvement was represented by reducing actual error by 10% and 20%. In other words, scientists used the actual and forecasted profiles to create artificial wind profiles to study different cases, one case with a forecasting error of 10% and another case with a prediction error of 20% (Lew et al., 2011).

Researchers quantified the potential benefit using the operating costs, energy not supplied, and wind power curtailment as the leading indicators. For a penetration level equal to 14%, if the forecasting error reduces from 10% to 20%, the financial benefits almost duplicate. The authors reported a similar conclusion for a penetration level of 24%. The authors found no specific relationship with the forecasting error for a penetration level of 14% concerning the operating reserve deficit because this level is too low. However, for a penetration level of 24% and 10% improvement in forecasting error, the reserve deficit reduces about 44%. For the same penetration level and 20% improvement in forecasting error, the reserve deficit reduces about 65% (Lew et al., 2011). On the one hand, there is no wind power curtailment for a penetration level of 14% because this level is low. On the other hand, if the penetration level increases to 24%, the situation is different. For 10% improvement on forecasting error, wind power curtailment reduces 4%, while if the improvement is 20%, wind power curtailment improves 6.4% (Lew et al., 2011).

Some authors have continuously improved the methods to quantify the effects of variability and uncertainty, specifically for the intra-hour operation or load-following conditions. A study performed with data from the Electric Reliability Council of Texas (ERCOT) revealed that variability cost reduces at a high capacity factor. Thus, DMs are encouraged to design and install wind farms in places of exceptional wind resources (Katzenstein & Apt, 2012).

Other models incorporate the power system structure to quantify the forecasting error value, incorporating the utility's perspective with higher accuracy. A model inspired by the US structure of Midcontinent, California, and New England Independent System Operators allows us to identify the role of the power mix in assessing the effects of wind power variability and uncertainty (Wang et al., 2016). Some European countries such as the UK, Belgium, and Spain have also studied this topic. Using information from the UK, a model incorporated the relationship between electricity prices and wind power capacity to investigate financial viability. Simulations suggested that prices could increase by around 6%. The accuracy of forecasting tools and the costs related to their development also influence economic profitability (Barthelmie et al., 2008).

Another study carried out using data from Belgium offers relevant conclusions regarding GHG emissions. Researchers concluded that the influence of wind power uncertainty on GHG emissions does not follow the same reasoning as the operating costs. Frequently, DMs do not include GHG emissions in the financial assessment of forecasting error. Under these circumstances, highly pollutant sources such as coal become economically viable and are intensively used in the generation mix. However, if GHG emission costs are adopted, the situation is different: gas-based units gain a more relevant role due to their operating capabilities and GHG emissions magnitudes, increasing expenses related to prediction error (Luickx et al., 2010).

Following this trend, scientists have incorporated the market structure, generation mix, and the associated GHG emission rates to quantify forecasting error impact. Several researchers have analyzed this topic using information from Spain, finding that pollutant emission reduction is significant even with forecasting error (Gonzalez-Aparicio & Zucker, 2015; Gutierrez-Martin et al., 2013).

Realizing all of these barriers to integrating renewable energies is helpful because they guide scientists to design the appropriate tools to deal with these problems. Researchers have suggested improving the traditional planning methodologies such as economic dispatch (ED) and unit commitment (UC) to increase the temporal resolution from minutes and hours to months and years. Thus, computational tools require integrating chronological simulation models to include an adjustable time step. They need to incorporate the characteristics of market structure and its processes, accurate solutions of stochastic unit commitment (SUC), and ED problems, including realistic operating constraints, contingencies, and likely scenarios for expanding the transmission system (Philbrick, 2011). Furthermore, planning methods need to analyze the effects of topology changes on load loss, avoiding the computational efforts associated with the power flow problem. Similarly, they need to establish a central or base case to evaluate the impact of perturbations on power system operation. Incorporating yearly analysis allows designers to include information on maintenance and repair processes and estimate pollutant emissions in the long term (Philbrick, 2011).

At this point, we have gained a general perspective over the practical benefits of adopting a forecasting system for environmental variables. The following section will discuss some of the essential techniques employed in the renewable energy industry.

5.2 Forecasting techniques

This section briefly describes the basic techniques employed in forecasting elements of some highly adopted renewable resources, such as wind speed and solar radiation, and other auxiliary variables, such as ambient temperature. We also comment on other methods related to electricity prices and demand prediction.

5.2.1 Wind speed and power forecasting

Inevitably, wind power forecasting has significant limitations because scientists do not have complete knowledge about the atmospheric processes involved in wind energy conversion. Meteorological processes related to air circulation between high-pressure zones are complex and depend on the impact of the boundary layer, temperature gradients, and terrain features, among many other factors. The imperfect knowledge about these factors generates systematic and random errors in the forecasting process (Pinson, 2013).

Research institutions such as the NOAA, the European Centre for Medium-Range Weather Forecasts (ECMWF), and the National Center for Environmental Prediction have experience in describing the weather state variables at a global scale. The thermal dynamics between the Earth's surface and the atmosphere create the intra-daily fluctuations of wind speed. The magnitude of such volatility on wind speed depends on the climatic characteristics of the place of interest and the corresponding season. Phenomena related to local weather, such as cloud motion and precipitation, strongly influence the wind speed variations that occur on a time interval of minutes or hours (Pinson, 2013).

On the one hand, generating companies and utilities trade electricity through a market. Forced by the electricity market structure, forecasting renewable power sources is reported up to between 12 and 36 h ahead in an hourly step. However, forecasting information between 10 min and 2 h is helpful for the ED and control processes. On the other hand, researchers express forecasting products depending on the mathematical approach employed by DMs. The DM uses the point forecast, which operates with some specific values along the forecasting horizon when the analysis is deterministic, solving the UC problem. If DMs incorporate uncertainty in the process, they require probabilistic forecasting, which offers more information about the expected prediction error. In this context, DMs base decisions on the solution to the SUC problem (Pinson, 2013).

Experts divide wind power forecasting techniques into two groups. The first group includes those methods that employ time series analysis (TSA), whereas the procedures in the second group use numerical weather prediction (NWP). However, some authors classify the forecasting methods as statistical, artificial intelligence (AI), and physical approaches. In addition, engineers and practitioners combine techniques of different groups to improve global performance. Another technique frequently used as a reference point is persistence analysis. Persistence consists of assuming that the wind speed during the next hour will be the same as the current one. This reasoning offers good results because the one-lag autocorrelation of wind speed is usually high (Foley et al., 2012).

Implementing TSA for forecasting purposes is not restricted to power system management. Researchers have been employing these techniques in other fields such as business and economics for a long time. Statisticians and applied scientists recognized the necessity of a theoretical framework to fulfill the requirements of both areas, obtaining more accurate predictions and consequently, more appropriate decisions (Fildes & Makridakis, 1995). The evolution of forecasting techniques has demonstrated the potential of combining several methods to increase the whole performance. Furthermore, scientists have found meaningful progress from the implementation of decomposition concepts (Webby & OConnor, 1996). The field of operational research has a close relationship with forecasting, and it has made significant contributions. Some specific applications such as inventory management and information sharing in the supply chain have demonstrated the successful relationship between these areas. Improvement of customer relationships through data mining and advanced computational techniques is another example of how these areas can operate in a consolidated manner (Fildes et al., 2008).

TSA is frequently implemented in a regressive form to predict the wind speed or power. Information from the past few steps is used to estimate the values in the next few steps. A representative example is the autoregressive moving average (ARMA) model, a widely used statistical method in many fields. In general, statistical techniques offer satisfactory results in forecasting monthly or annual wind speed (Foley et al., 2012).

Brown et al. (1984) wrote one of the earliest works using TSA to simulate and predict wind speed and power. The proposed technique applied the transformation and standardization process to prepare the wind speed time series for the ARMA-fitting procedure. Once the ARMA model is adjusted, this is evaluated regressively to obtain the transformed and standardized wind speed values 1 day ahead. After that, transformation and standardization processes are undone, and the wind speed forecasting is finally obtained. Wind speed is translated into power using the power curve of the wind farm. Additionally, the prediction intervals are incorporated in the proposed methodology.

Torres et al. (2005) performed an extensive analysis of TSA using an ARMA model over a high number of locations in Spain. The authors concluded that the TSA implemented were only valid for short-term forecasts. Furthermore, they did not observe a significant difference in the results with the geographical characteristics of the chosen locations. Erdem and Shi (2011) performed short-term forecasting of wind speed and its direction. To this end, they decomposed wind speed into its lateral and longitudinal components and represented them using an ARMA model. The authors also implemented an ARMA model for wind speed and another one for wind direction, and also used a vector autoregressive model. Using field measurements from a site in the US, the authors concluded that the model selection depends on the variable of interest (wind speed or direction).

As previously mentioned, techniques based on AI and neural network (NN) training, fuzzy systems, and support vector machines (SVMs) are also frequently employed (Foley et al., 2012). Kariniotakis et al. (1996) presented one of the earliest works in this field. The authors developed an algorithm based on the cross-validation approach to optimize the NN architecture of a recurrent neural network (RNN), which means carrying out the optimal selection of the number of past values, the learning rate, and the parameter of the activation function. The forecasting model was applied to manage a wind-diesel power system considering a forecasting horizon of 2 or 3 h with a 10-min step.

Cadenas and Rivera (2009) tested a wide range of NN architectures to forecast wind power in a Mexican location, considering an hourly time step. Investigators analyzed configurations with two and three layers and with three, four, and six neurons. The authors also performed a comparative analysis between an NN and an ARMA model (Cadenas & Rivera, 2007) for the monthly prediction of wind generation. According to their conclusions, both models offer good results, although the ARMA model presented lower statistical error. Finally, the researchers developed a combined ARMA-NN model (Cadenas & Rivera, 2010). First, the ARMA model is fitted, and the associated forecasting error is determined. Then, the NN is trained using the error signal previously obtained, catching the nonlinearities that the ARMA model could not recognize. The combined approach was tested in three different Mexican locations, demonstrating the capabilities of the hybrid model over the ARMA- and NN-based techniques.

In a general sense, determining the proper structure of an NN or an ARMA model is problematic because it is an intensive computation process. Flores et al. (2012) proposed a design technique based on a genetic algorithm (GA). The characteristics of the ARMA model or the NN are represented as a GA individual. The authors proposed a binary GA implementation so that considering or not considering a specific value in the past to perform the prediction can be indicated. The researchers tested their methodology using data from several Mexican regions. After an extensive study, they observed that the NN offered better results than the ARMA model.

Recently, NN has undergone a significant evolution due to the efforts of many scientists. Deep neural networks (DNN) (Lingras, 1998) are a result of such an effort. DNN considers several layers in its structure, improving its capacity to recognize complex characteristics of the training data. This type of NN has also been employed on forecasting tasks. Khodayar et al. (2017) developed a forecasting technique using a rough DNN with autoencoders to provide a robust prediction. Another advanced NN architecture recently applied in many fields is the long short-term memory (LSTM) (Hochreiter & Schmidhuber, 1997). In this regard, Zhou et al. (2019) developed a prediction methodology that combines the LSTM and the K-means clustering algorithm. The capabilities of LSTM are used to recognize the time-series correlation degree, while the K-means is used to define impact factors of wind generation. Additionally, the nonparametric kernel density estimation is incorporated to perform a probabilistic prediction.

Shahid et al. (2021) employed a GA to optimize the architecture of an LSTM. After evaluating the performance of the proposed approach using data from several European sites, the authors verified the quality of the forecasting products in terms of accuracy, reliability, and robustness.

Statistical methods are usually employed to predict wind speed and power in a horizon lower than 6 h. In contrast, engineers use forecasting techniques based on NWP to estimate hourly values in a higher horizon (Giebel et al., 2011). NWP employs computational techniques to understand the general circulation of the atmosphere and how the meteorological variables evolve. Behavior of crucial variables such as pressure and temperature are modeled using computational models of atmospheric dynamics (Marchuk, 1974). This approach has practical applications in many fields such as agriculture, not only renewable energy integration.

Richardson (2000) carefully analyzed the advantages of employing probabilistic environmental forecasting from an economic perspective, using a decision-analytic model. The scientist used the forecasting results obtained from the ECMWF and compared them with those obtained from an operational ensemble prediction system (EPS). The environmental variables analyzed are temperature and precipitation. EPS presents the forecasts and their associated probability. Then, depending on a probability threshold, the DM decides the actions to be taken. The threshold probability allows us to transform a probabilistic forecast into a deterministic one. The threshold parameter directly impacts the potential benefits of probabilistic forecasting. For the case of a temperature anomaly over Europe, the EPS approach offered potential benefits of between 10% and 20%. Based on these results, the researcher concluded that the information provided by the EPS is more valuable than that contained in a simple deterministic forecast.

Vitart et al. (2008) used the variable resolution EPS to create a forecasting system able to work from medium (15-day twice a day) to monthly (32-day once a week) horizon. This version offers relevant information for predicting extreme weather conditions at a small scale and accurate forecasting products at a large scale. According to the results reported, this model provides promising results indicating extreme events similar to the heatwave over Europe of 2003 and the wet summer over England of 2007.

Scientists have also shown interest in intra-seasonal prediction because DMs have recognized the potential benefits of its implementation in fields such as the humanitarian sector, public health, energy, agriculture, and water management, among many others. However, important issues need to be solved. Some issues are related to the models of the relationship between the atmosphere and ocean phenomena. This problem could introduce important errors. Another conflictive topic is the quantification of forecasting error. This problem refers to the fact that the performance of the environmental prediction systems depends on the initial atmospheric conditions (White et al., 2017).

Barbounis et al. (2006) developed a forecasting technique using information from a wind farm located in Greece through SKIRON, an atmospheric modeling system. The method uses three types of recurrent NN to analyze the wind speed and direction data, facilitating the forecasting of wind generation 72 h ahead. Lobo and Sanchez (2012) developed a forecasting technique based on wind farm aggregation to reduce the prediction error. The method analyzes the similarities between the information obtained from NWP and the historical data. The authors verified the effectiveness of their technique using data of a Spanish location, observing an improvement of between 4% and 8% compared to the forecasting tool SIPREOLICO.

Khalid and Savkin (2012) developed a model based on TSA and NWP for short-term prediction. The methodology collects information from several measuring installations, including wind speed and direction, and combines it with a synchronized signal obtained from NWP. The authors demonstrated the capabilities of their technique using data from an Australian wind farm, observing a better performance than persistent and Gray predictor models.

Chen et al. (2014) proposed a model that combines the reasoning behind NWP with probability theory. The technique combines a Gaussian process with NWP to forecast wind generation 24 h ahead. Essentially, the Gaussian process corrects the estimation of NWP. According to the conclusions presented by the authors, their methodology improves between 9% and 17% the results obtained from a NN. Xu et al. (2015) created a technique based on deleting the anomalous data from NWP. To this end, a data analyzer is implemented to recognize erroneous information of NWP. Then, the K-means clustering algorithm and an NN are combined hierarchically to perform the prediction. The NN is trained and tuned using the results of the data analyzer.

Buhan et al. (2016) presented a forecasting approach that incorporates measurements from a reference station and NWP. The approach employs the adaptive boosting algorithm to combine NWP information. Then, the results are clustered, and an NN-SVM model is used in each cluster to establish a relationship between the wind profiles and the reference station. The authors carried out a wind power prediction over the next 48 h successfully.

5.2.2 Solar irradiance forecasting

Solar irradiance is a critical concept in the assessment of solar energy systems. Global horizontal irradiance (GHI) represents the irradiance over a horizontal surface. Direct normal irradiance (DNI) is the irradiance that incises directly from the sun on the horizontal plane. Diffuse horizontal irradiance (DHI) remains scattered on the line between the sun and the position of the energy system. GHI has a close relationship with the sun position, DHI, and DNI, which is among the most critical variables in the operation of solar systems (Law et al., 2014).

Practitioners use GHI forecasting to assess the performance of solar photovoltaic (PV) and thermal systems. In systems with a concentrating infrastructure, the system concentrates DNI through a receiver. For this reason, DNI forecasting becomes crucial. DNI reduction depends on the clouds and aerosol concentration. The reduction in DNI could be between 30% and 100%, depending on the aerosol concentration. In the case of GHI, the decline could be 10% or even more. This phenomenon is the main reason to avoid estimating DNI from the GHI prediction (Law et al., 2014).

Scientists perform solar irradiance forecasting using TSA, AI, clear sky modeling (CSM), cloud motion vector (CMV), and NWP. CSM can estimate solar irradiance over a determined location on the Earth using the sun's position and the extraterrestrial solar irradiance. Scientists increase CSM precision by simulating some of the characteristics of the atmosphere, such as the aerosol optical depth. Global-to-diffuse or global-to-beam models are implemented in those cases where researchers cannot directly obtain DNI. These models are used to divide GHI into DNI and DHI using statistical techniques. In addition, CSM provides valuable information when researchers employ forecasting techniques based on parameters of atmospheric composition. Methods based on CMV use imaging instruments obtained from a satellite or employ a ground-based infrastructure to record information of the motion of the clouds. Usually, NWP uses radiative transfer models to perform GHI forecasting. An option is estimating DNI using a global-to-diffuse model and the relationship between the GHI prediction and the GHI value indicated by CSM (Law et al., 2014).

The technique adopted firmly depends on the forecasting horizon, which could be between minutes and days. Predictions between 10 min and 1 h, useful for short-term ramps and regulation, could be performed using ground-based CMV or TSA. In the case of a forecasting horizon between 1 and 5 h, valid for load-following applications, practitioners could use satellite-based CMV. Finally, predictions between 6 h and 10 days, employed in UC, are performed using NWP (Law et al., 2014).

Regarding the forecasting techniques based on TSA, David et al. (2016) developed a forecasting method based on the clear sky index estimated from the GHI. Specifically, the authors proposed combining the ARMA and the generalized autoregressive conditional heteroscedasticity (GARCH) models. The ARMA model performs the point forecasting, while the GARCH model estimates the corresponding prediction interval. The effectiveness of the ARMA-GARCH model was evaluated using data for six locations around the world, observing good results comparable with those obtained from other techniques such as SVM.

NN training is also widely used in solar irradiance forecasting. Ji and Chee (2011) proposed a prediction technique implemented in two stages. In the first stage, the non-stationary trend is removed from the time series. Then, the stationarity of the residuals is tested using the augmented Dickey-Fuller method. In the second stage, the residual time series is predicted using an ARMA model, while the solar radiation is finally forecast using the time-delay NN. Capizzi et al. (2012) combined the wavelet decomposition (WD) with an RNN to predict solar radiation 2 days ahead. The WD recognizes the correlation between solar radiation, wind speed, humidity, and temperature. The forecasting process is first carried out in the wavelet domain. Indeed, the RNN uses the wavelet coefficients estimated from the time series of the weather variables. Then the decomposition is undone using the inverse wavelet transform to obtain the forecasting product.

Liu et al. (2015) proposed incorporating the aerosol index, in addition to other meteorological variables. Specifically, time series of weather variables, PV power production, aerosol index, temperature, humidity, and wind speed are used to determine the day-ahead PV power forecasting training an NN with the backpropagation (BP) algorithm. In addition to NNs and SVMs, machine learning (ML) methods have enriched the forecasting theory. Indeed, other methods, such as regression tree, random forest, gradient boosting, among many others, have been successfully implemented for solar radiation prediction (Voyant et al., 2017). According to recent studies, ML can be classified as generalized, based on ensembling, and based on decomposition and clustering (Zhou et al., 2021).

As previously stated, researchers in many fields have paid attention to DNNs, including forecasting solar radiation. Following this trend, Huang and Kuo (2019) created the PVPNet model. PVPNet is a forecasting tool for day-ahead operation from deterministic and probabilistic perspectives. The methodology uses weather and operating data, such as PV power production, temperature, and solar radiation, incorporated into a DNN to obtain the PV power prediction.

Lai et al. (2021) employed a clustering method based on DNN to predict GHI an hour ahead. The clustering process allows us to identify those data portions with irregular patterns. Then, the feature attention deep forecasting (FADF) DNN is created for each cluster, obtaining their corresponding GHI forecast. Essentially, the FADF recognizes several features in the data and weights them during the forecasting process. According to the conclusions reported by the authors, their model improves the persistence model up to 12%, approximately.

Huang et al. (2021) combined many approaches to predict solar irradiation an hour ahead. The proposed technique includes WD, convolutional neural network (CNN), LSTM, and other methods, to recognize the key characteristics of several weather variables such as temperature, relative humidity, and wind speed. According to the reported results, the combined approach shown better results than BP, SVM, and RNN.

NWP also has a privileged position in solar radiation forecasting. In this regard, Voyant et al. (2021) combined information from NWP with an ARMA model and an NN. This approach overcomes the persistence model at 11.3% and the NN system at 3.5%. Deo and Şahin (2017) took advantage of the availability of satellite technology and proposed a methodology that employs moderate resolution imaging spectroradiometer (MODIS) land-surface temperature (LST). Specifically, the authors combined MODIS-LST with an NN trained using conjugate gradient and Levenberg-Marquard algorithms. Experiments in the prediction of monthly and seasonally forecasting have shown the superiority of this model over other statistical approaches such as the ARMA model.

5.2.3 Temperature, prices, and demand forecasting

Another variable of high relevance is the electricity price, forecasting of which could be performed using a multiagent simulation, a fundamental method, a reduced-form model, a technique of econometric analysis, or a methodology based on AI or NN training (Weron, 2014).

Multi-agent simulation models are also known as equilibrium or game-theoretic methods. These techniques simulate the interaction between generating companies, which are represented as agents, and how they influence the market clearing process. These techniques incorporate the bidding strategies of each participant, which is a crucial feature in liberalized electricity markets (Weron, 2014).

Fundamental methods are also known as structural techniques. These methodologies perform the forecasting process considering significant physical and economic parameters of the price evolution. The relationship between physical variables such as electricity demand and environmental conditions is represented and independently forecasted using time series analysis or NN training. Then, their influence on electricity price is finally estimated (Weron, 2014).

Reduced-form models analyze the statistical evolution of prices with the time for risk assessment and management. These models represent the probability distribution and correlation of electricity prices with other variables in the long term (Weron, 2014). At first, engineers performed electricity demand using time series analysis and similar econometric models. Then, researchers employed these techniques on price forecasting by only changing the electricity demand by price. Similarly, AI and NN training consists of performing the price forecast by training an NN with historical price data (Weron, 2014). With the massive integration of renewable energies, scientists have developed probabilistic techniques for electricity price forecasting. Some approaches can estimate reasonably good profiles, including the probability of a price spike in the long term (Ziel & Steinert, 2018).

Ambient temperature is an environmental variable that affects the performance of some energy conversion systems, such as PV generators, and influences how people consume energy. Investigators have been motivated by the necessity of improving energy efficiency. Thus, they have developed forecasting models oriented on the optimal operation of heating, ventilation and, air conditioning systems, taking advantage of the theoretical concepts already known. Following this trend, Zhang, Tang, and Li (2014) proposed a prediction model using the relationship between ambient temperature, solar irradiance, clouds, ground heat radiation, and heat convection. Under real conditions, the model has shown good performance for subhourly and hourly time intervals.

The relationship between ambient temperature and electricity demand has been analyzed statistically. Using field data, Valenzuela et al. (2000) demonstrated the crucial role of temperature forecasting and hourly load correlation on the quality of the power system cost prediction. In addition, they confirmed the dependence of production costs uncertainty on the electricity demand forecasting error. Similarly, Wang et al. (2018) demonstrated the relationship between temperature and gas consumption by applying an extensive statistical analysis. In this regard, electricity and gas demand prediction requires ambient temperature information to improve its performance. Mestekemper et al. (2013) proposed a forecasting model based on periodic autoregression and dynamic factor modeling that integrates environmental information. The effectiveness of this approach was estimated using data from Germany and Australia, concluding that the inclusion of weather information substantially enhances the quality of the forecasting product.

The incorporation of new NN structures and training algorithms has motivated using AI to forecast a broad set of variables. ML techniques have recently gained interest due to their promising capabilities to predict the ambient temperature and other environmental variables regionally and globally. ML facilitates processing large datasets, accelerating the computing process and increasing its accuracy. However, special care should be taken during the designing and training process (Cifuentes et al., 2020).

Successful implementation of ML techniques and NN training depends on having a representative amount of data-points, and on the optimization techniques selected for the training process and its parameter setting. In addition, the training process requires a long computational time, which can be reduced by using parallel computing implemented on graphics processing units (GPU) (Boukabara et al., 2019). Once the NN has been trained, the computing time reduces drastically.

Despite its characteristics, ML results need to be carefully analyzed by scientists and experts to guarantee the feasibility and applicability of the forecasting product. Specific knowledge about the problem under study, selecting the appropriate ML architecture, selecting and normalizing data points, selecting the proper training algorithm and its parameters, and accuracy assessment require a high level of expertise in multiple areas (Boukabara et al., 2019).

Integration of weather information remotely obtained to an NWP system offers an excellent opportunity to increase the quality of the current forecasting products. However, there are problems that need to be solved; some of them are related to processing a large amount of data, improvements on the currently used models and their applications, and user requirements (Boukabara et al., 2019).

The first barrier is related to the fact that NWP systems have problems dealing with abundant and diverse data types. The amount of data from satellites has been increasing, and consequently, integrating this information in real-time has significant limitations related to hardware and latency restrictions. In addition, new types and formats for data representation require new manners to take advantage of this information. Due to this barrier, NWP systems use only between 1% and 3% of the available satellite information (Boukabara et al., 2019).

The problem of processing a set of information with a massive number of datapoints will continue because smart sensors and the internet of things will increase data sources. This fact allows scientists to improve the temporal and special resolutions of meteorological forecasting. Furthermore, the technological development that humanity is currently undergoing provides smaller satellites that are commercially affordable (Boukabara et al., 2019). The deficiencies of current computation systems and the limitations on the latency of communications directly affect short-term forecasting. Thus, improvements required to deal with these issues consist of enhancing the data assimilation system of satellite information, including quality control and bias error correction, and increasing the quality of the meteorological forecasts, improving the users' decision-making weather data (Boukabara et al., 2019).

Scientists have also recognized the advantages of incorporating ML techniques to join remotely sensed data with NWP tools. ML methods increase computational efficiency because evaluating the ML model is often faster than evaluating the physical model. Employing a large dataset during the NN training process results in an accurate simulation of environmental phenomena. The flexibility of the whole model increases because physical variables not modeled actually could be included using ML. Implementing ML methodologies has been simplified because modern computing languages already include subroutines to facilitate its programming (Boukabara et al., 2019).

However, significant concerns have been raised regarding ML. One of them is related to the reliability of the results when nonfrequent events are evaluated. In these cases, the data available for NN training is not abundant, and consequently, the learning process becomes complicated. Another issue is related to the mathematical representation of physical restrictions. Other topics currently being discussed are related to the uncertainty representation, the maturity of the available ML models, and their integration with the existing standards and procedures (Boukabara et al., 2019).

In order to deal with some of these problems, Jeong et al. (2021) developed a prediction model for ambient temperature using a DNN, which offers forecasting products at a regional scale. The forecasting system is provided with a data assimilation system. It combines a bidirectional LSTM with a CNN. This combination results in a system able to provide temperature prediction 14 days ahead. Concerning electricity price forecasting, Jasiński (2021) carried out research to understand its intrinsic characteristics and the relationship with the ambient temperature; results using a DNN show that forecasting error reduces up to 15.3%.

5.3 Uncertainty management

The impact of forecasting error on renewable energy systems is a crucial barrier to be overcome. These circumstances force statisticians and engineers to characterize the forecasting error in terms of its probability density function (PDF) and design effective techniques to incorporate such PDFs in the optimization problem required to determine the optimal operating point of the energy system (ED, UC, and SUC).

Concerning the PDF of forecasting error, the Gaussian distribution was widely adopted in the early days. However, many other distributions have been adopted as the interest in renewable energies has grown.

Bludszuweit et al. (2008) analyzed wind generation time series and, employing the persistence technique, proposed using the Beta distribution. The authors observed that forecasting error is fat-tailed, and it could have a behavior similar to a Gaussian distribution depending on the value of the kurtosis. Giorgi et al. (2011) considered other prediction techniques such as the ARMA model and NN training for different forecasting horizons between 1 and 24 h. In addition, they estimated the probability that forecasting error occurs between ±10% and ±20% intervals. Regarding the ±10% interval, the probability is between 48.8% and 78.9%, depending on the technique and forecasting horizon. In the case of ±20%, the probability is between 75.4% and 91.55%.

An extensive analysis of forecasting error using data from NREL and other organizations with persistence forecasting helped Tewari et al. (2011) to conclude that a mixed distribution offers satisfactory results to represent wind power uncertainty. Hodge and Milligan (2011) analyzed data from the ERCOT, concluding that the forecasting horizon has an influence on the probability distribution of prediction error. From a statistical analysis, the authors suggested using the Cauchy-Lorentz distribution.

Zhang et al. (2013) performed a statistical analysis using data from a windfarm in China and concluded that the versatile distribution could be an excellent option to model forecasting error. The versatile distribution offers generality due to its adjustable parameters. In addition, it can be analytically treated due to its mathematical form. Bruninx and Delarue (2014) studied a Belgian windfarm. After a deep analysis, the authors observed that the probability distribution of wind generation has a heavy tail and proposed using the Lévy α-stable distribution.

Following similar reasoning to that with EPS, DMs use scenario generation and reduction methods to incorporate uncertainty into the decision-making process, such as SUC. Regarding this topic, Ma et al. (2013) created a methodology for scenario generation assuming a *t* location-scale distribution for wind generation. Specifically, the technique employs a multivariate normal distribution with an inverse transform sampling to generate many likely scenarios. Wind generation time series is used to estimate the covariance matrix of the multivariate normal distribution. The method's effectiveness was illustrated using data from a windfarm in Ireland, observing that the variability modeled by the scenario set can be modified using the covariance characteristics.

Chen et al. (2018) proposed using a generative adversarial network—a combination of two DNNs—for scenario generation. The proposed approach is a data-driven technique that recognizes patterns in temporal and spatial dimensions. Furthermore, the effects of environmental events and chronological aspects can be included in the scenario generation process. Using information from NREL, the authors demonstrated the capabilities of this methodology to generate scenarios for PV generation. Similarly, Nuño et al. (2018) developed a technique for scenario generation of PV power production. The methodology performs a geographical analysis, aggregates power curves, performs a data preprocessing, and finally, performs the random simulation. The stochastic model uses a multivariate version of the ARMA model to consider the temporal and spatial characteristics.

Zhang, Kang, et al. (2014) proposed a statistical model using copula theory, another way to model stochastic dependence. The authors demonstrated the effectiveness of this approach using wind generation and its corresponding day-ahead prediction. Similarly, Becker (2018) used the C-vine and D-vine structures of copula theory to generate scenarios. According to the results observed from the analysis of a windfarm in Belgium, D-vine has shown excellent results.

The following section will discuss basic concepts related to the forecasting process using an NN.

5.4 Using a neural network for forecasting purposes

Previously, in Chapter 3, we explained the procedure related to NN training using a real-coded GA. From this section, we extend this content by treating the problem of short-term forecasting of several variables of interest such as electricity price and demand, ambient temperature, and renewable resources. Specifically, we evaluate the benefits of combining the classical BP algorithm with the already discussed real-coded GA to improve the optimization results and computational efficiency. Usually, investigators use an NN to model the complex behavior within a dataset, which is frequently presented using the input–output format. The training process consists of finding out the NN matrices so that the relationship between the input and output data points is represented with minimal error.

The BP algorithm was designed to solve this optimization problem to train a feedforward neural network (FFNN). A drawback of this method is that the error signals can blow up or vanish; however, more advanced networks such as the LSTM (Hochreiter & Schmidhuber, 1997) have already addressed this issue. In this chapter, we center our attention on the short-term forecasting problem and the performance of the combination of BP and GA. In the following subsections, we describe the type and specifications of the NN considered and how to build the input–output data format required for NN training.

5.4.1 Neural network type and specifications

Usually, the input-out dataset is represented using two different matrices with T rows. Thus, we have T points to train the network. The features of the input and output datasets can be represented by the number of columns of each matrix (M for the input matrix and Q for the output matrix). In general, we use the indices m ($m = 1, \ldots, M$) and q ($q = 1, \ldots, Q$) to represent the features of the input and output datasets, respectively.

Specifically, in the forecasting problem, we take information from the past and use the NN to predict the future one step ahead. Using the general formulation, we take M past-values of the time series to predict the next one. On the one hand, the number of

$$
\boldsymbol{u}_{(t)} = \begin{bmatrix} \dot{u}_{(t,1)} \\ \dot{u}_{(t,2)} \\ \vdots \\ \dot{u}_{(t,m)} \\ \dot{u}_{(t,m+1)} \\ \vdots \\ \dot{u}_{(t,M-1)} \\ \dot{u}_{(t,M)} \end{bmatrix}
\begin{array}{l} \leftarrow \text{Variable at } M^{\text{th}} \text{ hour} \\ \leftarrow \text{Variable at } (M\text{-}1)^{\text{th}} \text{ hour} \\ \\ \leftarrow \text{Variable at } m^{\text{th}} \text{ hour} \\ \leftarrow \text{Variable at } (m+1)^{\text{th}} \text{ hour} \\ \\ \leftarrow \text{Variable at } 2^{\text{nd}} \text{ hour} \\ \leftarrow \text{Variable at } 1^{\text{st}} \text{ hour} \end{array}
$$

Fig. 5.1 Structure of the NN input data: Relationship between the input-data vector and the chronological time series.

features of the input dataset (M) represents the number of values we evaluate in the NN to predict the next step. On the other hand, we are only predicting one step, so that the number of features of the output dataset is one ($Q = 1$).

In this context, a determined input ($\boldsymbol{u}_{(t)}$) and output ($\boldsymbol{w}_{(t)}$) data-point (t) is represented using Eqs. (5.1) and (5.2), respectively.

$$
\boldsymbol{u}_{(t)} = \begin{bmatrix} u_{(t,\,1)} \\ \vdots \\ u_{(t,\,m)} \\ \vdots \\ u_{(t,\,M)} \end{bmatrix} \tag{5.1}
$$

$$
\boldsymbol{w}_{(t)} = \left[w_{(t,\,1)} \right] \tag{5.2}
$$

Figs. 5.1 and 5.2 present the structure of a typical input ($\boldsymbol{u}_{(t)}$) and output ($\boldsymbol{w}_{(t)}$) data-point, respectively. The vector $\boldsymbol{u}_{(t)}$ takes information from the actual and past moments, while the vector $\boldsymbol{w}_{(t)}$ stores the prediction one step ahead. Thus, we can identify information from the past and future simultaneously. In this context, we can identify the actual moment as the element $u_{(t,\,M)}$ in Fig. 5.1. Arbitrarily, we have defined this as the first hour of the past, or in other words, the value of the variable at the first hour in the past. Then, we can complete the remainder of the vector by taking the data chronologically.

Regarding the vector $\boldsymbol{w}_{(t)}$ shown in Fig. 5.2, which takes information from the future by evaluating the NN, it is only composed of the value of the variable of interest in the future, or in other words, the variable at the first hour in the future.

$$
\boldsymbol{w}_{(t)} = \left[\dot{w}_{(t,1)} \right] \leftarrow \text{Variable at } 1^{\text{st}} \text{ hour}
$$

Fig. 5.2 Structure of the NN output data: The evaluation of the NN offers a scaled estimation of the variable prediction one step ahead.

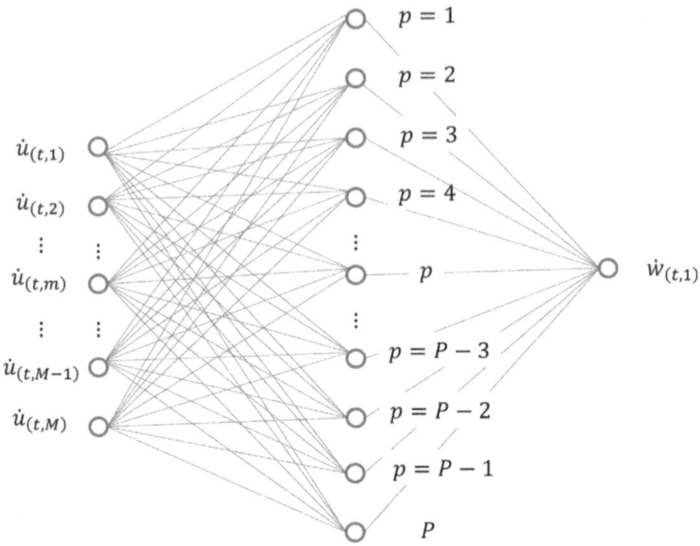

Fig. 5.3 Structure of a simple FFNN: This is the NN with several inputs and one output used for prediction one step ahead.

Fig. 5.3 presents the structure of FFNN used to predict a value one step ahead. As can be observed, values of the actual and past history are evaluated by the NN to estimate the value of the time series one step ahead.

Eqs. (5.3) and (5.4) show the expressions used for the NN evaluation. As stated before, the matrices $\boldsymbol{W}_{\mathbf{hx}}$, $\boldsymbol{b}_{\mathbf{h}}$, $\boldsymbol{W}_{\mathbf{yh}}$, $\boldsymbol{b}_{\mathbf{y}}$ have to be estimated to minimize the mismatch between the input and output data-points.

$$\boldsymbol{h}_{(t)} = \text{sigmoid}\left(\boldsymbol{W}_{\mathbf{hx}} \times \boldsymbol{u}_{(t)} + \boldsymbol{b}_{\mathbf{h}}\right) \tag{5.3}$$

$$\boldsymbol{w}_{(t)} = \text{sigmoid}\left(\boldsymbol{W}_{\mathbf{yh}} \times \boldsymbol{h}_{(t)} + \boldsymbol{b}_{\mathbf{y}}\right) \tag{5.4}$$

The matrix $\boldsymbol{\mathcal{L}}_{\mathbf{NN}}$, defined in Eq. (5.5), for the specific case of the forecasting problem, allows us to store the number of rows and columns of NN matrices. Thus, the first row of $\boldsymbol{\mathcal{L}}_{\mathbf{NN}}$ stores the number of rows and columns of the matrix $\boldsymbol{W}_{\mathbf{hx}}$. The second row stores the number of rows and columns of the matrix $\boldsymbol{b}_{\mathbf{h}}$. The third row stores the number of rows and columns of the matrix $\boldsymbol{W}_{\mathbf{yh}}$, and the fourth row stores the number of rows and columns of the matrix $\boldsymbol{b}_{\mathbf{y}}$. As the NN has only one output, the number of features of the output dataset (Q) has been set to one.

$$\boldsymbol{\mathcal{L}}_{\mathbf{NN}} = \begin{bmatrix} P & M \\ P & 1 \\ 1 & P \\ 1 & 1 \end{bmatrix} \tag{5.5}$$

The matrix $\mathcal{L}_{\mathbf{NN}}$ is particularly useful during the retrieving process required for the evaluation of a determined individual ($\mathcal{L}_{\mathbf{IND}}$) of the GA.

5.4.2 Data preparation for neural network training

First of all, we need to perform a scaling process to express the time series data in the interval [0, 1]. Before NN training, the data needs to be prepared in the appropriate format. In this work, we have divided the entire time series into two groups. The first group is the section of the raw data used for training, and the second group is the section of the raw data used for testing. We identify the current or actual moment, the past, and the future in the training portion, as shown in Fig. 5.4. The information in the present and past is evaluated in the FFNN. The result would be an estimation of the first step in the future.

Then, the first value of the future dataset is moved to the present, and the value that used to be the present one becomes the first value of the past; this procedure represents the time elapsed.

Regarding the testing data, we labeled this as *beyond the future* to differentiate it from the future part of the training dataset. If the procedure previously explained is applied sequentially to all the available data, we would obtain the required training data in the input–output format. The input data is represented by the matrix $\boldsymbol{U}$ of Eq. (5.6) and the output vector $\boldsymbol{W}$ of Eq. (5.7).

$$\boldsymbol{U} = \begin{bmatrix} \text{transpose}\left(\boldsymbol{u}_{(1)}\right) \\ \vdots \\ \text{transpose}\left(\boldsymbol{u}_{(t)}\right) \\ \vdots \\ \text{transpose}\left(\boldsymbol{u}_{(T)}\right) \end{bmatrix} \tag{5.6}$$

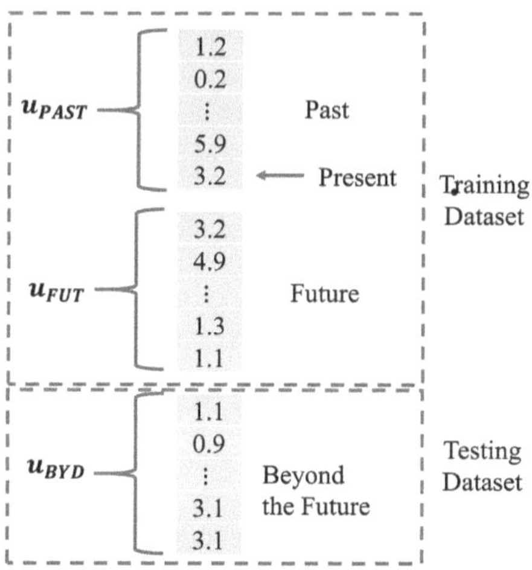

Fig. 5.4 Categorization of time series for forecasting purposes: Dividing the chronological data to create the training and testing datasets.

$$\mathbf{W} = \begin{bmatrix} w_{(1)} \\ \vdots \\ w_{(t)} \\ \vdots \\ w_{(T)} \end{bmatrix} \tag{5.7}$$

The specific method for data preparation is explained as follows:

Step 1: Initialize the NN input dataset by creating a matrix ($\boldsymbol{U}$) with T rows and M columns.

Step 2: Calculate the number of days on the future dataset ($\boldsymbol{u}_{\mathbf{FUT}}$). To this end, we divide the number of hours of the future dataset to the number of hours of forecasting horizon ($H = 24$). The result is stored on the variable D.

Step 3: Copy the past values stored in the variable $\boldsymbol{u}_{\mathbf{PAST}}$ to the history vector ($\boldsymbol{u}_{\mathbf{HIST}}$).

Step 4: Set $u \leftarrow 0$.

Step 5: Analyze the first day of the forecasting horizon by setting $d \leftarrow 1$.

Step 6: If $d \leq D$, go to Step 7, else end.

Step 7: Calculate the number of elements in the history vector ($\boldsymbol{u}_{\mathbf{HIST}}$) and store the result in the variable T_A.

Step 8: Define an auxiliary vector ($\boldsymbol{u}_{\mathbf{AUX}}$) as equal to the elements of the future vector ($\boldsymbol{u}_{\mathbf{FUT}}$) from $(1 + u)$ to $(H + u)$. In other words, set $\boldsymbol{u}_{\mathbf{AUX}} = \boldsymbol{u}_{\mathbf{FUT}}((1 + u) : (H + u))$.

Step 9: Analyze the first hour of the day d by setting $h \leftarrow 1$.

Step 10: If $h \leq H$, go to Step 11, else go to Step 15.

Step 11: Calculate the number of elements in the history vector ($\boldsymbol{u}_{\mathbf{HIST}}$) and store the result in the variable T_B.

Step 12: Complete the row $u + h$ of the input matrix ($\boldsymbol{U}$) with the transposed elements of the history vector ($\boldsymbol{u}_{\mathbf{HIST}}$) from the element $T_B - M + 1$ to the element T_B. In other words, set $\boldsymbol{U}_{(u+h,:)} \leftarrow \text{transpose}(\boldsymbol{u}_{\mathbf{HIST}}((T_B - M + 1) : T_B))$.

Step 13: Complete the element $T_A + h$ of the history vector ($\boldsymbol{u}_{\mathbf{HIST}}$) with the element h of the auxiliary vector ($\boldsymbol{u}_{\mathbf{AUX}}$). In other words, $\boldsymbol{u}_{\mathbf{HIST}}\ (T_A + h) \leftarrow \boldsymbol{u}_{\mathbf{AUX}}(h)$.

Step 14: Set $h \leftarrow h + 1$ and go to Step 10.

Step 15: Set $u \leftarrow u + H$.

Step 16: Set $d \leftarrow d + 1$ and go to Step 6.

The NN output dataset is represented by the matrix $\boldsymbol{W}$, which is equal to the vector u_{FUT}.

5.4.3 Combining backpropagation and genetic algorithms

Previously, in Chapter 3, we observed that training an NN with a real-coded GA requires extensive computational resources to get a good result. Depending on the problem under study, the heuristic optimization techniques can be outperformed by specialized algorithms (Dorigo et al., 1996), in our case, the BP and Levenberg–Marquardt (LM) algorithms.

Scientists frequently combine different optimization algorithms to enhance global performance. In this chapter, the classical BP algorithm and the GA are combined. Initially, the BP algorithm is applied to find a high-quality solution to the training optimization problem. Then, this solution is included in the initial population of the GA. This approach allows us to take advantage of the starting point offered by the BP algorithm, which can be found in a relatively short time, and the searching power of GA, which needs more computational resources and time. Thus, the optimization efforts performed by GA are focused on searching for a high-quality solution guided by the results obtained from the implementation of the classical BP algorithm (Dorigo & Gambardella, 1997).

The BP algorithm has been implemented by following the procedure explained in Rojas (1996). This is a publically available material, so it will not be repeated in this chapter. Fig. 5.5 shows the structure of the population, where the matrices $\boldsymbol{W}_{\mathbf{hx}}$, $\boldsymbol{b}_{\mathbf{h}}$, $\boldsymbol{W}_{\mathbf{yh}}$, and $\boldsymbol{b}_{\mathbf{y}}$, obtained from the application of the BP algorithm, have been incorporated in the last individual (I). Regarding the remainder of the population, they are supposed to be randomly initialized.

Regarding the training method, we employ the real-coded GA proposed by Chuang et al. (2015, 2016), minimizing the root mean square error (RMSE). (y_{obj} is the objective function obtained from the RMSE, x_{pop} is the population vector of Fig. 5.5 with I elements). This method consists of three stages. The first stage is the ranking selection (RS), the second stage is the direction-based crossover (DBX), and the third stage is the dynamic random mutation (DRM). The procedures are set by means of three

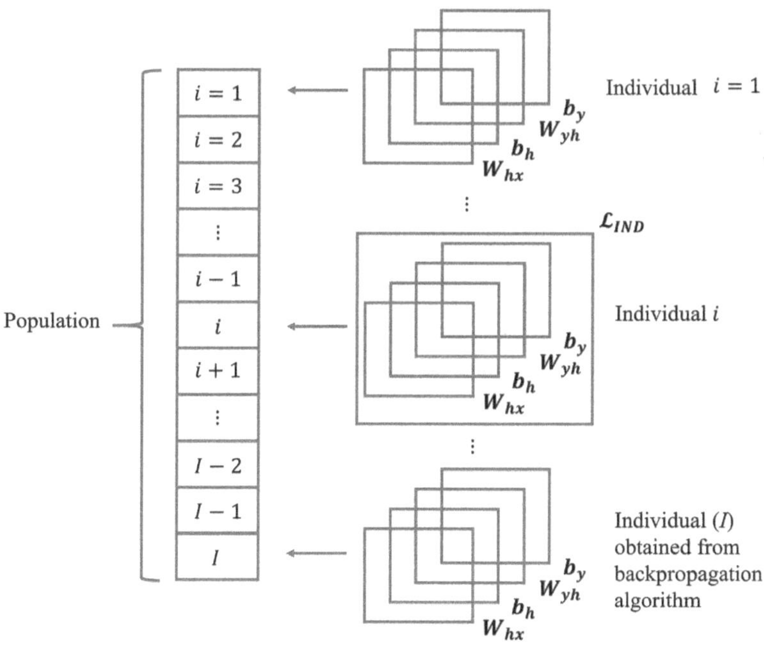

Fig. 5.5 Structure of the population: Creating the population from individuals that are three-dimensional matrices.

parameters adjusted by the decision-maker, the crossover rate (p_c), the mutation rate (p_m), and the proportional parameter (p_r).

During RS, we define a group of individuals with the best performance in terms of fitness values. The size of the group is determined by using the proportional parameter. The RS stage is implemented by following these steps. First, the individuals are sorted according to their fitness. From this process, the order in which individuals should be clustered is determined and stored in a list ($\mathcal{L}_{\mathbf{A}}$). Second, the group of individuals with the worst performance is sized by using the proportional parameter. The size of the worst-performance group would be floor($I \times p_r$). Third, the group of individuals with the best performance is determined. To this end, we select the first I − floor($I \times p_r$) elements of the list $\mathcal{L}_{\mathbf{A}}$. The results are stored in another list ($\mathcal{L}_{\mathbf{B}}$). The individuals that do not belong to this list can be discarded. Fourth, we select the first floor($I \times p_r$) elements of the list $\mathcal{L}_{\mathbf{A}}$ and save them in another list ($\mathcal{L}_{\mathbf{C}}$). The list $\mathcal{L}_{\mathbf{C}}$ contains the individuals used to replace those with the worst performance. Fifth, create a new list ($\mathcal{L}_{\mathbf{D}}$) by concatenating the lists $\mathcal{L}_{\mathbf{B}}$ and $\mathcal{L}_{\mathbf{C}}$ previously built. Finally, sort the individuals in the list $\mathcal{L}_{\mathbf{D}}$ according to their fitness and save the results in another list ($\mathcal{L}_{\mathbf{E}}$), this is the result obtained from the RS procedure.

DBX divides the group of individuals obtained from RS into two parts. The first part is the group of high-ranking individuals, and the second part is those individuals with a lower ranking. During DBX, individuals from both parts are crossed sequentially. Using the list of individuals ($\mathcal{L}_{\mathbf{E}}$) obtained from RS procedure, DBX process is implemented as follows:

Step 1: Set $u \leftarrow$ floor(I/2).

Step 2: Start the sequential crossing of pairs of individuals. Go to Step 2.1.

Step 2.1: Set $n \leftarrow 1$.

Step 2.2: If $n \leq$ floor(I/2), go to Step 2.3, else end.

Step 2.3: Set $u \leftarrow u + 1$.

Step 2.4: Select the individual n of $\mathcal{L}_{\mathbf{E}}\left(\mathcal{L}_{E(n)}\right)$, it will be called $x^a_{\text{pop}}(x^a_{\text{pop}} \leftarrow x_{\text{pop}}\left(\mathcal{L}_{E(n)}\right))$.

Step 2.5: Select the individual u of $\mathcal{L}_{\mathbf{E}}\left(\mathcal{L}_{E(u)}\right)$, it will be called $x^b_{\text{pop}}(x^b_{\text{pop}} \leftarrow x_{\text{pop}}\left(\mathcal{L}_{E(u)}\right))$.

Step 2.6: If $RAND > 1 - p_c$ and $(\max\{y_{\text{obj}}(i)\ \forall\ i = 1,\ldots,I\} - \min\{y_{\text{obj}}\ \forall\ i = 1,\ldots,I\}) > \varepsilon$, calculate the variable S_c using Eq. (5.8). $RAND$ is a uniform random number in the interval [0, 1], and ε is a small number (tolerance).

$$S_c = \frac{\left| y_{\text{obj}}\left(x^a_{\text{pop}}\right) - y_{\text{obj}}\left(x^b_{\text{pop}}\right) \right|}{\max\left\{y_{\text{obj}}(i) \forall i = 1, \ldots, I\right\} - \min\left\{y_{\text{obj}}(i) \forall i = 1, \ldots, I\right\}} \tag{5.8}$$

Step 2.7: For each dimension $g = 1, 2, 3, 4$, calculate the element of the corrective matrix ($\Delta x_{\text{pop}}(g)$) using Eq. (5.9). The corrective matrix is a three-index matrix.

$$\Delta x_{\text{pop}(g)} = \begin{cases} 0; RAND < 0.5 \\ x^a_{\text{pop}(g)} - x^b_{\text{pop}(g)}; RAND \geq 0.5 \end{cases} \tag{5.9}$$

Step 2.8: Calculate the offspring of individuals and using Eq. (5.10).

$$\begin{aligned} x^a_{\text{pop}} &\leftarrow x^a_{\text{pop}} + S_c \Delta x_{\text{pop}} \\ x^b_{\text{pop}} &\leftarrow x^b_{\text{pop}} + S_c \Delta x_{\text{pop}} \end{aligned} \tag{5.10}$$

Step 2.9: Perform DBX in the next group of individuals by setting $n \leftarrow n + 1$ and going to Step 2.2.

DRM enhances the diversity of the individuals in the population. As we are not considering any limit on the values of the matrices $\boldsymbol{W}_{\mathbf{hx}}$, $\boldsymbol{b}_{\mathbf{h}}$, $\boldsymbol{W}_{\mathbf{yh}}$, and $\boldsymbol{b}_{\mathbf{y}}$, the original algorithm has been modified. DRM for NN training can be implemented as follows:

Step 1: Considering the generation under study (k), calculate the variable S_m using Eq. (5.11), where K is the total number of generations ($k = 1, \ldots, K$).

$$S_m = \left(\frac{1-k}{K}\right)^2 \tag{5.11}$$

Step 2: Apply DRM over the population. Go to Step 2.1.

Step 2.1: Set $i \leftarrow 1$.

Step 2.2: If $i \leq I$, go to Step 2.3, else end.

Step 2.3: If $RAND < p_m$, go to Step 2.4.

Step 2.4: Set $x_{\text{pop}}(i) \leftarrow x_{\text{pop}}(i) + S_m RANG$, where $RANG$ is a random number from a Gaussian distribution with mean 0 and standard deviation 1. The rows and columns of $RANG$ are given in Eq. (5.5).

Step 2.5: Set $i \leftarrow i + 1$ and go to Step 2.2.

To retain those promising individuals, the GA implemented is an elitist one. To this end, we first select a position in the population randomly, and then we insert the best individual in this position.

5.4.4 Forecasting with a trained neural network

Once the NN has been trained using the BP-GA methodology previously explained. We can use it to predict the variable of interest several steps ahead. Due to the electricity market structure and other human activities, we have fixed the forecasting horizon to 24 h. The forecasting procedure used in this chapter is described as follows:

Step 1: Calculate the number of elements in the future vector ($\boldsymbol{u}_{\mathbf{FUT}}$) and store the result in the variable T_A.

Step 2: Define the history vector ($\boldsymbol{u}_{\mathbf{HIST}}$) as equal to the elements of the future vector from $(T_A - M + 1)$ to T_A. In other words, set $\boldsymbol{u}_{\mathbf{HIST}} = \boldsymbol{u}_{\mathbf{FUT}}((T_A - M + 1) : T_A)$.

Step 3: Calculate the number of days on the beyond the future dataset ($\boldsymbol{u}_{\mathbf{BYD}}$). To this end, we divide the number of hours of the beyond the future dataset to the number of hours of forecasting horizon ($H = 24$). The result is stored on the variable D.

Step 4: Create an auxiliary matrix ($\boldsymbol{\mathcal{L}}_{\mathbf{AUX}}$) with H rows and D columns.

Step 5: Set $u \leftarrow 0$.

Step 6: Analyze the first day of the forecasting horizon by setting $d \leftarrow 1$.

Step 7: If $d \leq D$, go to Step 7, else go to Step 23.

Step 8: Calculate the number of elements in the history vector ($\boldsymbol{u}_{\mathbf{HIST}}$) and store the result in the variable T_A.

Step 9: Create an auxiliary vector ($\boldsymbol{u}_{\mathbf{AUX}}$). Copy the values of the history vector ($\boldsymbol{u}_{\mathbf{HIST}}$) to the one just created. This means, set $\boldsymbol{u}_{\mathbf{AUX}} \leftarrow \boldsymbol{u}_{\mathbf{HIST}}$.

Step 10: Create an auxiliary vector ($\boldsymbol{v}_{\mathbf{AUX}}$) with H elements.

Step 11: Analyze the first hour of the day d by setting $h \leftarrow 1$.

Step 12: If $h \leq H$, go to Step 13, else go to Step 19.

Step 13: Calculate the number of elements in the history vector ($\boldsymbol{u}_{\mathbf{HIST}}$) and store the result in the variable T_B.

Step 14: Create an auxiliary vector ($\boldsymbol{w}_{\mathbf{AUX}}$) to store the transposed elements of the history vector ($\boldsymbol{u}_{\mathbf{HIST}}$) from the element $T_B - M + 1$ to the element T_B. In other words, set $\boldsymbol{w}_{\mathbf{AUX}} \leftarrow \text{transpose}(\boldsymbol{u}_{\mathbf{HIST}}((T_B - M + 1) : T_B))$.

Step 15: Evaluate the already trained NN using the vector $\boldsymbol{w}_{\mathbf{AUX}}$. Then, store the result in the element $T_A + h$ of the vector $\boldsymbol{u}_{\mathbf{AUX}}(\boldsymbol{u}_{\mathbf{AUX}}(T_A + h))$.

Step 16: Use the auxiliary vector ($\boldsymbol{v}_{\mathbf{AUX}}$). Set the element h of the vector $\boldsymbol{v}_{\mathbf{AUX}}$ to the element $T_A + h$ of the vector $\boldsymbol{u}_{\mathbf{AUX}}$. In other words, set $\boldsymbol{v}_{\mathbf{AUX}}\ (h) \leftarrow \boldsymbol{u}_{\mathbf{AUX}}\ (T_A + h)$.

Step 17: Set $\boldsymbol{u}_{\mathbf{HIST}} \leftarrow \boldsymbol{u}_{\mathbf{AUX}}$.

Step 18: Set $h \leftarrow h + 1$ and go to Step 12.

Step 19: Create an auxiliary vector ($\boldsymbol{z}_{\mathbf{AUX}}$) to store the elements of the beyond the future vector ($\boldsymbol{u}_{\mathbf{BYD}}$) from $(1 + u)$ to $(H + u)$. In other words, set $\boldsymbol{z}_{\mathbf{AUX}} \leftarrow \boldsymbol{u}_{\mathbf{BYD}}\ ((1 + u) : (H + u))$.

Step 20: Complete the elements of the history vector ($\boldsymbol{u}_{\mathbf{HIST}}$) between $T_B - H + 1$ and T_B. In other words, set $\boldsymbol{u}_{\mathbf{HIST}}((T_B - H + 1) : T_B) \leftarrow \boldsymbol{z}_{\mathbf{AUX}}$.

Step 21: Set the column d of the auxiliary matrix $\boldsymbol{\mathcal{L}}_{\mathbf{AUX}}(\boldsymbol{\mathcal{L}}_{\mathbf{AUX}}(:, d))$ to $\boldsymbol{v}_{\mathbf{AUX}}$. In other words, $\boldsymbol{\mathcal{L}}_{\mathbf{AUX}}(:, d) \leftarrow \boldsymbol{v}_{\mathbf{AUX}}$.

Step 22: Set $u \leftarrow u + H$ and $d \leftarrow d + 1$. Then, go to Step 7.

Step 23: Create an auxiliary vector ($\boldsymbol{s}_{\mathbf{AUX}}$) to save the reshaped values of the matrix $\boldsymbol{\mathcal{L}}_{\mathbf{AUX}}$. In other words, express the forecasted values of the matrix $\boldsymbol{\mathcal{L}}_{\mathbf{AUX}}$ as a vector ($\boldsymbol{s}_{\mathbf{AUX}}$) with $H \times D$ elements.

The values of the vector $\boldsymbol{S}_{\mathbf{AUX}}$ and $\boldsymbol{u}_{\mathbf{BYD}}$ correspond to the forecasted and actual values, respectively. The DM can use them to calculate the RMSE index to evaluate the performance of the forecasting system.

5.5 Scenario reduction using a genetic algorithm

Implementation of scenario generation frequently requires a reduction procedure to select only the most promising scenarios. The reduction process could be carried out using a clustering algorithm such as the K-means previously mentioned. However, in this work, we are going to apply an integer-coded GA to solve the clustering problem. A representative example of this approach was developed by Maulik and Bandyopadhyay (2000).

We assume that a large set of scenarios ($s = 1, \ldots, S$) is available, already generated using any of the techniques previously mentioned. Then, our goal is to select a few of them ($v = 1, \ldots, V$), minimizing the information loss. Indeed, this is a mono-objective optimization problem. In this regard, we assume that the number of scenarios is large ($S \rightarrow \infty$), and the number of scenarios, or the number of clusters, in the reduced set is small ($V \to 0$). This assumption is practical and increases the difficulty of the optimization problem.

Integer-coded GA implementation is similar to the binary-coded one explained in Chapter 2. We will center our attention on explaining the calculation of the objective function to this optimization problem. Eqs. (5.12)–(5.16) present the structure of a typical individual ($i = 1, \ldots, I$), including its inherent limitations and objective function. The element $x_{pop}(i, v)$ is an integer number between $x_{pop}^{min}(v) = 1 \forall v = 1, \ldots, V$ and $x_{pop}^{max}(v) = S \forall v = 1, \ldots, V$.

$$\boldsymbol{x}_{\mathbf{pop}}(i) = \left[x_{pop}(i, 1) \ \cdots \ x_{pop}(i, v) \ \cdots \ x_{pop}(i, V)\right] \tag{5.12}$$

$$x_{pop}^{min}(v) \le x_{pop}(i, v) \le x_{pop}^{max}(v) \forall 1 \le v \le V \tag{5.13}$$

$$\boldsymbol{x}_{\mathbf{pop}}^{\mathbf{min}} = \left[x_{pop}^{min}(1) \ \cdots \ x_{pop}^{min}(v) \ \cdots \ x_{pop}^{min}(V)\right] \tag{5.14}$$

$$\boldsymbol{x}_{\mathbf{pop}}^{\mathbf{max}} = \left[x_{pop}^{max}(1) \ \cdots \ x_{pop}^{max}(v) \ \cdots \ x_{pop}^{max}(V)\right] \tag{5.15}$$

$$\boldsymbol{y}_{\mathbf{obj}}\left(\boldsymbol{x}_{\mathbf{pop}}(i)\right) = \boldsymbol{y}_{\mathbf{obj}}(i) = \left[y_{obj}(i, 1)\right] \tag{5.16}$$

Estimating the objective function ($\boldsymbol{y}_{\mathbf{obj}}(i)$) of a determined individual ($i = 1, \ldots, I$) starts by determining the cluster ($v = 1, \ldots, V$) to which each scenario ($s = 1, \ldots, S$) belongs. To this end, create a vector with V elements ($\boldsymbol{\alpha}_{\mathbf{AUX}}$) and another vector with S elements ($\boldsymbol{\beta}_{\mathbf{AUX}}$). Then apply the algorithm explained as follows:

Step 1: Analyze the first scenario by setting $s \leftarrow 1$.

Step 2: If $s \le S$, go to Step 3, else end.

Step 3: Analyze the first cluster by setting $v \leftarrow 1$.

Step 4: If $v \le V$, go to Step 5, else go to Step 9.

Step 5: Identify the cluster v as $\boldsymbol{x}_{\mathbf{cls}} = \boldsymbol{x}_{\mathbf{scn}}\left(:, x_{pop}(i, v)\right)$.

Step 6: If $x_{pop}(i, v) \ne s$, go to Step 7, else set $\boldsymbol{\alpha}_{\mathbf{AUX}}(v) \leftarrow \infty$.

Step 7: $\alpha_{AUX}(v) \leftarrow \sqrt{\sum_s (x_{scn}(:, s) - x_{cls})^2}$ (Euclidean distance).

Step 8: Set $s \leftarrow s + 1$ and go to Step 4.

Step 9: Calculate the index of the minimum value in $\boldsymbol{\alpha}_{\mathbf{AUX}}$. Then, store the result in the element s of the vector $\boldsymbol{\beta}_{\mathbf{AUX}}$.

Step 10: Set $s \leftarrow s + 1$ and go to Step 2.

Now, we need to determine the number of scenarios per cluster. To this end, create a vector ($\boldsymbol{\gamma}_{\mathbf{AUX}}$) with V elements. Then, apply the procedure described as follows:

Step 1: Set $v \leftarrow 1$.

Step 2: If $v \le V$, go so Step 3, else end.

Step 3: Set $s \leftarrow 1$.

Step 4: If $s \le S$, go to Step 5, else go to Step 8.

Step 5: If $v = \boldsymbol{\beta}_{\mathbf{AUX}}(s)$, go to Step 6, else go to Step 7.

Step 6: Set $\boldsymbol{\gamma}_{\mathbf{AUX}}(v) \leftarrow \boldsymbol{\gamma}_{\mathbf{AUX}}(v) + 1$.

Step 7: Set $s \leftarrow s + 1$ and go to Step 4.

Step 8: Set $v \leftarrow v + 1$ and go to Step 3.

Finally, we estimate the objective function of individuals i by applying the algorithm described as follows (van der Merwe & Engelbrecht, 2003):

Step 1: Set $h \leftarrow 0$.
Step 2: Set $v \leftarrow 1$.
Step 3: If $v \leq V$, go to Step 4, else go to Step 13.
Step 4: Identify the cluster v as $\boldsymbol{x}_{\mathbf{cls}} = \boldsymbol{x}_{\mathbf{scn}}\left(:, x_{\text{pop}}(i, v)\right)$.
Step 5: Set $u \leftarrow 0$.
Step 6: Set $s \leftarrow 1$.
Step 7: If $s \leq S$, go to Step 8, else go to Step 11.
Step 8: If $v = \boldsymbol{\beta}_{\mathbf{AUX}}(s)$, go to Step 9, else go to Step 10.
Step 9: Set $u \leftarrow u + \sqrt{\sum_s \left(\boldsymbol{x}_{\mathbf{scn}}(:, s) - \boldsymbol{x}_{\mathbf{cls}}\right)^2}$.
Step 10: Set $s \leftarrow s + 1$ and go to Step 7.
Step 11: Set $h \leftarrow h + {}^{u}/_{\gamma_{\text{AUX}}(v)}$.
Step 12: Set $v \leftarrow v + 1$ and go Step 3.
Step 13: Calculate the objective function of individual as $\boldsymbol{y}_{\mathbf{obj}}(i) \leftarrow h/V$.

Now we have a general perspective of the benefits of adopting a forecasting system and its implementation using an NN, we will illustrate the previous prediction method considering ambient temperature, wind speed, power, and electricity prices. Finally, another case will show the scenario generation and reduction approach aforementioned.

5.6 Case studies

This section illustrates the relationship between operational research and forecasting (Fildes et al., 2008) using as an example the temperature, wind speed and power, and electricity price. In other words, we perform NN training combining the classical BP and GA algorithms. Then, the trained NNs are used to forecast the variables of interest, determining the RMSE index value. In addition to the BP-GA method, the FFNN has been trained using the LM algorithm (Marquardt, 1963) to have a reference point. Thus, we can assess the capabilities of the BP-GA implemented in this work based on another widely accepted method. We have implemented the optimization algorithm in a computer provided with i7-3630QM CPU at 2.40 GHz, 8GB of RAM, and a 64-bit operating system. The computing language used was MATLAB®.

As previously explained in Section 5.4.3, the classical BP algorithm and GA are combined for NN training. Thus, we first train the FFNN using the classical BP, considering a learning rate equal to 0.001 and 5000 epochs. After that, we implement the real-coded GA explained in Section 5.4, considering a proportional parameter equal to 0.3 ($p_r = 0.3$), a crossover rate equal to 0.95 ($p_c = 0.95$), and a mutation rate equal to 0.1 ($p_m = 0.1$). Finally, the population size was set to 50 ($I = 50$), and the number of generations was set to 100. The reference training method was the LM algorithm, considering an NN with a hyperbolic tangent sigmoid transfer function and a linear activation function. The training process was performed over 100 epochs. The remainder characteristics are the same as the previously described NN for the classical BP-GA method.

5.6.1 Daily temperature forecasting

Temperature forecasting is illustrated using an hourly time series measured in Zaragoza, Spain (Agencia Estatal de Meteorología (AEMET), 2021) in 2005. In order to prepare the data for the NN training process, it was categorized as past, future, and beyond the future data following directions explained in Section 5.4.2. Thus, the set of past and present values ($\boldsymbol{u}_{\mathbf{PAST}}$) has 24 elements, starting with the first of January, and the set of future values ($\boldsymbol{u}_{\mathbf{FUT}}$) has 1416 elements. Finally, the set beyond the future ($\boldsymbol{u}_{\mathbf{BYD}}$) used for testing purposes has 744 elements. Concerning the NN structure, it has five hidden units ($P = 5$) and 24 input variables ($M = 24$). In this regard, we use 24 values of the present and past to predict the variable at the next hour. The NN structure is an FFNN with a sigmoid activation function, as shown in Eqs. (5.3) and (5.4).

Fig. 5.6 shows the evolution of the training process for the classical BP and the GA separately. After 5000 epochs of the classical BP algorithm, the resulting RMSE is 0.037411. Then, the GA takes this individual and performs a search, finding a solution

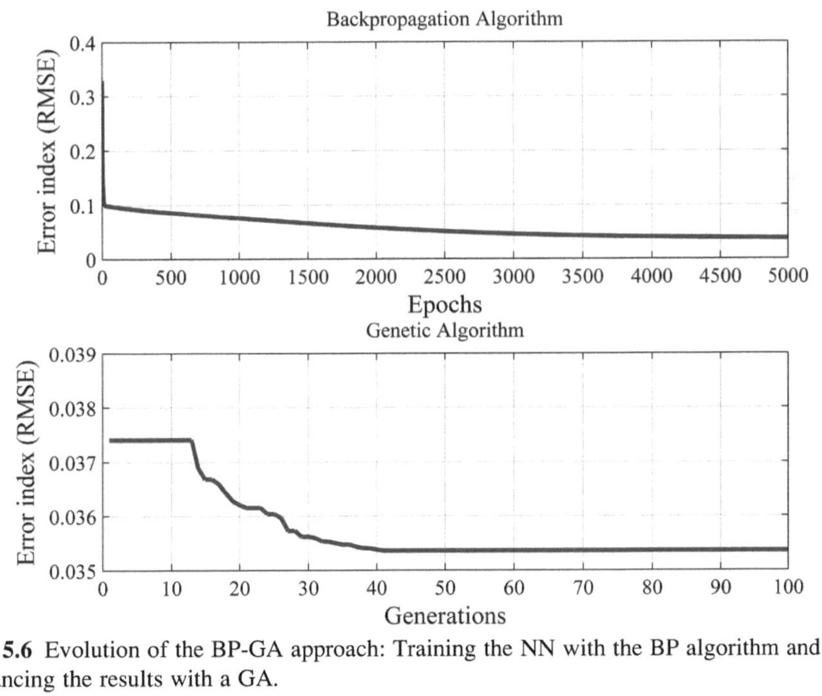

Fig. 5.6 Evolution of the BP-GA approach: Training the NN with the BP algorithm and then enhancing the results with a GA.

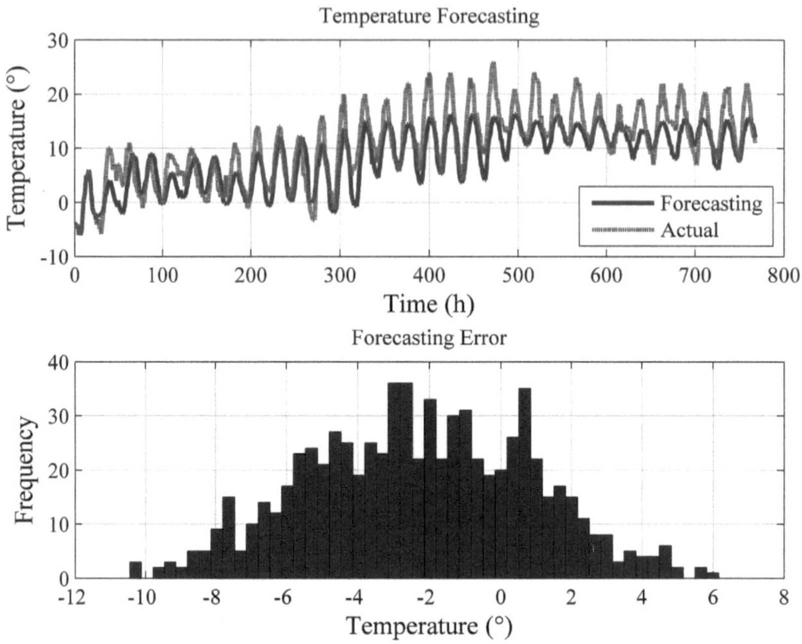

Fig. 5.7 Temperature forecasting and its error: Daily forecasting of the ambient temperature.

with RMSE equal to 0.035358. Thus, the GA improves the results of the classical BP by 5.48%. The computational time required was 10.593 min.

Fig. 5.7 shows the day-ahead temperature forecasting over the testing dataset and the PDF of the error. According to these results, the RMSE is 3.884 degree. Regarding the reference FFNN, trained with the LM algorithm, the RMSE is 3.771 degree and the computational time required was 0.479 min.

5.6.2 Short-term wind speed and power forecasting

Wind speed and power forecasting have been illustrated using hourly measurements of the Valkenburg weather station in the Netherlands (Koninklijk Nederlands Meteorologisch Instituut (KNMI), 2021). This section only considered a forecasting horizon of up to 6 h because this is the range in which the forecasting product is reliable and valuable in practice. A higher horizon requires NWP, a technique beyond this work scope (Giebel et al., 2011). Specifically, the forecasting process was implemented using the algorithm of Section 5.4.2 but considering $H = 1, 2, 3, 4$, and 6.

The entire dataset consists of hourly wind speed measurements of the years 1983–2005. This information was divided into months in order to preserve the seasonal variations. Then, the forecasting procedure was carried out monthly for the range of prediction horizons previously mentioned. For each month, the set $\boldsymbol{u}_{\mathbf{PAST}}$ was built

Table 5.1 Number of elements in the testing dataset (wind speed).

Month	Elements of *uBYD*	Month	Elements of *uBYD*
1	8184	7	8184
2	6528	8	8184
3	8184	9	7632
4	7632	10	8184
5	8184	11	7632
6	7632	12	8184

considering 168 elements, and the set $u_{\mathbf{FUT}}$ was built considering 8760 values. However, the elements of the testing dataset ($u_{\mathbf{BYD}}$) were selected according to Table 5.1.

Regarding the NN structure, it has 15 hidden units ($P = 15$) and 12 inputs ($M = 12$). The remainder of the characteristics and parameters are equal to those previously reported. Once wind speed has been forecasted, the corresponding power was calculated using the typical power curve (Giorsetto & Utsurogi, 1983) of a 1-MW turbine with cut-in, rated, and cut-off speed of 3, 12, and 25 m/s, respectively.

Figs. 5.8–5.19 show the RMSE values of wind speed and power forecasting for each month and the forecasting horizon considered. In the legend, the term BP-GA refers to the combined methodology explained in this chapter, while the term BP refers to the NN with the reference training method (LM algorithm), considering a

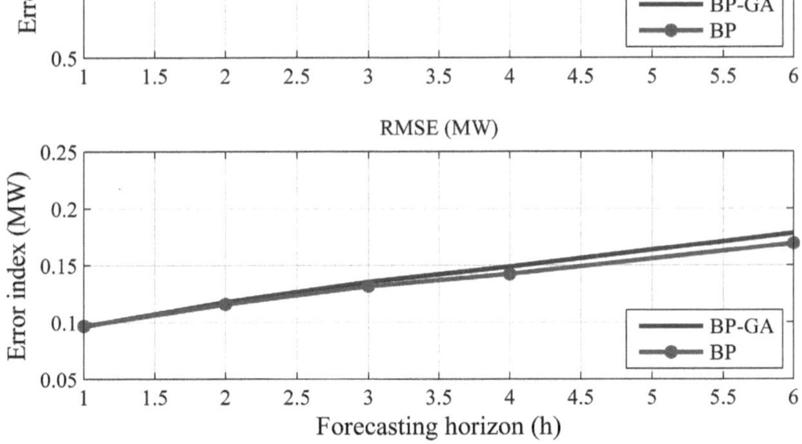

Fig. 5.8 Wind speed and power prediction: Short-term wind speed and power—January.

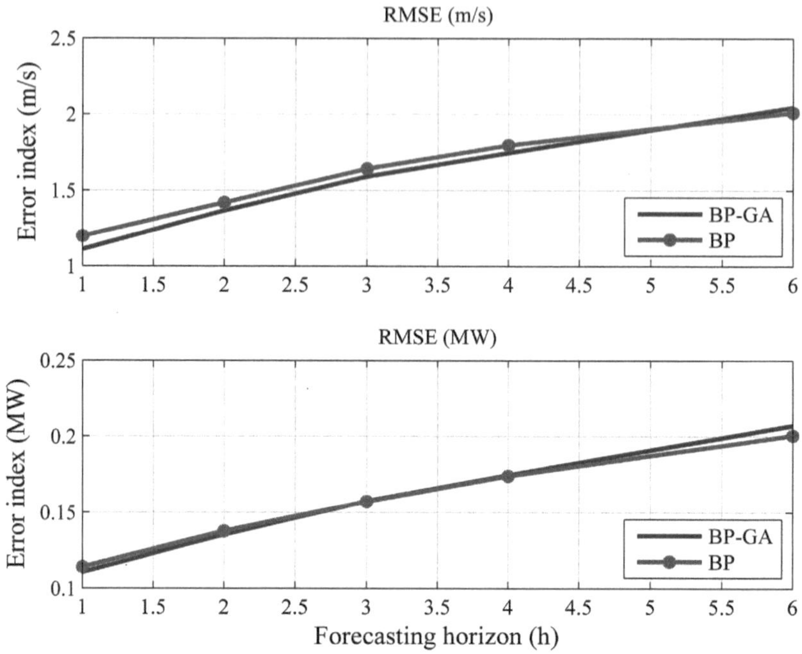

Fig. 5.9 Wind speed and power prediction: Short-term wind speed and power forecasting—February.

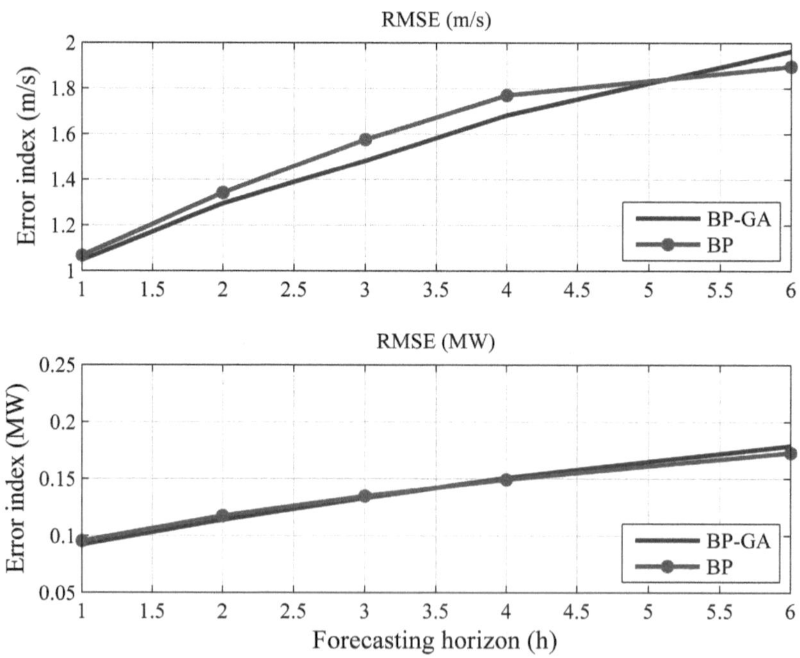

Fig. 5.10 Wind speed and power prediction: Short-term wind speed and power forecasting—March.

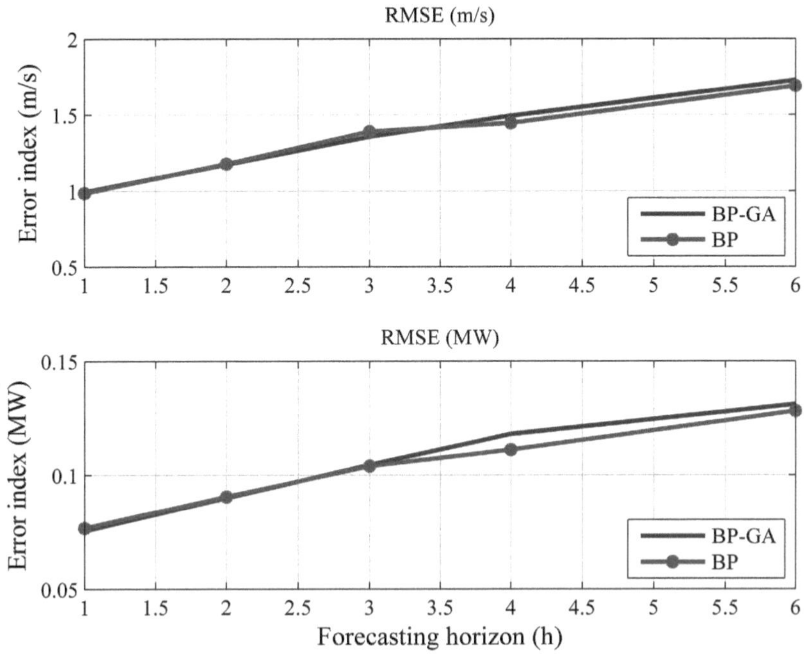

Fig. 5.11 Wind speed and power prediction: Short-term wind speed and power forecasting—April.

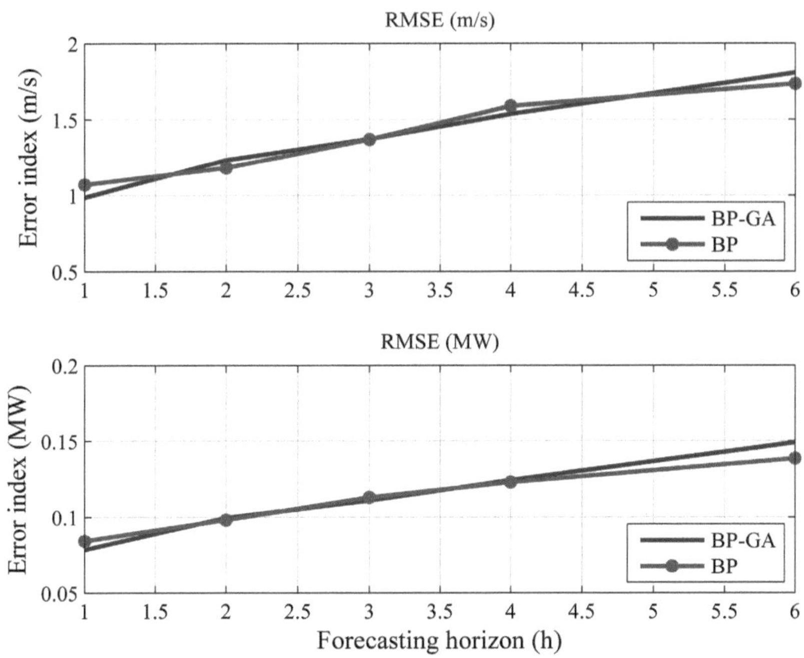

Fig. 5.12 Wind speed and power prediction: Short-term wind speed and power forecasting—May.

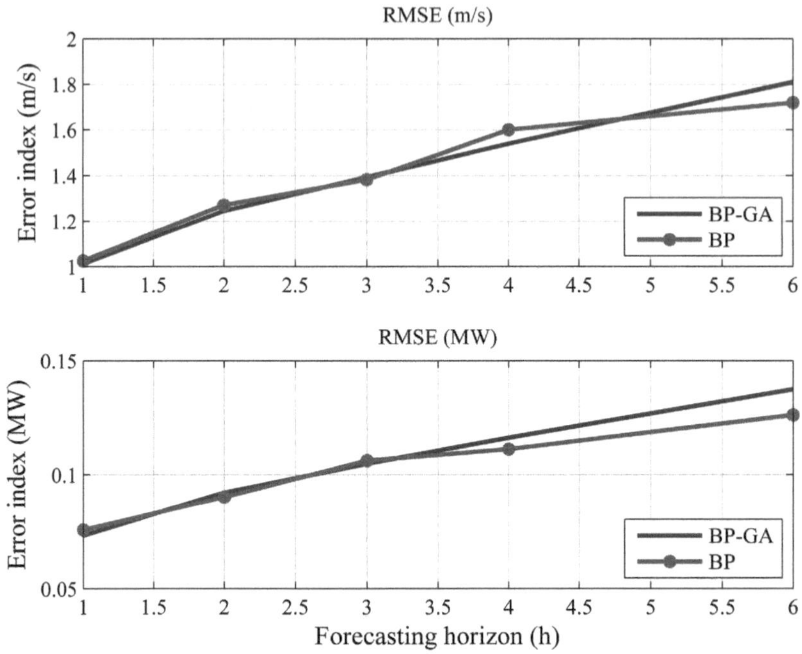

Fig. 5.13 Wind speed and power prediction: Short-term wind speed and power forecasting—June.

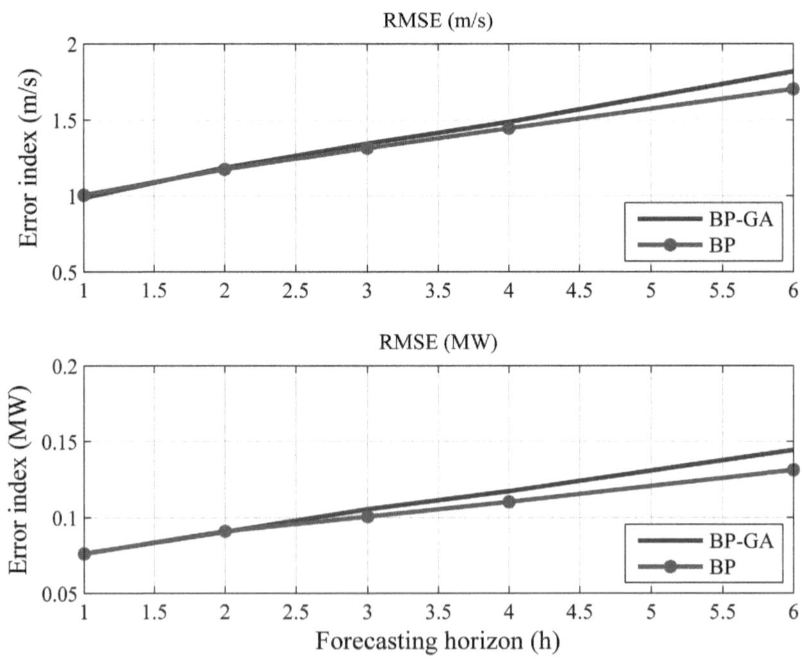

Fig. 5.14 Wind speed and power prediction: Short-term wind speed and power forecasting—July.

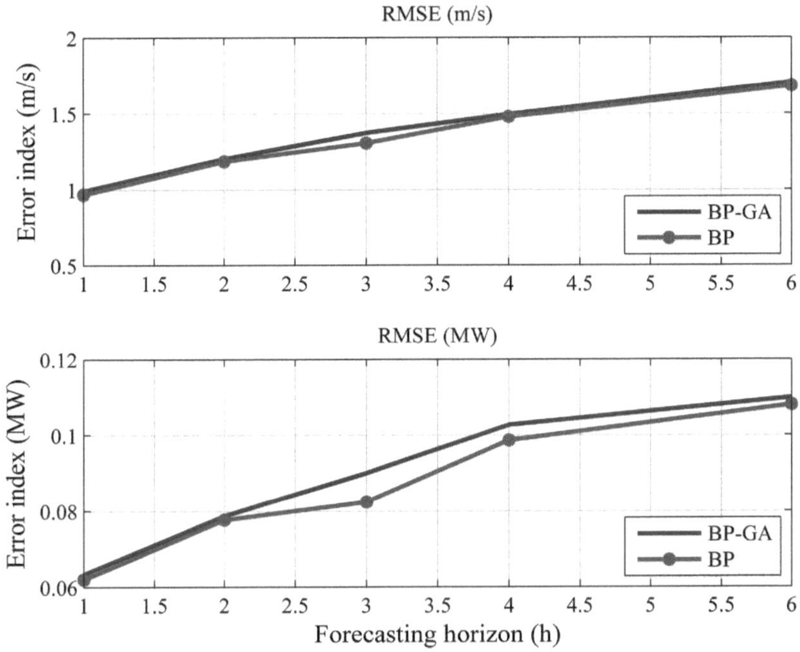

Fig. 5.15 Wind speed and power prediction: Short-term wind speed and power forecasting—August.

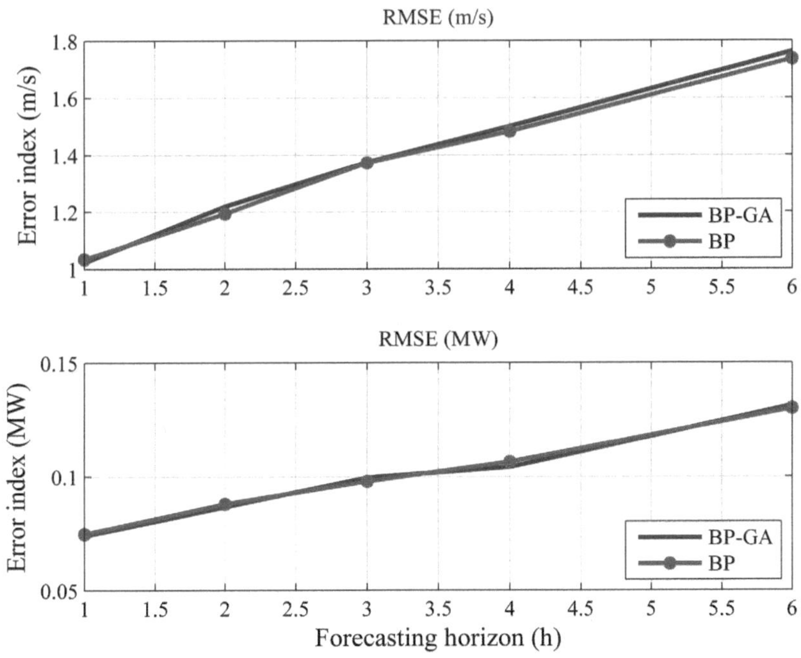

Fig. 5.16 Wind speed and power prediction: Short-term wind speed and power forecasting—September.

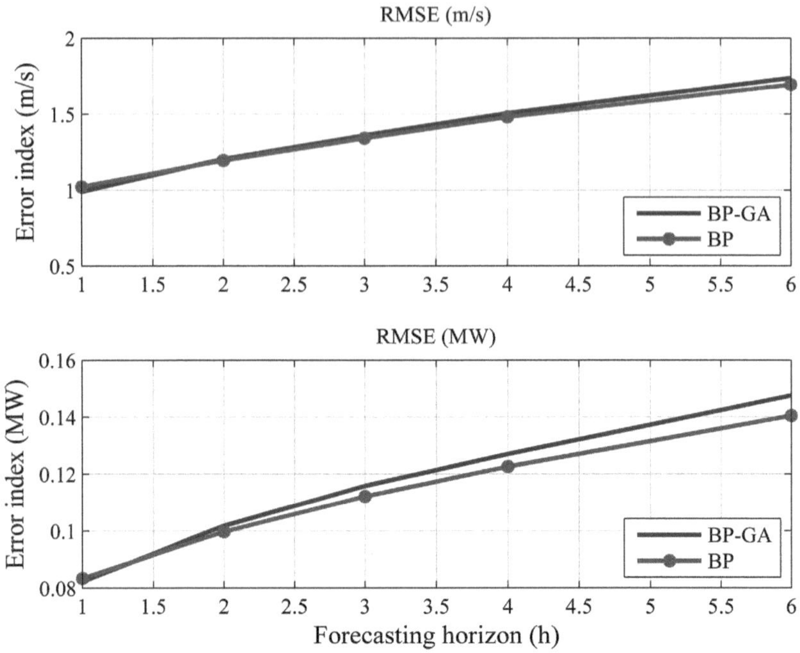

Fig. 5.17 Wind speed and power prediction: Short-term wind speed and power forecasting—October.

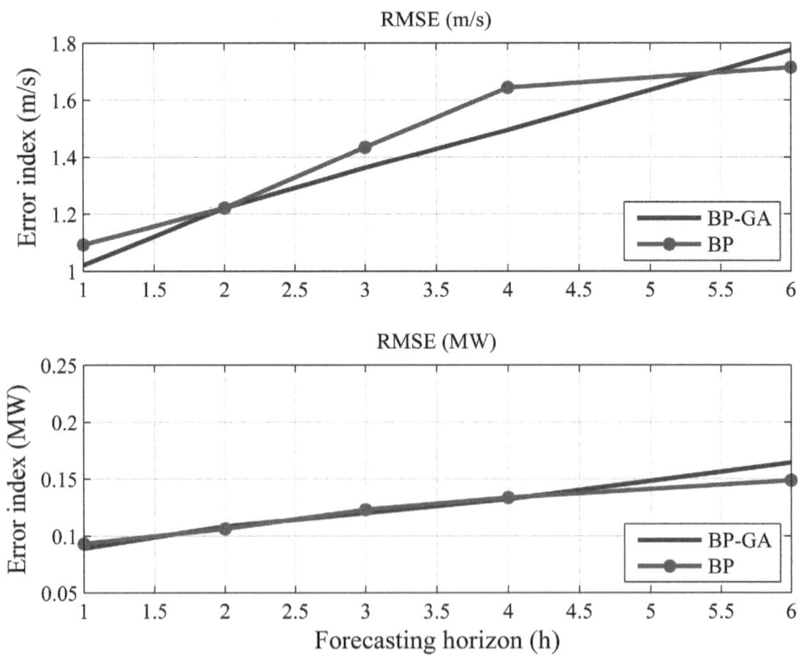

Fig. 5.18 Wind speed and power prediction: Short-term wind speed and power forecasting—November.

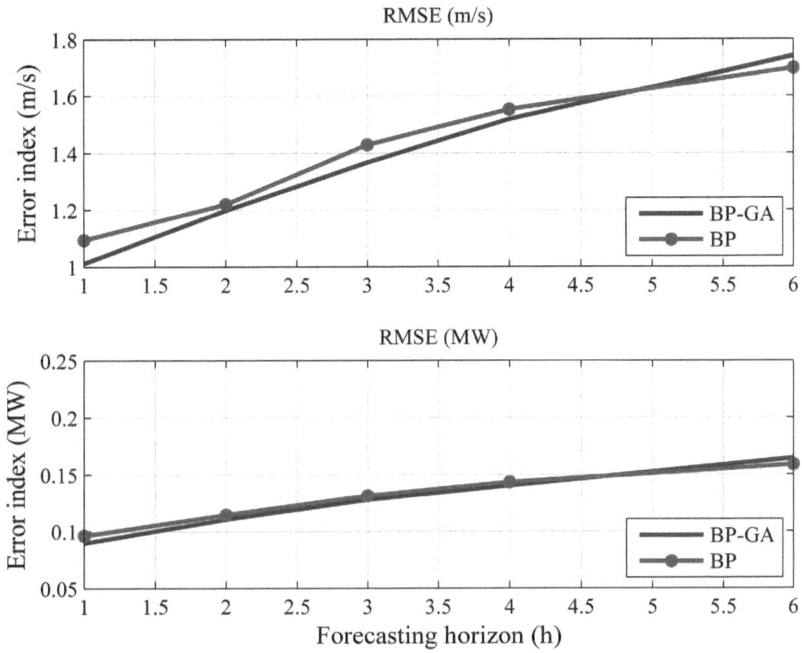

Fig. 5.19 Wind speed and power prediction: Short-term wind speed and power forecasting—December.

hyperbolic tangent sigmoid transfer function and a linear activation function. Regarding the computational time, the combination of classical BP and GA requires around 72 min, while the reference method requires 4.2 min, approximately.

5.6.3 Daily electricity price forecasting

Day-ahead electricity price forecasting is illustrated using information from the Spanish market (Operador del Mercado Ibérico de Energía (OMIE), 2021) between 2010 and 2018. The hourly data was organized on a monthly basis so that the seasonal characteristics are preserved. The number of elements in the set $\boldsymbol{u}_{\mathbf{PAST}}$ is 168, and the number of elements in the sets $\boldsymbol{u}_{\mathbf{FUT}}$ and $\boldsymbol{u}_{\mathbf{BYD}}$ are reported in Table 5.2.

The NN structure has five hidden units ($P = 5$), and it takes 24 h of the time series to perform prediction, which means $M = 24$. The settings for the combined BP-GA model and the reference NN structure trained with the LM algorithm are as previously reported. Fig. 5.20 shows the RMSE index for both techniques. The GA initialized with the classical BP algorithm is shown in the legend as BP-GA, while the NN trained by the LM algorithm is shown in the legend as BP.

Table 5.2 Number of elements in the testing dataset (electricity price).

Month	Elements of *uFUT*	Elements of *uBYD*
1	5784	744
2	5256	672
3	5784	744
4	5592	720
5	5784	744
6	5592	720
7	5784	744
8	5784	744
9	5592	720
10	5784	744
11	5592	720
12	5784	744

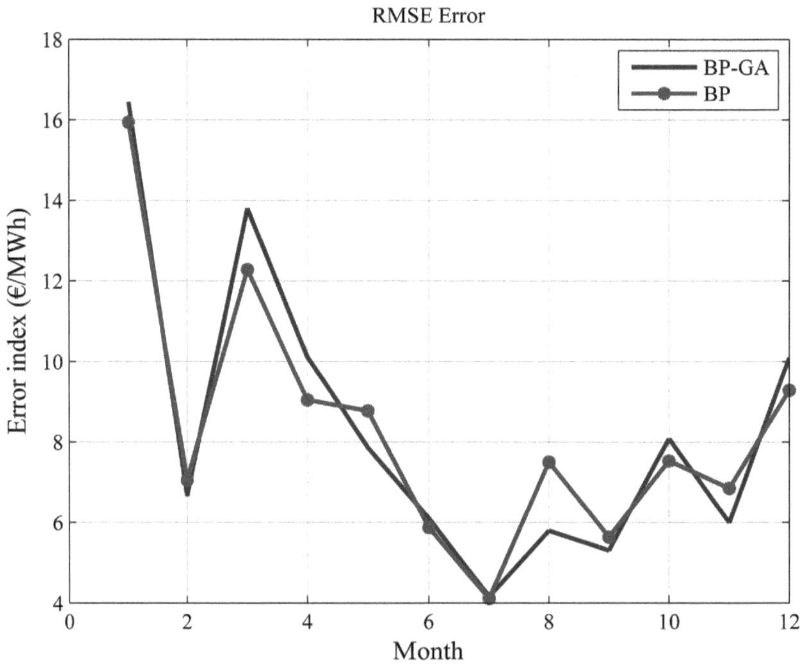

Fig. 5.20 Electricity price prediction over a year: Electricity price forecasting error.

Regarding the computational time, implementation of the classical BP-GA requires 44.8 min, approximately, while the reference NN training approach takes 0.6 min. According to the observed results for all the variables involved (temperature, wind speed and power, and electricity prices), we can verify that the forecasting errors obtained using the combined BP-GA approach are comparable to those obtained from the LM algorithm. However, the computational time required by the BP-GA is much higher than that required by the specialized algorithm.

5.6.4 Scenario generation and reduction

Solving the optimization problem of data clustering using an integer-coded GA is illustrated in this section. The wind power scenarios to be clustered are assumed to be already available. In this case study, we have considered those shown in Fig. 5.21 (right) in gray. They are 2000 scenarios, which follow the trend of the profile shown in Fig. 5.21 (left). We need to reduce the set of 2000 scenarios to only 50 to perform a stochastic optimization in a further study. To this end, we implement the algorithm described in Section 5.5 with $S = 2000$ and $V = 50$. The optimization process was carried out considering a population size of 50 individuals, 100 generations, a crossover rate of 0.95, and a mutation rate of 0.1. The results of the data clustering are shown in Fig. 5.21 (right) in black, while the evolution of the GA is shown in Fig. 5.22. The computational time required was 33.8 min.

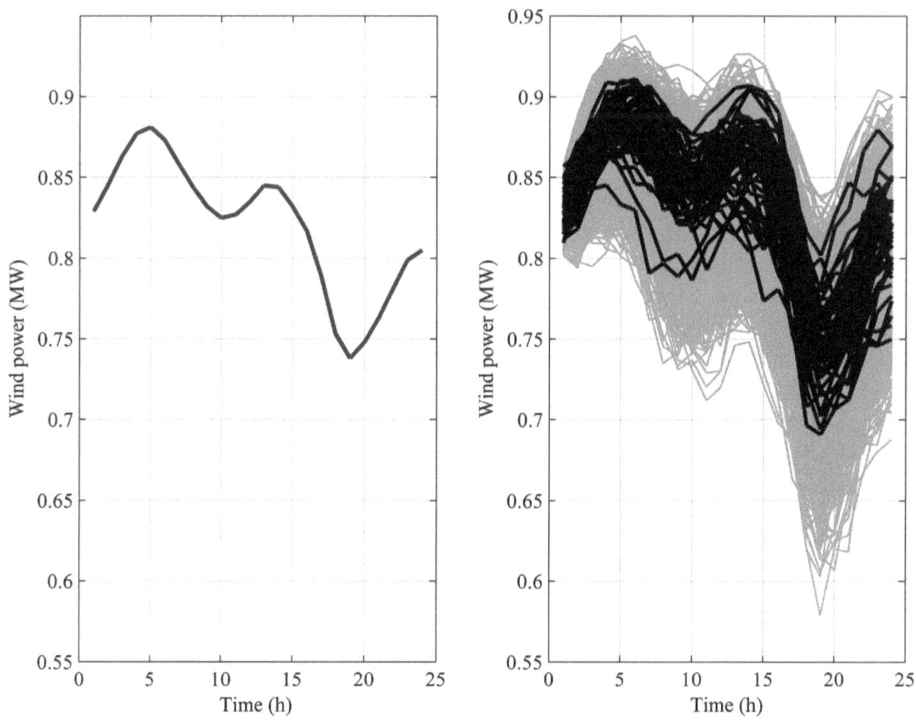

Fig. 5.21 Wind power scenario reduction: Original wind-power profile (*left*) and the associated scenarios (*right*).

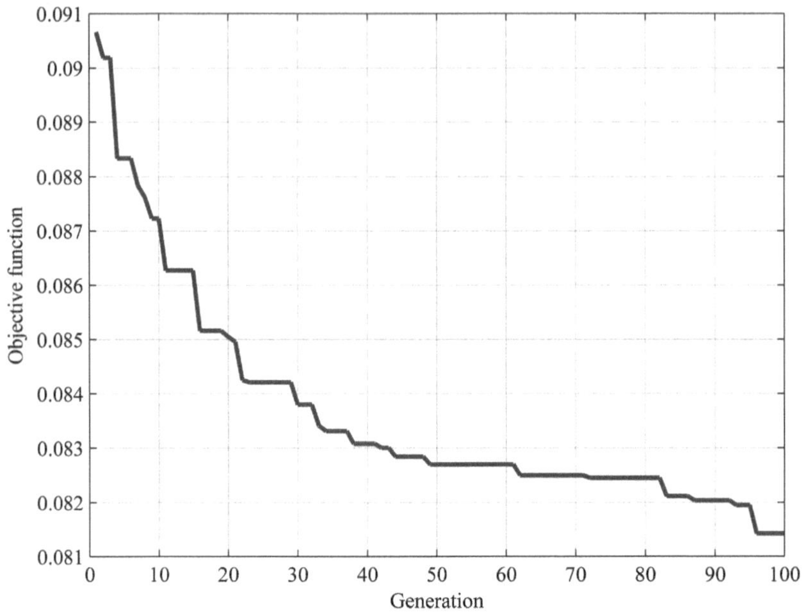

Fig. 5.22 Progress of scenario reduction of wind power scenarios by a GA: Evolution of GA implemented for scenario reduction.

5.7 Conclusions

This chapter presented the preliminary concepts related to the standard prediction methods frequently applied in power systems operation. A comprehensive discussion of the benefits associated with an accurate forecasting system was clearly laid out. Then, commonly used techniques for wind and solar energies, ambient temperature, electricity prices, and load demand were described, including TSA, NWP, and AI. In addition to these topics, the representation of forecasting error using PDFs and scenario generation and reduction procedures were also explained. Thus, the reader should have gained a broad idea of the daily management of a typical energy system. Several case studies for temperature, wind speed and power, and electricity prices were presented for illustrative purposes. Specifically, the NN training algorithm explained in Chapter 3 was initialized using the results of the classical BP algorithm. Thus, the GA's computational effort and searching capabilities are used to improve the quality of an available solution.

The results were compared to those obtained from the LM algorithm, observing similar results. However, the combined BP-GA method has a high computational burden, which requires substantial processing time. This problem can be solved by implementing the BP-GA model in a GPU.

The problem of data clustering was also analyzed in this chapter. An integer-coded GA was implemented to select a small and representative group of scenarios from a large set. The clustering technique was centered on the scenario reduction methodology of wind power scenarios.

References

Agencia Estatal de Meteorología (AEMET). (2021). http://www.aemet.es/en/portada.

Albadi, M. H., & El-Saadany, E. F. (2010). Overview of wind power intermittency impacts on power systems. *Electric Power Systems Research*, *80*, 627–632. https://doi.org/10.1016/j.epsr.2009.10.035.

Baker, K., Hug, G., & Li, X. (2017). Energy storage sizing taking into account forecast uncertainties and receding horizon operation. *IEEE Transactions on Sustainable Energy*, *8*(1), 331–340. https://doi.org/10.1109/TSTE.2016.2599074.

Barbounis, T. G., Theocharis, J. B., Alexiadis, M. C., & Dokopoulos, P. S. (2006). Long-term wind speed and power forecasting using local recurrent network models. *IEEE Transactions on Energy Conversion*, *21*(1), 273–284. https://doi.org/10.1109/TEC.2005.847954.

Barthelmie, R. J., Murray, F., & Pryor, S. C. (2008). The economic benefit of short-term forecasting for wind energy in the UK electricity market. *Energy Policy*, *36*(5), 1687–1696. https://doi.org/10.1016/j.enpol.2008.01.027.

Becker, R. (2018). Generation of time-coupled wind power infeed scenarios using pair-copula construction. *IEEE Transactions on Sustainable Energy*, *9*(3), 1298–1306. https://doi.org/10.1109/TSTE.2017.2782089.

Bludszuweit, H., Domínguez-Navarro, J. A., & Llombart, A. (2008). Statistical analysis of wind power forecast error. *IEEE Transactions on Power Systems*, *23*(3), 983–991. https://doi.org/10.1109/TPWRS.2008.922526.

Boukabara, S. A., Krasnopolsky, V., Stewart, J. Q., Maddy, E. S., Shahroudi, N., & Hoffman, R. N. (2019). Leveraging modern artificial intelligence for remote sensing and NWP: Benefits and challenges. *Bulletin of the American Meteorological Society*, *100*(12), ES473–ES491. https://doi.org/10.1175/BAMS-D-18-0324.1.

Brown, B. G., Katz, R. W., & Murphy, A. H. (1984). Time series models to simulate and forecast wind speed and wind power. *Journal of Climate and Applied Meteorology*, *23*(8), 1184–1195. https://doi.org/10.1175/1520-0450(1984)023<1184:TSMTSA>2.0.CO;2.

Bruninx, K., & Delarue, E. (2014). A statistical description of the error on wind power forecasts for probabilistic reserve sizing. *IEEE Transactions on Sustainable Energy*, *5*(3), 995–1002. https://doi.org/10.1109/TSTE.2014.2320193.

Buhan, S., Özkazanç, Y., & Çadırcı, I. (2016). Wind pattern recognition and reference wind mast data correlations with NWP for improved wind-electric power forecasts. *IEEE Transactions on Industrial Informatics*, *12*(3), 991–1004. https://doi.org/10.1109/TII.2016.2543004.

Cadenas, E., & Rivera, W. (2007). Wind speed forecasting in the south coast of Oaxaca, Mexico. *Renewable Energy*, *32*, 2116–2128. https://doi.org/10.1016/j.renene.2006.10.005.

Cadenas, E., & Rivera, W. (2009). Short term wind speed forecasting in La Venta, Oaxaca, Mexico, using artificial neural networks. *Renewable Energy*, *34*, 274–278. https://doi.org/10.1016/j.renene.2008.03.014.

Cadenas, E., & Rivera, W. (2010). Wind speed forecasting in three different regions of Mexico, using a hybrid ARIMA-ANN model. *Renewable Energy*, *35*, 2732–2738. https://doi.org/10.1016/j.renene.2010.04.022.

Capizzi, G., Napoli, C., & Bonano, F. (2012). Innovative second-generation wavelets construction with recurrent neural networks for solar radiation forecasting. *IEEE Transactions on Neural Networks and Learning Systems*, *23*(11), 1805–1815. https://doi.org/10.1109/TNNLS.2012.2216546.

Chen, N., Qian, Z., Nabney, I. T., & Meng, X. (2014). Wind power forecasts using Gaussian processes and numerical weather prediction. *IEEE Transactions on Power Systems*, *29*(2), 656–665. https://doi.org/10.1109/TPWRS.2013.2282366.

Chen, Y., Wang, Y., Kirschen, D., & Zhang, B. (2018). Model-free renewable scenario generation using generative adversarial networks. *IEEE Transactions on Power Systems*, *33*(3), 3265–3275. https://doi.org/10.1109/PESGM40551.2019.8974096.

Chuang, Y. C., Chen, C. T., & Hwang, C. (2015). A real-coded genetic algorithm with a direction-based crossover operator. *Information Sciences*, *305*, 320–348. https://doi.org/10.1016/j.ins.2015.01.026.

Chuang, Y. C., Chen, C. T., & Hwang, C. (2016). A simple and efficient real-coded genetic algorithm for constrained optimization. *Applied Soft Computing*, *38*, 87–105. https://doi.org/10.1016/j.asoc.2015.09.036.

Cifuentes, J., Marulanda, G., Bello, A., & Reneses, J. (2020). Air temperature forecasting using machine learning techniques: A review. *Energies*, *13*(6), 1–28. https://doi.org/10.3390/en13164215.

David, M., Ramahatana, F., Trombe, P. J., & Lauret, P. (2016). Probabilistic forecasting of the solar irradiance with recursive ARMA and GARCH models. *Solar Energy*, *133*, 55–72. https://doi.org/10.1016/j.solener.2016.03.064.

Deo, R. C., & Şahin, M. (2017). Forecasting long-term global solar radiation with an ANN algorithm coupled with satellite-derived (MODIS) land surface temperature (LST) for regional locations in Queensland. *Renewable and Sustainable Energy Reviews*, *72*, 828–848. https://doi.org/10.1016/j.rser.2017.01.114.

Dorigo, M., & Gambardella, L. M. (1997). Ant colonies for the travelling salesman problem. *Biosystems*, *43*(2), 73–81. https://doi.org/10.1016/S0303-2647(97)01708-5.

Dorigo, M., Maniezzo, V., & Colorni, A. (1996). Ant system: Optimization by a colony of cooperating agents. *IEEE Transactions on Systems, Man, and Cybernetics-Part B: Cybernetics*, *26*(1), 29–41. https://doi.org/10.1109/3477.484436.

Erdem, E., & Shi, J. (2011). ARMA based approaches for forecasting the tuple of wind speed and direction. *Applied Energy*, *88*(4), 1405–1414. https://doi.org/10.1016/j.apenergy.2010.10.031.

Fildes, R., & Makridakis, S. (1995). The impact of empirical accuracy studies on time-series analysis and forecasting. *International Statistical Review*, *63*(3), 289–308. https://doi.org/10.2307/1403481.

Fildes, R., Nikolopoulos, K., Crone, S. F., & Syntetos, A. A. (2008). Forecasting and operational research: A review. *Journal of the Operational Research Society*, *59*(9), 1150–1172. https://doi.org/10.1057/palgrave.jors.2602597.

Flores, J. J., Graff, M., & Rodriguez, H. (2012). Evolutive design of ARMA and ANN models for time series forecasting. *Renewable Energy*, *44*, 225–230. https://doi.org/10.1016/j.renene.2012.01.084.

Foley, A. M., Leahy, P. G., Marvuglia, A., & McKeogh, E. J. (2012). Current methods and advances in forecasting of wind power generation. *Renewable Energy*, *37*, 1–8. https://doi.org/10.1016/j.renene.2011.05.033.

Giebel, G., Brownsword, R., Kariniotakis, G., Denhard, M., & Draxl, C. (2011). *The state-of-the-art in short-term prediction of wind power. A literature overview*. ANEMOS. https://orbit.dtu.dk/en/publications/the-state-of-the-art-in-short-term-prediction-of-wind-power-a-lit.

Giorgi, M. G. D., Ficarella, A., & Tarantino, M. (2011). Error analysis of short term wind power prediction models. *Applied Energy*, *88*, 1298–1311. https://doi.org/10.1016/j.apenergy.2010.10.035.

Giorsetto, P., & Utsurogi, K. F. (1983). Development of a new procedure for reliability modelling of wind turbine generators. *IEEE Transactions on Power Apparatus and Systems*, *PAS-102*(1), 134–143. https://doi.org/10.1109/TPAS.1983.318006.

Gonzalez-Aparicio, I., & Zucker, A. (2015). Impact of wind power uncertainty forecasting on the market integration of wind energy in Spain. *Applied Energy*, *159*, 334–349. https://doi.org/10.1016/j.apenergy.2015.08.104.

Gutierrez-Martin, F., Da Silva-Alvarez, R. A., & Montoro-Pintado, P. (2013). Effects of wind intermittency on reduction of CO2 emissions: The case of the Spanish power system. *Energy*, *61*, 108–117. https://doi.org/10.1016/j.energy.2013.01.057.

Hansen, J. W. (2002). Realizing the potential benefits of climate prediction to agriculture: Issues, approaches, challenges. *Agricultural Systems*, *74*(3), 309–330. https://doi.org/10.1016/S0308-521X(02)00043-4.

Hochreiter, S., & Schmidhuber, J. (1997). Long short-term memory. *Neural Computation*, *9*(8), 1735–1780. https://doi.org/10.1162/neco.1997.9.8.1735.

Hodge, B. M., & Milligan, M. (2011). Wind power forecasting error distributions over multiple timescales. In *2011 IEEE power and energy society general meeting*. Detroit, MI, USA: IEEE. https://doi.org/10.1109/PES.2011.6039388.

Huang, C. J., & Kuo, P. H. (2019). Multiple-input deep convolutional neural network model for short-term photovoltaic power forecasting. *IEEE Access*, *7*, 74822–74834. https://doi.org/10.1109/ACCESS.2019.2921238.

Huang, X., Li, Q., Tai, Y., Chen, Z., Zhang, J., Shi, J., Gao, B., & Liu, W. (2021). Hybrid deep neural model for hourly solar irradiance forecasting. *Renewable Energy*, *171*, 1041–1060. https://doi.org/10.1016/j.renene.2021.02.161.

Jasiński, T. (2021). Use of new variables based on air temperature for forecasting day-ahead spot electricity prices using deep neural networks: A new approach. *Energy*, *213*. https://doi.org/10.1016/j.energy.2020.118784, 118784.

Jeong, S., Park, I., Kim, H. S., Song, C. H., & Kim, H. K. (2021). Temperature prediction based on bidirectional long short-term memory and convolutional neural network combining observed and numerical forecast data. *Sensors*, *21*(3), 1–20. https://doi.org/10.3390/s21030941.

Ji, W., & Chee, K. C. (2011). Prediction of hourly solar radiation using a novel hybrid model of ARMA and TDNN. *Solar Energy*, *85*(5), 808–817. https://doi.org/10.1016/j.solener.2011.01.013.

Jones, J. W., Hansen, J. W., Royce, F. S., & Messina, C. D. (2000). Potential benefits of climate forecasting to agriculture. *Agriculture Ecosystems and Environment*, *82*(1–3), 169–184. https://doi.org/10.1016/S0308-521X(02)00043-4.

Jonghe, C. D., Hobbs, B. F., & Belmans, R. (2014). Value of price responsive load for wind integration in unit commitment. *IEEE Transactions on Power Systems*, *29*(2), 675–685. https://doi.org/10.1109/TPWRS.2013.2283516.

Jonsson, T., Pinson, P., & Madsen, H. (2010). On the market impact of wind energy forecasts. *Energy Economics*, *32*, 313–320. https://doi.org/10.1016/j.eneco.2009.10.018.

Kariniotakis, G. N., Stavrakakis, G. S., & Nogaret, E. F. (1996). Wind power forecasting using advanced neural networks models. *IEEE Transactions on Energy Conversion*, *11*(4), 762–767. https://doi.org/10.1109/60.556376.

Katzenstein, W., & Apt, J. (2012). The cost of wind power variability. *Energy Policy*, *51*, 233–243. https://doi.org/10.1016/j.enpol.2012.07.032.

Khalid, M., & Savkin, A. V. (2012). A method for short-term wind power prediction with multiple observation points. *IEEE Transactions on Power Systems*, *27*(2), 579–586. https://doi.org/10.1109/TPWRS.2011.2160295.

Khodayar, M., Kaynak, O., & Khodayar, M. E. (2017). Rough deep neural architecture for short-term wind speed forecasting. *IEEE Transactions on Industrial Informatics*, *13*(6), 2770–2779. https://doi.org/10.1109/TII.2017.2730846.

Koninklijk Nederlands Meteorologisch Instituut (KNMI). (2021). https://www.knmi.nl/over-het-knmi/about.

Lai, C. S., Zhong, C., Pan, K., Ng, W. W. Y., & Lai, L. L. (2021). A deep learning based hybrid method for hourly solar radiation forecasting. *Expert Systems with Applications, 177*. https://doi.org/10.1016/j.eswa.2021.114941, 114941.

Law, E. W., Prasad, A. A., Kay, M., & Taylor, R. A. (2014). Direct normal irradiance forecasting and its application to concentrated solar thermal output forecasting—A review. *Solar Energy*, *108*, 287–307. https://doi.org/10.1016/j.solener.2014.07.008.

Lerner, J., Grundmeyer, M., & Garvert, M. (2009). The importance of wind forecasting. *Renewable Energy Focus*, *10*(2), 64–66. https://doi.org/10.1016/S1755-0084(09)70092-4.

Lew, D., Milligan, M., Jordan, G., & Piwko, R. (2011). The value of wind power forecasting. *National Renewable Energy Laboratory*.. https://www.nrel.gov/docs/fy11osti/50814.pdf.

Lingras, P. (1998). Comparison of neofuzzy and rough neural networks. *Information Sciences*, *110*, 207–215. https://doi.org/10.1016/S0020-0255(97)10045-7.

Liu, J., Fang, W., Zhang, X., & Yang, C. (2015). An improved photovoltaic power forecasting model with the assistance of aerosol index data. *IEEE Transactions on Sustainable Energy*, *6*(2), 434–442. https://doi.org/10.1109/TSTE.2014.2381224.

Lobo, M. G., & Sanchez, I. (2012). Regional wind power forecasting based on smoothing techniques, with application to the Spanish peninsular system. *IEEE Transactions on Power Systems*, *27*(4), 1990–1997. https://doi.org/10.1109/TPWRS.2012.2189418.

Luickx, P. J., Delarue, E. D., & D'haeseleer, W. D. (2010). Impact of large amounts of wind power on the operation of an electricity generation system: Belgian case study. *Renewable and Sustainable Energy Reviews*, *14*, 2019–2028. https://doi.org/10.1016/j.rser.2010.03.018.

Ma, X. Y., Sun, Y. Z., & Fang, H. L. (2013). Scenario generation of wind power based on statistical uncertainty and variability. *IEEE Transactions on Sustainable Energy*, *4*(4), 894–904. https://doi.org/10.1109/TSTE.2013.2256807.

Marchuk, G. (1974). *Numerical methods in weather prediction* (1st ed.). Academic Press. https://doi.org/10.1016/B978-0-12-470650-7.X5001-4.

Marquardt, D. W. (1963). An algorithm for least-squares estimation of nonlinear parameters. *Journal of the Society for Industrial and Applied Mathematics*, *11*(2), 431–441. https://www.jstor.org/stable/2098941.

Maulik, U., & Bandyopadhyay, S. (2000). Genetic algorithm-based clustering technique. *Pattern Recognition*, *33*, 1455–1465. https://doi.org/10.1016/S0031-3203(99)00137-5.

Meinke, H., & Stone, R. (2005). Seasonal and inter-annual climate forecasting: The new tool for increasing preparedness to climate variability and change in agricultural planning and operations. *Climatic Change*, *70*(1–2), 221–253. https://doi.org/10.1007/1-4020-4166-7_11.

Mestekemper, T., Kauermann, G., & Smith, M. S. (2013). A comparison of periodic autoregressive and dynamic factor models in intraday energy demand forecasting. *International Journal of Forecasting*, *29*(1), 1–12. https://doi.org/10.1016/j.ijforecast.2012.03.003.

Nuño, E., Koivisto, M., Cutululis, N. A., & Sørensen, P. (2018). On the simulation of aggregated solar PV forecast errors. *IEEE Transactions on Sustainable Energy*, *9*(4), 1889–1898. https://doi.org/10.1109/TSTE.2018.2818727.

Operador del Mercado Ibérico de Energía (OMIE). (2021). https://www.omie.es/en.

Orwig, K., Clark, C., Cline, J., Benjamin, S., Wilczak, J., Marquis, M., Finley, C., Stern, A., & Freedman, J. (2012). Enhanced short-term wind power forecasting and value to grid operations. *National Renewable Energy Laboratory*. https://www.nrel.gov/docs/fy12osti/56258.pdf.

Philbrick, C. R. (2011). Wind integration and the need for advanced decision support tools. In *2011 IEEE power and energy society general meeting*. Detroit, MI, USA: IEEE. https://doi.org/10.1109/PES.2011.6039685.

Pinson, P. (2013). Wind energy: Forecasting challenges for its operational management. *Statistical Science*, *28*(4), 564–585. https://www.jstor.org/stable/43288436.

Richardson, D. S. (2000). Skill and relative economic value of the ECMWF ensemble prediction system. *Quarterly Journal of the Royal Meteorological Society*, *126*(563), 649–667. https://doi.org/10.1002/qj.49712656313.

Rojas, R. (1996). The backpropagation algorithm. In *Neural networks: A systematic introduction* (pp. 151–184). Springer-Verlag. https://page.mi.fu-berlin.de/rojas/neural/chapter/K7.pdf.

Sahin, C., Shahidehpour, M., & Erkmen, I. (2013). Allocation of hourly reserve versus demand response for security-constrained scheduling of stochastic wind energy. *IEEE Transactions on Sustainable Energy*, *4*(1), 219–228. https://doi.org/10.1109/TSTE.2012.2213849.

Shahid, F., Zameer, A., & Muneeb, M. (2021). A novel genetic LSTM model for wind power forecast. *Energy*, *223*. https://doi.org/10.1016/j.energy.2021.120069, 120069.

Tewari, S., Geyer, C. J., & Mohan, N. (2011). A statistical model for wind power forecast error and its application to the estimation of penalties in liberalized markets. *IEEE Transactions on Power Systems*, *26*(4), 2031–2039. https://doi.org/10.1109/TPWRS.2011.2141159.

Torres, J. L., Garcia, A., De Blas, M., & De Francisco, A. (2005). Forecast of hourly average wind speed with ARMA models in Navarre (Spain). *Solar Energy*, *79*(1), 65–77. https://doi.org/10.1016/j.solener.2004.09.013.

Valenzuela, J., Mazumdar, M., & Kapoor, A. (2000). Influence of temperature and load forecast uncertainty on estimates of power generation production costs. *IEEE Transactions on Power Systems*, *15*(2), 668–674. https://doi.org/10.1109/59.867157.

van der Merwe, D. W., & Engelbrecht, A. P. (2003). Data clustering using particle swarm optimization. In *The 2003 congress on evolutionary computation*. Canberra, ACT, Australia: IEEE. https://doi.org/10.1109/CEC.2003.1299577.

Vitart, F., Buizza, R., Balmaseda, M. A., Balsamo, G., Bidlot, J. R., Bonet, A., Fuentes, M., Hofstadler, A., Molteni, F., & Palmer, T. N. (2008). The new VarEPS-monthly forecasting system: A first step towards seamless prediction. *Quarterly Journal of the Royal Meteorological Society*, *134*(636), 1789–1799. https://doi.org/10.1002/qj.322.

Voyant, C., Muselli, M., Paoli, C., & Nivet, M. L. (2021). Numerical weather prediction (NWP) and hybrid ARMA/ANN model to predict global radiation. *Energy*, *39*(1), 341–355. https://doi.org/10.1016/j.energy.2012.01.006.

Voyant, C., Notton, G., Kalogirou, S., Nivet, M. L., Paoli, C., Motte, F., & Fouilloy, A. (2017). Machine learning methods for solar radiation forecasting: A review. *Renewable Energy*, *105*, 569–582. https://doi.org/10.1016/j.renene.2016.12.095.

Wang, H., Gu, C., Zhang, X., Li, F., & Gu, L. (2018). Identifying the correlation between ambient temperature and gas consumption in a local energy system. *CSEE Journal of Power and Energy Systems*, *4*(4), 479–486. https://doi.org/10.17775/CSEEJPES.2017.00260.

Wang, Q., Wu, H., Florita, A. R., Martinez-Anido, C. B., & Hodge, B.-M. (2016). The value of improved wind power forecasting: Grid flexibility quantification, ramp capability analysis, and impacts of electricity market operation timescales. *Applied Energy*, *184*, 696–713. https://doi.org/10.1016/j.apenergy.2016.11.016.

Webby, R., & OConnor, M. (1996). Judgemental and statistical time series forecasting: A review of the literature. *International Journal of Forecasting*, *12*(1), 91–118. https://doi.org/10.1016/0169-2070(95)00644-3.

Weron, R. (2014). Electricity price forecasting: A review of the state-of-the-art with a look into the future. *International Journal of Forecasting*, *30*, 1030–1081. https://doi.org/10.1016/j.ijforecast.2014.08.008.

White, C. J., Carlsen, H., Robertson, A. W., Klein, R. J. T., Lazo, J. K., Kumar, A., Vitart, F., de Perez, E. C., Ray, A. J., Murray, V., Bharwani, S., MacLeod, D., James, R., Fleming, L., Morse, A. P., Eggen, B., Graham, R., Kjellstrom, E., Becker, E., & Zebiak, S. E. (2017). Potential applications of subseasonal-to-seasonal (S2S) prediction. *Meteorological Applications*, *24*(3), 315–325. https://doi.org/10.1002/met.1654.

Wu, H., Shahidehpour, M., Alabdulwahab, A., & Abusorrah, A. (2015). Thermal generation flexibility with ramping costs and hourly demand response in stochastic security-constrained scheduling of variable energy sources. *IEEE Transactions on Power Systems*, *30*(6), 2955–2964. https://doi.org/10.1109/TPWRS.2014.2369473.

Xu, Q., He, D., Zhang, N., Kang, C., Xia, Q., Bai, J., & Huang, J. (2015). A short-term wind power forecasting approach with adjustment of numerical weather prediction input by data mining. *IEEE Transactions on Sustainable Energy*, *6*(4), 1283–1291. https://doi.org/10.1109/TSTE.2015.2429586.

Zhang, N., Kang, C., Xia, Q., & Liang, J. (2014). Modeling conditional forecast error for wind power in generation scheduling. *IEEE Transactions on Power Systems*, *29*(3), 1316–1324. https://doi.org/10.1109/TPWRS.2013.2287766.

Zhang, Z. S., Sun, Y. Z., Gao, D. W., Lin, J., & Cheng, L. (2013). A versatile probability distribution model for wind power forecast errors and its application in economic dispatch. *IEEE Transactions on Power Systems*, *28*(3), 3114–3125. https://doi.org/10.1109/TPWRS.2013.2249596.

Zhang, X., Tang, S. C., & Li, G. (2014). Development of an ambient air temperature prediction model. *Energy and Buildings*, *73*, 166–170. https://doi.org/10.1016/j.enbuild.2014.01.006.

Zhou, Y., Liu, Y., Wang, D., Liu, X., & Wang, Y. (2021). A review on global solar radiation prediction with machine learning models in a comprehensive perspective. *Energy Conversion and Management*, *235*. https://doi.org/10.1016/j.enconman.2021.113960, 113960.

Zhou, B., Ma, X., Luo, Y., & Yang, D. (2019). Wind power prediction based on LSTM networks and nonparametric kernel density estimation. *IEEE Access*, *7*, 165279–165292. https://doi.org/10.1109/ACCESS.2019.2952555.

Zhu, Y., Toth, Z., Wobus, R., Richardson, D., & Mylne, K. (2002). The economic value of ensemble-based weather forecasts. *Bulletin of the American Meteorological Society*, *83*(1), 73–83. https://doi.org/10.1175/1520-0477(2002)083<0073:TEVOEB>2.3.CO;2.

Ziel, F., & Steinert, R. (2018). Probabilistic mid- and long-term electricity price forecasting. *Renewable and Sustainable Energy Reviews*, *94*, 251–266. https://doi.org/10.1016/j.rser.2018.05.038.

Optimization of renewable energy systems by genetic algorithms

6

6.1 Introduction

iHOGA and MHOGA (improved hybrid optimization by genetic algorithms and megawatt hybrid optimization by genetic algorithms) (Dufo-López, 2022) are two versions of the hybrid optimization by genetic algorithms (HOGA) software, developed in C++ by researchers of the University of Zaragoza (Spain) for the simulation and optimization of electric power generation systems based on renewable energies.

iHOGA software is for systems from a few W up to 5 MW power, while MHOGA is for MW power systems, without any limit. The software can model systems with electrical energy consumption load (DC and/or AC) and/or hydrogen load, as well as consumption of water from tanks or reservoirs previously pumped.

It can include different components of AC or DC coupled (Figs. 6.1 and 6.2): photovoltaic (PV) generator (including bifacial and concentrating PV, CPV), wind turbine, hydroelectric turbine (pumped hydro storage, PHS, can be considered), backup generator (diesel, gasoline, biogas, hydrogen, etc.), inverter or inverter-charger, battery energy storage system (BESS, lead-acid, Li-ion, or others), components of hydrogen (electrolyzer, hydrogen tank and fuel cell), and thermoelectric generator (TEG, Seebeck effect). Auxiliary components such as battery charge controller or PV inverter are also defined in iHOGA low-power systems.

iHOGA and MHOGA can simulate and optimize off-grid systems or systems connected to the AC grid, with or without their own load consumption, and different cases of net metering or net billing can be defined (monthly or annual, with or without rolling credit, considering hourly periods, etc.).

In systems with load consumption, the optimization is the minimization of the total net present cost of the system (NPC). In power generation systems (without load or with low load), the optimization can be the maximization of the total net present value of the system (NPV) or the minimization of the levelized cost of energy (LCOE), calculating also the internal rate of return (IRR). Economic calculations are very accurate (different inflation rates can be defined for different costs) and can be done in any currency.

The software includes multiperiod simulation and optimization (considering the degradation of the PV modules, wind turbines, and batteries during the years, the increase of the load consumption during the years, etc.), multiobjective optimization, simulation in time steps from 1 min to 1 h, sensitivity analysis, probability analysis, etc. It uses very detailed models of the components, obtaining detailed simulations and very accurate estimates of the operation of the system. It includes control strategies that can be optimized for off-grid systems or for grid-connected systems (batteries energy management and/or PHS energy management for the charge/discharge,

Genetic Optimization Techniques for Sizing and Management of Modern Power Systems
https://doi.org/10.1016/B978-0-12-823889-9.00007-2

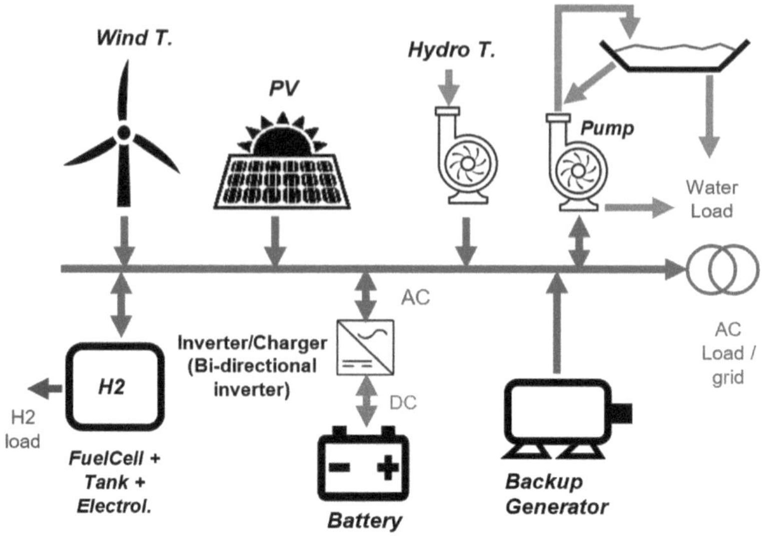

Fig. 6.1 iHOGA/MHOGA components, AC coupled high power system.

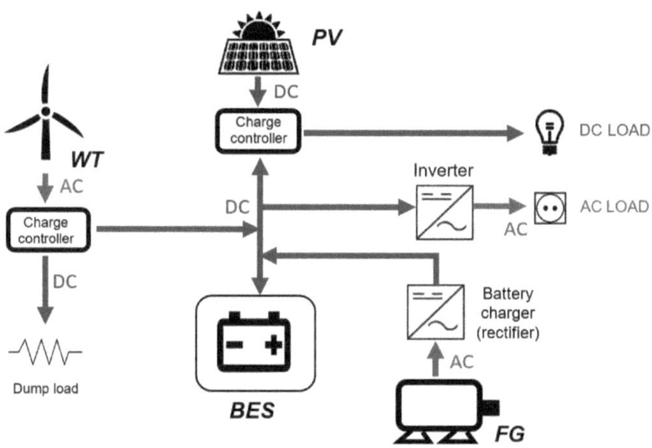

Fig. 6.2 iHOGA components, typical DC coupled low power system.

depending on the purchasing or selling electricity price of each time step). It includes an extensive database of the different components and it allows the automatic download from NASA's website of irradiation, wind, and temperature data, and also the download of hourly data from PVGIS (European Union) or Renewables Ninja websites.

iHOGA and MHOGA use advanced models to accurately estimate batteries' lifetimes (batteries are generally the most expensive components and the ones that require most replacements). This implies that the estimation of the NPC or NPV is obtained in a much more realistic way than other software tools, being able to compare correctly between different combinations of components and control strategies and obtaining the real optimal solution. It includes advanced optimization models (genetic algorithms (GA)), which implies the possibility of obtaining the optimum system using very low computational time.

There are several software tools to simulate and optimize hybrid renewable systems (Al-falahi et al., 2017; Bernal-Agustín & Dufo-López, 2009c; Sinha & Chandel, 2014). Unlike other commercial hybrid optimization software tools, iHOGA and MHOGA use advanced models in the simulations: variable inverter efficiency (dependent on the inverter output power), variable electrolyzer and fuel cell efficiencies, advanced battery lifetime models, advanced grid-connected storage control (batteries, hydrogen, and/or PHS), advanced grid options (hourly electricity purchase price, hourly sell price, hourly periods and optimization for the contracted power, net metering and net billing, and other features), probability analysis, etc. Fig. 6.3 shows the main screen of iHOGA software in a project of high power with PV, wind, and BESS; in this example, hydro turbine, backup generator, and hydrogen components were not considered.

Although the advantages of hybrid systems vs single energy systems are clear in many cases, the optimal calculation of a hybrid system is very complex, due to the variability of the availability of renewable resources (solar irradiation, wind, hydro resource), as well as the energy load. In addition, some components of the system have

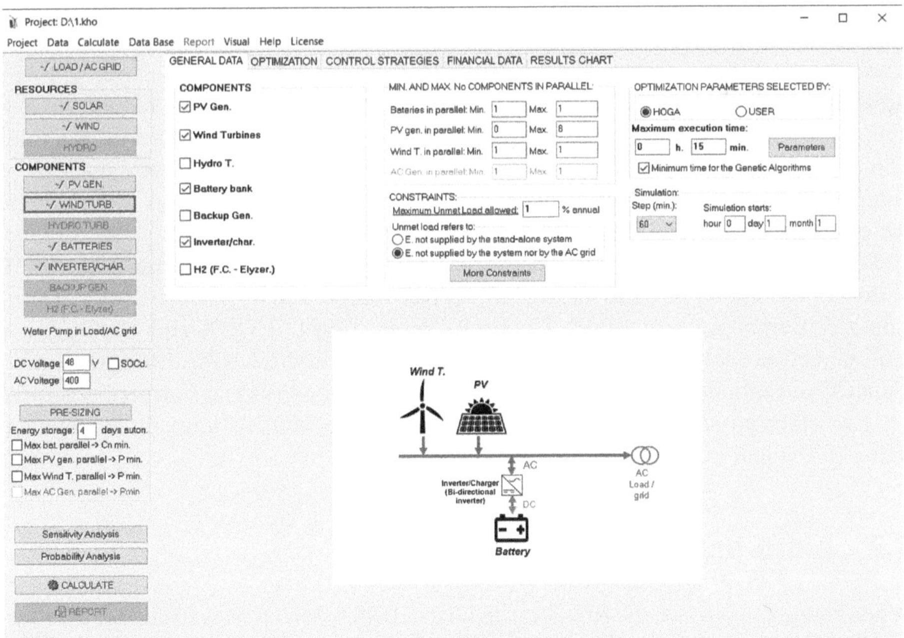

Fig. 6.3 iHOGA software main screen, high power project with PV, wind turbines, and BESS.

nonlinear characteristics, and the number of variables influencing the optimization is very high. The optimization of a hybrid system has a large number of possible solutions (combinations of PV modules/generators, wind turbines, hydraulic turbine, BESS, backup generator, fuel cell, electrolyzer, inverter/charger, and TEG, or some of them, and also control strategy variables), therefore, the optimization is a very complicated task involving classical mathematical techniques (for example, with mixed-integer programming).

None of the commercial renewable software tools use advanced optimization algorithms, except for iHOGA and MHOGA. Other programs (e.g., HOMER) perform the optimization by testing all possible combinations. One problem with that methodology is that if the number of possible combinations is very high, the calculation time increases enormously and becomes inadmissible. In addition, the control strategies are too simple, and no in-depth optimization is possible, as it is in iHOGA/MHOGA. Other advantages of iHOGA/MHOGA are in the models used, more precise than in HOMER, in the components as well as in the economic calculations, besides functionalities like presizing, optimization of the slope of the photovoltaic modules, probability analysis, advanced battery aging models, multiobjective optimization, etc.

iHOGA and MHOGA use GA (Bernal-Agustín & Dufo-López, 2009a; Dufo-López et al., 2007; Dufo-López & Bernal-Agustín, 2005), which are used to perform the optimization, for the combination of the components (main GA) and also for the control strategy (secondary GA). GA makes it possible to obtain satisfactory solutions when applied to problems with a great level of complexity, and does so with very low calculation times. If the number of possible combinations is very high, using a population of roughly 0.003% of the possible combinations, with 15 generations, i.e., evaluating approximately 0.04% of the possible combinations, there is a very high probability of obtaining the optimal combination or a solution very close to the optimal one (Bernal-Agustín & Dufo-López, 2009a). That is, the optimization time can be reduced to less than 0.1% with great probability of obtaining the optimal solution or a very close solution. The software also has the possibility to evaluate all possible combinations of components and control variables of the strategy, i.e., optimization according to the enumerative method, without GA.

For multiobjective optimization (Bernal-Agustín et al., 2006; Bernal-Agustín & Dufo-López, 2009b; Coello et al., 2002; Dufo-López et al., 2011, 2016a; Dufo-López & Bernal-Agustín, 2008; Zitzler & Thiele, 1999), iHOGA and MHOGA use multiobjective evolutionary algorithms (MOEA) to minimize NPC, life cycle equivalent CO_2 emissions, and/or unmet load by the stand-alone system. Furthermore, there are two other possible objectives to maximize by the MOEA: human development index (HDI) and job creation.

6.2 Main options

When creating a project, the first thing to do in iHOGA/MHOGA is to choose the main options. For the simulation of each combination of components and control strategies we can choose:

- Simulation of just the first year and extrapolate the results.
- Simulation multiperiod: simulate all the years of the system lifetime, considering the annual increase in load, the decrease in generation (PV, wind, hydro), the reduction in the battery capacity due to its degradation, the annual change in the electricity price and in the hydrogen price, the annual variation in resources, and/or the annual variation in O&M costs.

In addition, we must choose the type of project considering the economic optimization:

- Minimization of the NPC: typical for off-grid systems or grid-connected systems with high load.
- Maximization of the NPV: typical for grid-connected generators with low load or without any load, where all the energy or almost all is sold to the AC grid. Green hydrogen generating systems (grid-connected renewable generators that export electricity to the AC grid and also generate hydrogen in electrolyzers) can be considered. In these cases, we can select maximization of NPV or minimization of LCOE.

6.3 Input data

Load, AC grid purchase/sell electricity prices, irradiation resources (irradiation, wind speed, temperature), and components data must be set. In addition, data for the control strategies, constraints, financial data, and other data must be defined.

6.3.1 Load

The load can be AC, DC, hydrogen (for external use), and/or water stored in tank, previously pumped. Each type of load can be defined in hourly values for the typical day of each month (defining a daily, hourly, and minute-step variability), or we can use preset load profiles, or we can import the load from file (the values of a whole year in steps of 1 h or less, down to 1-min time steps). Fig. 6.4 shows the load screen of an example with the load profile of 100 kWh/day average (residential).

6.3.2 Electricity and hydrogen prices

In the case where there is an AC grid (grid-connected system), we must define the electricity prices for the electricity injected to the AC grid (exported to the grid, that is, sold energy) and for the electricity imported from the AC grid (purchased). We can define fixed prices or hourly prices: time-of-use (TOU) tariffs or real-time pricing (RTP) for the electricity purchased, for the electricity sold, for the access charges, and for the back-up charges (in this case). Different annual inflation values can be defined for the electricity bought and for the electricity sold. Fig. 6.5 shows the screen where the electricity bought to the AC grid is defined (in this example, in hourly periods, a TOU tariff).

The AC grid contracted power must also be defined, with a fixed value or by hourly periods; the limit power to the electricity sold or purchased can be the contracted power or it can be calculated from the registered values during the simulation (and

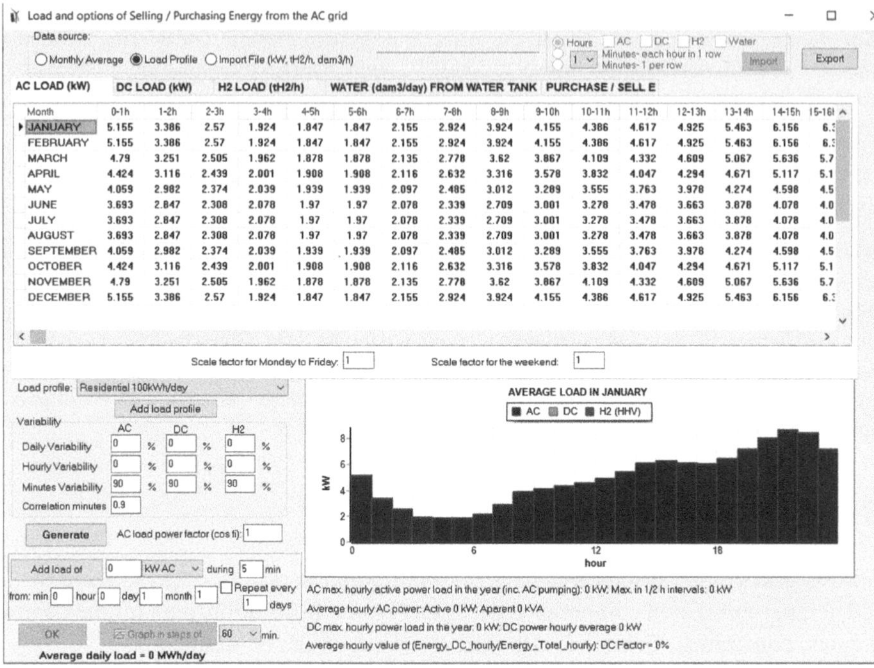

Fig. 6.4 iHOGA software, load data screen (residential AC load of 100 kWh/day average selected).

the payment will depend on the maximum register); it can even be optimized by the software.

The hydrogen price and its annual inflation (when generated hydrogen is not used by our system and it is sold to be used externally) must also be set.

6.3.3 Resources

Solar irradiation, wind speed, temperature, and water flow are the resources that must be defined for the PV generator, wind turbine, and hydro turbine. Monthly average values or 1-year basis hourly (or in lower steps, down to 1-min time steps) imported values can be used. In addition, irradiation, wind speed, and temperature can be downloaded in hourly values of a whole year from different databases: NASA (NASA Prediction of Worldwide Energy Resources, 2022), PVGIS European Commission, 2020), or Renewables Ninja (Pfenninger & Staffell, 2022). The data in hourly or minute time steps can also be exported from the software. A local database of all the world (22-year average monthly average values of NASA) can be installed, which includes Weibull shape factor of the wind speed series.

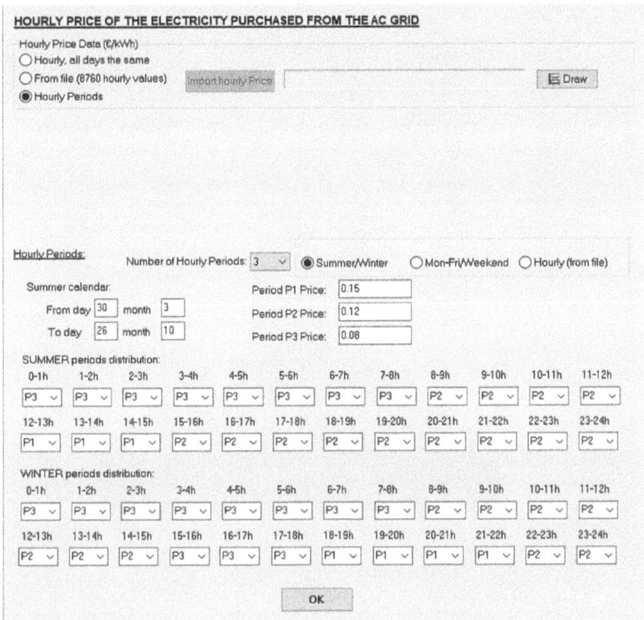

Fig. 6.5 iHOGA/MHOGA software, electricity purchased from the AC grid considering hourly periods (TOU tariff).

6.3.3.1 *Solar data*

Latitude and longitude must be set (directly or clicking in the world map) and also, the tilt inclination angle and azimuth angle of the PV generator must be defined (Fig. 6.6). Two different surfaces can be defined for the PV generator, with a percentage of the total size of the PV generator for each one (for example, 65% of the size of the PV generator at 30 degree tilt and −90 degree azimuth, and 35% at 30 and 90 degrees azimuth). Solar tracking can also be defined (no tracking, horizontal axis, vertical axis, or both axes). Shadows over the PV generator are defined with the elevation of the obstacles (degree) vs the azimuth (degree), as well as the reduction factor of the direct irradiation.

Irradiation data can be defined in horizontal surface monthly average daily values, or it can be imported in horizontal or inclined surface data, in hourly data (or in lower time steps, down to 1 min) for a whole year. In addition, monthly average horizontal surface data or hourly horizontal surface data can be downloaded from NASA (a local database is also available with the software, using the NASA average data during 22 years, for all the world); whereas, hourly irradiation over the tilted surface (for any azimuth) can be downloaded from PVGIS or from Renewable Ninja. MHOGA also allows importation of the PV generation from a real PV generator in hourly or minute steps and its use as an input for the PV generation.

When using horizontal surface monthly average irradiation data (or when importing or downloading from NASA horizontal surface hourly irradiation), the

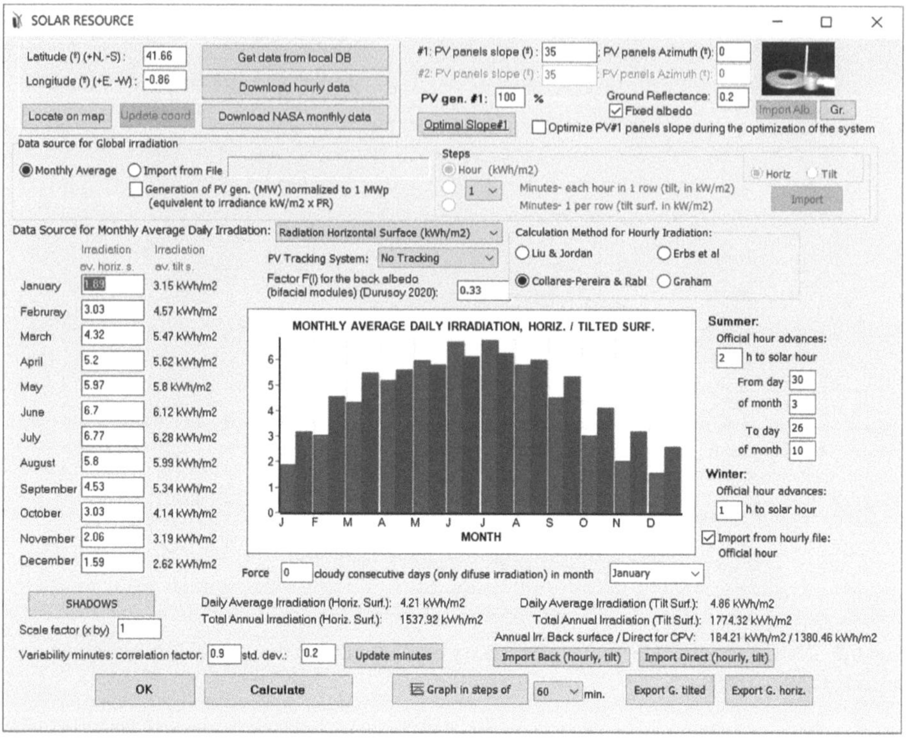

Fig. 6.6 iHOGA/MHOGA software, solar resource screen.

hourly irradiance on the inclined surface can be calculated by the method of Graham and Hollands (1990), which includes statistical variability, or through the model of Liu and Jordan (1960), Hay and Davies (1978), and Rietveld (1978), by using different correlations: Liu and Jordan (1960), Collares Pereira and Rabl (1979), or Erbs et al. (1982). iHOGA and MHOGA can calculate the optimal tilt angle for the PV generator (considering the azimuth, the horizontal irradiation, and the load), obtaining the optimal tilt angles to maximize the PV generation for each month, the optimal tilt angle to maximize the minimum ratio PV generation/load during the year, and the optimal tilt angle to maximize the PV generation during the whole year. The slope of the PV generator can also be a variable to be optimized during the optimization of the system.

Once the hourly irradiation data have been obtained, the global irradiance $G_{(t)}$ (kW/m^2) over the tilted surface of the PV generator in 1-min steps during a whole year is obtained by using a first-order autoregressive function based on the work of Aguiar and Collares-Pereira (1992). Fig. 6.7 shows the 1-min step irradiation for a specific day.

First global and diffuse minute irradiance data between hours ($G_{\mathrm{hours}(t)}$ and $D_{\mathrm{hours}(t)}$) are obtained by means of linear interpolation between hourly values. Then a first order autoregressive (AR) function ($\gamma_{(t)}$) is obtained (Eq. 6.1), which is used to calculate $G_{(t)}$ (Eq. 6.2), that cannot be lower than the diffuse component:

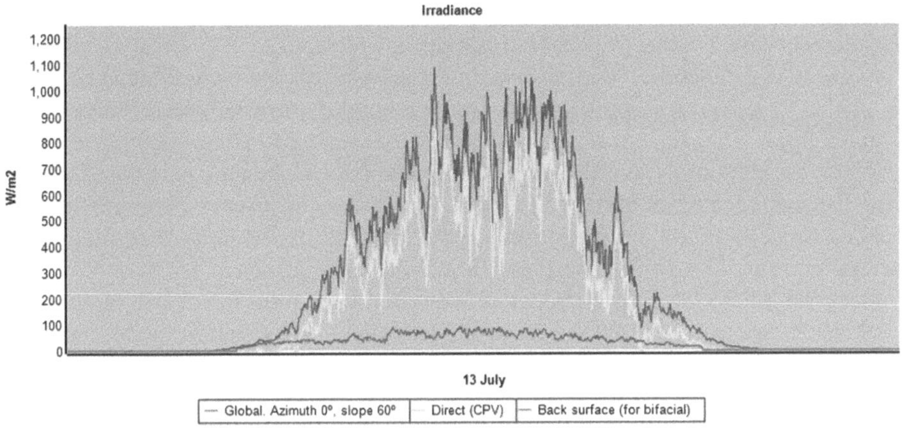

Fig. 6.7 iHOGA/MHOGA 1-min step irradiation obtained for a specific day (global over tilted surface, direct over tilted surface, and global over back surface).

$$\gamma_{(t)} = \phi_{1G}\, \gamma_{(t-\Delta t)} + \mathcal{N}\left(0, \sigma_G \sqrt{1 - \left(\phi_{1G}{}^2\right)}\right) \tag{6.1}$$

$$G_{(t)} = \max\left(D_{\text{hours}(t)}, G_{\text{hours}(t)}\left(1 + \gamma_{(t)}\right)\right) \tag{6.2}$$

where ϕ_{1G} is the correlation factor, σ_G is the standard deviation and $\mathcal{N}(a, b)$ is a random number following normal distribution with mean a and standard deviation b.

The direct irradiance $B_{(t)}$ (kW/m^2) over the tilted surface of the PV generator (needed in the case where we consider concentrating PV) in 1-min steps during a whole year is obtained by using a similar AR function. The irradiation over the back surface of the PV modules (needed in the case where we consider bifacial PV) is calculated using the methodology shown by Durusoy et al. (2020), expanded to any azimuth and tracking method.

6.3.3.2 Wind data

As for irradiation, monthly average values or hourly (or lower minute steps, down to 1 min) wind speed data can be used. The anemometer height (z_{anem}) must be set because the software must know the height of the measured wind data $w_{(t)}$; if it is different from the hub height (z_{hub}) of the wind turbines, the wind data will be converted from the anemometer height to the hub height, obtaining $w_{\text{hub}(t)}$, considering the surface roughness length z_0 (which must also be set), according to Eq. (6.3):

$$W_{\text{hub}(t)} = w_{(t)} \left[\frac{\ln\left(\frac{z_{\text{hub}}}{z_0}\right)}{\ln\left(\frac{z_{\text{anem}}}{z_0}\right)}\right] \tag{6.3}$$

Wind speed data can also be downloaded from the same databases used for the irradiation. Monthly average wind speed can be downloaded at 10 m or at 50 m height from the NASA database. We can also download hourly data from Renewable Ninja (at any height), from PVGIS (data only for 10 m height), or from NASA (only at 10 or 50 m height).

When using monthly average data, hourly wind speed data are calculated considering the shape factor (Weibull distribution, defined by the user or obtained from the local database) and the correlation factor ϕ_W (defined by the user) according to the method presented by Dufo-López and Bernal-Agustín (2012).

Once the hourly wind speed data has been calculated or downloaded, minute-step wind speed data are then obtained using a first order AR function model using the same correlation factor ϕ_W and a standard deviation defined by the user. In the right-hand part of the screen, the probability density function (PDF) of the wind speed series (red color distribution) is shown, as well as the PDF Weibull distribution curve (green curve) for the Weibull shape factor that best fits the generated series. Fig. 6.8 shows the software wind speed screen while Fig. 6.9 shows an example of the generated wind speed for 20 days in October.

In MHOGA software, we can define wind turbines or a wind farm with 16 power vs wind speed curves, one for each wind direction sector (from 0 to 337.5 degrees in steps of 22.5 degree). In this case, we must import the wind speed and also wind direction.

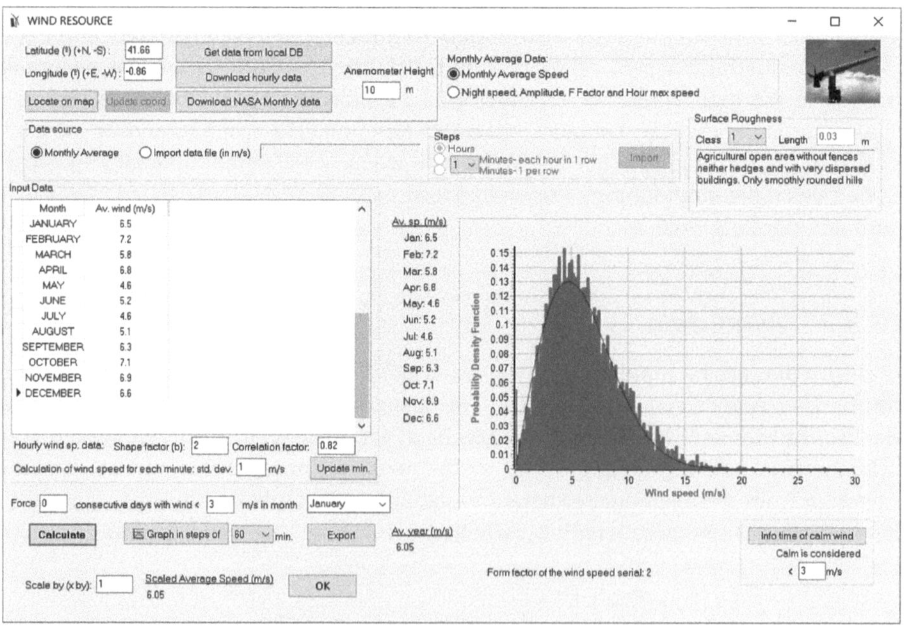

Fig. 6.8 iHOGA/MHOGA software, wind resource screen.

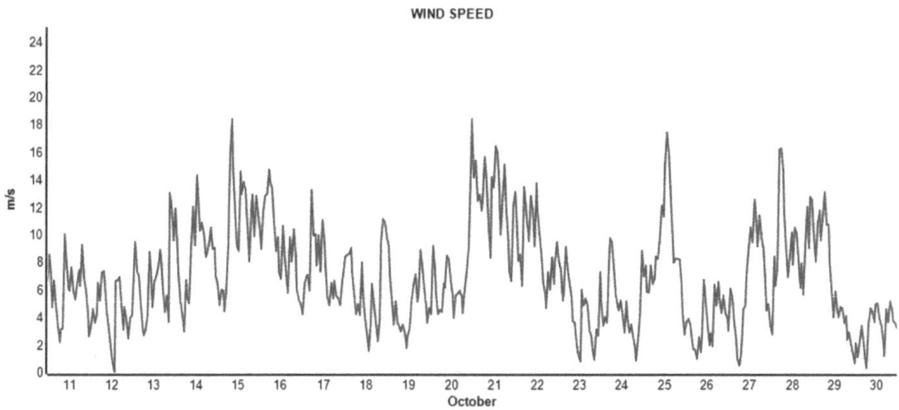

Fig. 6.9 iHOGA/MHOGA software, example of wind speed obtained for several days.

6.3.3.3 Hydro resource data

Total head H (the difference in elevation between the head water level and the tail water level, expressed in m) and the pressure losses L in the power canal and draft tube (m) must be set. iHOGA and MHOGA use these data to calculate the available head $H' = H - L$. Losses in penstock water, in percentage, LP (%), are used to obtain the net head $H' = H' \cdot (100 - LP)$. Data sources for the water flow can be in monthly average values (including daily and hourly flow variability) or in imported hourly values. Minute-step data will be the same as the hourly data (no variation is considered into each hour). An estimate for the maximum power available, based on head and water flow, is obtained (to be considered when defining hydro turbines. Fig. 6.10 shows the hydro resource screen.

6.3.3.4 Ambient temperature data

Ambient temperature affects PV generation and wind turbine generation. Monthly average or hourly temperature data can be used (imported or downloaded from the same databases as irradiation and wind speed). When using monthly average temperature data, the hourly temperature is calculated according to the model proposed by Erbs et al. (1983) using irradiation data.

Temperature also affects the performance and the aging of the batteries; however, the ambient temperature of the batteries is commonly different from the ambient temperature of the location, as batteries are usually inside buildings and, therefore, their ambient temperatures are less extreme than outside. That is why the ambient temperature for the batteries is usually defined with different data from the ambient temperature for PV and for wind turbines.

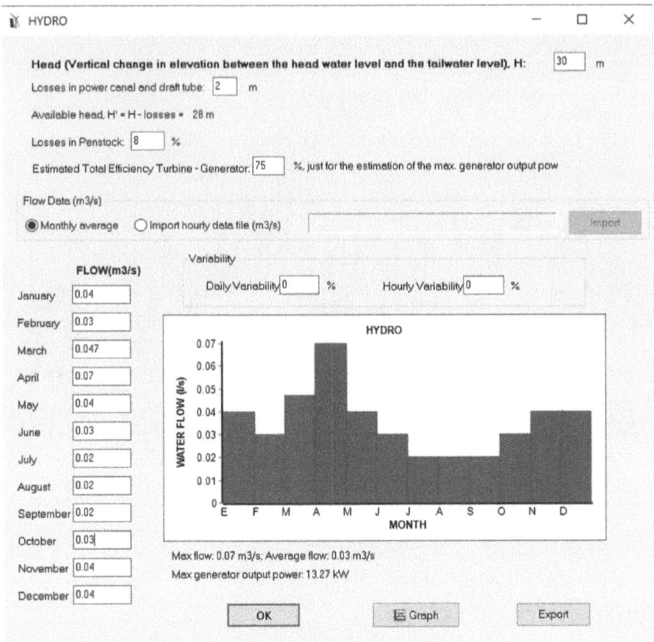

Fig. 6.10 iHOGA/MHOGA software, hydro resource screen.

6.3.3.5 TEG temperature data

In iHOGA low-power systems, if the system includes thermoelectric generator (TEG), the temperatures of the hot source (from the combustion exhaust gas) and of the cold source (where the cold side of the TEG is connected) are needed. We suppose that the energy supply for the TEG (that is, the hot source temperature) is free of charge (no cost related to the hot source), as it is the residual exhaust gas of the combustion that was obtained in other process (for example for heating or cooking).

For each time step (1 h or less, down to 1 min) of the year, data of the temperature of the two sources (hot source and cold source) must be introduced. It can also be obtained from monthly-average data.

6.3.4 Components data

The system can consider several units (types) of all or some of the different components: PV generators or modules, wind turbines, hydro turbines, batteries, inverters, inverter/chargers, backup generators, TEG (only in iHOGA), and/or hydrogen components (electrolyzers, hydrogen tank, and fuel cells). The user decides which kind of components can be considered for the system, and he/she can define its own components (taking into account its local availability, prices, etc.) or use the components of the software database (which can be edited).

6.3.4.1 PV modules/generators

In low-power systems (iHOGA), the user can define (or use from the database) different PV modules, which will be considered in the different combinations of components of the system. In high-power systems (iHOGA or MHOGA), the software uses PV generators of different sizes.

The PV modules/generators parameters needed for the simulations are the following: nominal voltage (only for low-power systems in iHOGA, used to calculate the number of PV modules in serial as nominal DC bus voltage divided by this value); short-circuit current (only for low-power systems in iHOGA); peak nominal power under standard test conditions (STC, which must be defined); acquisition cost (CAPEX, which must include the mounting structure cost, etc.); operation and maintenance (O&M, OPEX) annual cost; expected lifespan; normal operating temperature of the cell (NOCT); coefficient of variation of the power with the temperature, (%/°C); bifaciality factor (in bifacial PV modules, this is the ratio of the nominal efficiency at the back side, with respect to the nominal efficiency of the front side, that is, the ratio of the rear power with respect to the front power of the PV module, measured at STC); emissions of CO_2 in manufacturing, transport, and recycling; and weight. In addition, the modules/generators can be defined as normal PV or concentrating PV (CPV). If the PV generator is connected to the AC bus by its own inverter, its data must be set (power, efficiency) and its cost must be included in the CAPEX cost of the PV modules/generators. To take into account the degradation of the modules, losses in wires, dirt in modules, etc. (but not the losses in the inverter, which are considered apart), a loss factor ($f_{\text{PV\_loss}}$) must also be set.

In AC coupled systems, PV grid inverters include maximum power point trackers (MPPT) for maximizing the power production of the PV generator, however, in low-power DC coupled systems, MPPT is not usually included in the PV charge controllers. We must set whether the system includes MPPT or not.

In low-power systems (iHOGA) without MPPT, with a global irradiance over the PV surface $G_{(t)}$ (kW/m^2), the PV module of P_N nominal peak power (usually defined at standard conditions, STC) (Wp) and I_{sc} shortcut current (A) injects a DC current into the inverter or into the battery charger (depending on the system configuration) during time t, $I_{PV(t)}$ (A), which can be calculated using Eq. (6.4) (Dufo-López et al., 2014):

$$I_{PV(t)} = I_{sc}\left(G_{(t)}/G_{\text{ref}}\right) \tag{6.4}$$

where G_{ref} is the global irradiance under the conditions in which the PV was tested (1 kW/m^2 for STC).

As the PV generator is composed of N_{PV_p} strings in parallel, the output power of the PV generator (W) will be calculated in Eq. (6.5):

$$P_{PV(t)} = N_{PV\_p} \times I_{PV(t)} \times V_{DC(t)} \times f_{\text{PV\_loss}} \tag{6.5}$$

where $V_{DC(t)}$ (V) is the PV generator DC voltage during time step t.

If the inverter (or the battery charger) includes MPPT, the PV modules output will be the maximum power point of the $I = f(V)$ curve; therefore, the output power of the PV generator is shown in Eq. (6.6):

$$P_{PV(t)} = N_{PV\_p} \times N_{PV\_s} \times P_N \times \left(G_{(t)}/G_{\text{ref}}\right) \times f_{\text{PV\_loss}} \times \left[1 + (\alpha/100) \times \left(T_{c(t)} - T_{c\_\text{ref}}\right)\right] \tag{6.6}$$

where N_{PV_s} is the number of PV modules in serial, α is the power temperature coefficient of the PV module (%/°C), $T_{\text{c\_ref}}$ is the cell temperature under the conditions in which the PV was tested (25 °C for STC), and $T_{c(t)}$ (°C) is the PV cell temperature during time t, which is obtained using Eq. (6.7):

$$T_{c(t)} = T_{a(t)} + \left(\frac{NOCT - 20}{0.8}\right) \times G_{(t)} \tag{6.7}$$

where $NOCT$ is the nominal operation cell temperature (°C) and T_a is the ambient temperature (°C).

If the PV modules are bifacial, the power generated by the rear surface is calculated as follows, which will be added to the power generated by the PV front surface, as shown in Eq. (6.8):

$$P_{\text{PV\_back}(t)} = N_{PV\_p} \times N_{PV\_s} \times P_N \times \left(G_{\text{back}(t)}/G_{\text{ref}}\right) \times Bif \times f_{\text{PV\_loss}} \times \left[1 + (\alpha/100) \times \left(T_{c(t)} - T_{\text{c\_ref}}\right)\right] \tag{6.8}$$

where $G_{\text{back}(t)}$ is the global irradiance over the rear surface of the PV modules and Bif is the bifaciality factor of the modules.

If the generator is a concentrating PV (CPV), its internal cell temperature is calculated by Eq. (6.9) (Leloux et al., 2014):

$$T_{c(t)} = T_{a(t)} + B_{(t)} \times 1000 \times \left(\frac{\eta_{\text{opt}} \times 0.75}{\mu_{(t)}}\right) + 5 \tag{6.9}$$

where $B_{(t)}$ is the direct irradiance over the surface of the CPV (kW/m^2), η_{opt} is the optics efficiency and $\mu_{(t)}$ (W/m^2/K) is a function of the coolant, of the exchange surface, and of the wind speed, given by Eq. (6.10):

$$\mu_{(t)} = \mu + 2.5 \times ws_{(t)} \tag{6.10}$$

where μ is the thermal conductivity in natural convection (W/m^2/K) and $ws_{(t)}$ is the wind speed (m/s) at the height of the CPV.

After calculating the internal temperature of the cell, the power generated by the CPV generator of nominal peak power P_N is calculated by Eq. (6.11) (where $T_{\text{c\_ref}}$ is usually 20 °C and B_{ref} is the direct irradiance under the conditions in which the CPV was tested, commonly 0.9 kW/m^2):

$$P_{CPV(t)} = N_{PV\_p} \times N_{PV\_s} \times P_N \times \left({}^{B_{(t)}}/_{B_{\text{ref}}} \right) \times f_{\text{PV\_loss}} \times \left[1 + (\alpha/100) \times \left(T_{c(t)} - T_{\text{c\_ref}} \right) \right] \quad (6.11)$$

6.3.4.2 Wind turbines

Each wind turbine considered (from the database or defined by the user) has its own data: cost (CAPEX), lifespan, O&M cost (OPEX), hub height, etc., including the output curve vs wind speed, which is usually given by the manufacturer at standard conditions at sea level (air density of $\rho_0 = 1.225\,\text{kg/m}^3$ corresponding to temperature 288.15 K and 101.325 kPa). Fig. 6.11 shows the wind turbines screen.

During each time step, the standard output power vs wind speed curve is transformed to the actual air density $\rho_{(t)}$ by multiplying the curve by $\rho_{(t)}/\rho_0$. The air density in the international standard atmosphere up to an altitude of 5000 m is obtained by using Eq. (6.12) (Manwell et al., 2009):

$$\rho_{(t)} = 3.4837 \times \left(\frac{101.29 - 0.011837 \times h + \left(4.793 \times 10^{-7}\right) \times h^2}{T_{(t)}} \right) \quad (6.12)$$

where $T_{(t)}$ is the temperature during time t (K) and h the altitude of the location (m).

In the simulation, during each time step t, the output power will be obtained considering the wind speed in the wind turbine hub $w_{\text{hub}(t)}$ (m/s) at time step t and the actual output power curve of the wind turbine (for altitude h and temperature $T_{(t)}$).

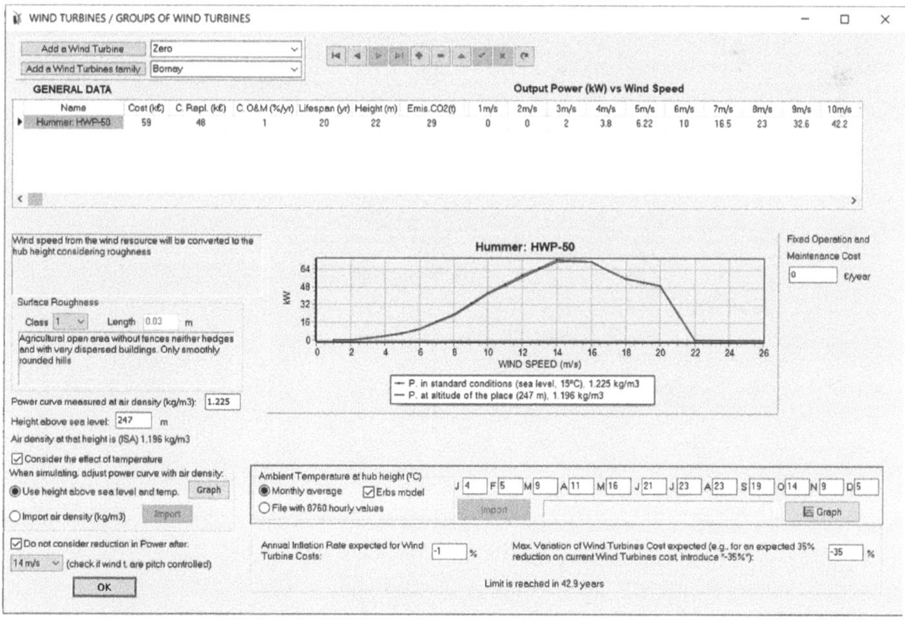

Fig. 6.11 iHOGA software, wind turbines screen.

In MHOGA software, we can define wind turbines or a wind farm with 16 power vs wind speed curves, one for each wind direction sector (from 0 to 337.5 degrees in steps of 22.5 degree). This way, the wind generation will depend not only on the wind speed but also on the wind direction (interpolating in the curves for both inputs).

6.3.4.3 Hydro turbines

Each hydro turbine considered (from the database or defined by the user) has its own data: cost (CAPEX), lifespan, O&M cost (OPEX), maximum flow, minimum and maximum pressure height, and efficiency vs percentage of maximum flow. Fig. 6.12 shows the hydro turbines screen.

Pumped hydro storage (PHS) can be defined, by means of a reversible pump-turbine machine or by means of two different machines.

6.3.4.4 Battery energy system storage (BESS)

In off-grid applications, lead-acid battery is still the prevalent technology (Krieger et al., 2013), but with the reduced prices of Li-ion batteries, they have increasingly been installed in recent years. In grid-connected systems, Li-ion batteries are the most widely used BESS.

In iHOGA and MHOGA software, the following battery data shown by the manufacturer must be introduced (Fig. 6.13): nominal capacity, nominal voltage (used to calculate the number of batteries in serial as nominal DC bus voltage divided by this

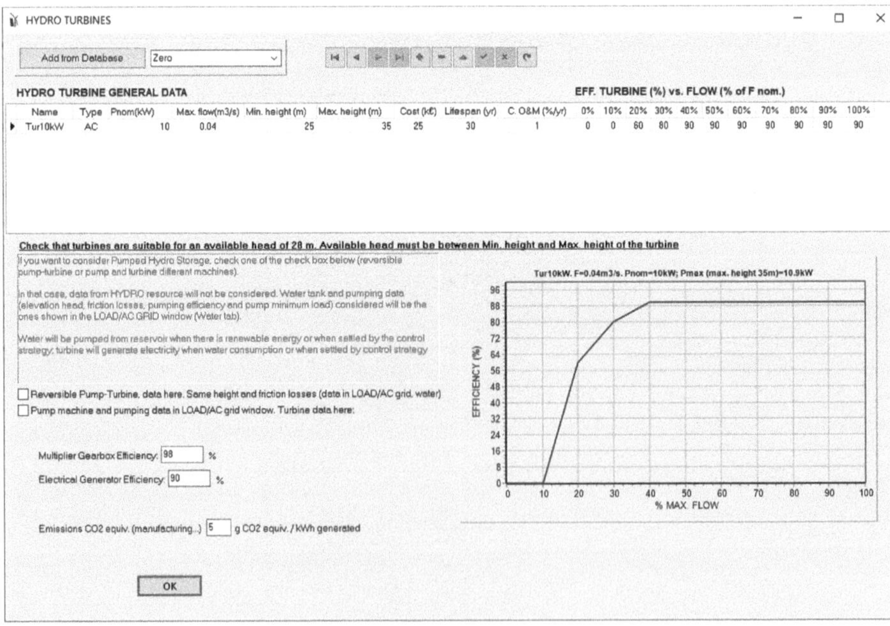

Fig. 6.12 iHOGA software, hydro turbines screen.

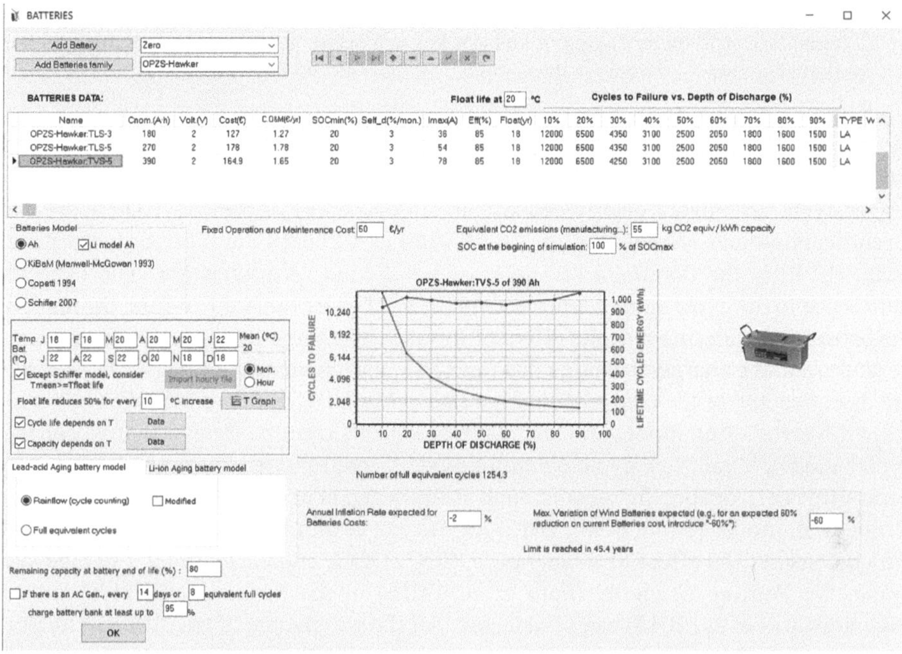

Fig. 6.13 iHOGA software, BESS screen.

value), acquisition cost (CAPEX) and O&M cost (OPEX), minimum state of charge (SOC), self-discharge coefficient, maximum current, roundtrip efficiency, floating lifetime (at a specific temperature, usually 20 °C), and cycles to failure vs depth of discharge.

The accurate estimation of battery lifetime is very important, as the BESS is usually the most expensive component, especially in off-grid systems. If the estimation of the battery lifetime (aging) is not done correctly, the estimated system NPC will be very different from the real NPC, as the replacement costs of the BESS will be wrongly considered. Unlike other simulation and optimization software tools, iHOGA and MHOGA use models to accurately predict the battery lifetime, for lead-acid and for Li-ion batteries.

Regarding lead-acid batteries, there are two main types: flat plate or tubular plate. Flat plate can be flooded (liquid electrolyte, vented lead-acid VLA) or sealed (valve regulated lead-acid VRLA, with enclosed electrolyte and, therefore, maintenance-free). In tubular plate types, they can also be flooded (OPzS) or gel sealed (OpzV). iHOGA/MHOGA can use different models to estimate the lead-acid battery lifetime:

(a) The "equivalent full cycles model," calculating the full charge (Ah throughput) cycled by the battery since the beginning and comparing to the manufacturers' standard test data (the real working conditions depth of discharge DOD, current, temperature, etc., are usually quite different from standard tests, but the effect of the temperature can be considered by the software as it models the dependence of cycles to failure vs temperature).

(b) The "rainflow cycle counting model" (Downing & Socie, 1982), which also considers the effect of the depth of discharge.

(c) Schiffer et al. (2007) weighted Ah throughput aging model, which is a complex but very accurate lead-acid battery aging model (Dufo-López et al., 2014), using weighting factors for the calculation of the charge throughput over the battery lifetime to model the lost capacity due to the different aging mechanisms (corrosion, acid stratification, sulphation and degradation of the active material).

Considering Li-ion batteries, their advantages compared with lead-acid batteries are: higher cycle life, higher energy density, and higher energy efficiency. There are different technologies (Armand et al., 2020; Zubi et al., 2018), and lithium iron phosphate (LFP) is the dominant technology in this kind of application. The capacity fade is due to the cycle aging (dependent to the total charge in ampere-hour throughput the battery, considering also the effect of the state of charge SOC, the DOD, the temperature, and the charge/discharge rate) and also due to calendar aging (when batteries are disconnected or the charge/discharge rate is low). iHOGA and MHOGA can consider different aging models for the cycle aging of Li-ion batteries: "equivalent full cycles model," "rainflow cycle counting model" (both models can consider the temperature effect), or specific models, such as the Wang et al. model (Wang et al., 2011), Groot et al. model (Groot et al., 2015), or Saxena et al. model (Saxena et al., 2016), which consider the effect of temperature, SOC, DOD, and current rate. For calendar aging, the Petit et al. model (Petit et al., 2016) or the Swierczynski et al. model (Swierczynski et al., 2015) can be selected. All the parameters of the different models can be modified to fit with the particular battery considered.

In iHOGA software, when considering low power systems, the battery charge controller must also be defined (in the case that it is not included into the inverter/charger).

6.3.4.5 Inverters and inverter/chargers

When there is AC load, the inverter is needed to convert from DC to AC; in AC coupled systems, the PV generator needs the inverter and also the BESS needs an inverter and a charger (rectifier, AC/DC converter) or an inverter/charger. Data of output nominal apparent power, lifetime, acquisition cost (CAPEX), maximum charge current, minimum and maximum operating DC voltage, and maximum input power form the renewable sources must be set. Furthermore, the charger efficiency and the inverter efficiency vs the output power must be defined. Many researchers and software tools consider a constant value for the inverter efficiency, however iHOGA and MHOGA consider its real dependence on the output power. In systems where there are several hours in the day with low PV generation and/or with low load, the low inverter efficiency at low output power (Fig. 6.14, red curve) can have a great effect on the performance of the system.

6.3.4.6 Backup generator

Data of output nominal apparent power, acquisition (CAPEX) and O&M (OPEX) costs, lifetime, minimum output power (to avoid premature failure due to a poor combustion process), fuel type, cost and annual inflation, emissions, and fuel consumption parameters for each generator must be set. The fuel consumption for each time step is calculated in Eq. (6.13):

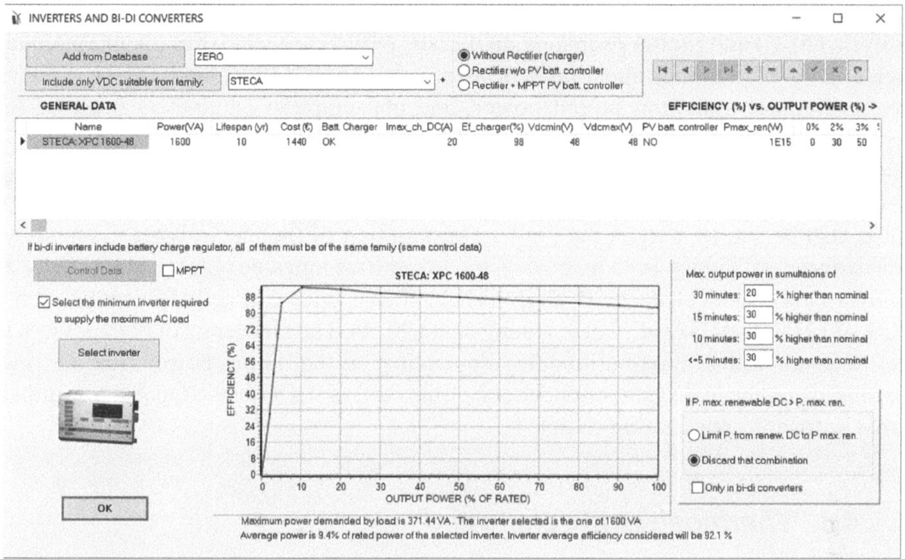

Fig. 6.14 iHOGA software, inverter/chargers screen.

$$F_{(t)}^{FG} = \left[c_1^F \left(\frac{P_{(t)}^{FG}}{1000} \right) + c_2^F \left(\frac{P_{\text{rated}}^{FG}}{1000} \right) \right] \Delta t \tag{6.13}$$

where c_1^F and c_2^F (L/kWh) are the fuel consumption parameters, $P_{(t)}^{FG}$ is the output power of the backup generator during time t (W), P_{rated}^{FG} (W) is the nominal rated power, and Δt is the time step measured in h. Skarstein and Ulhen (1989) suggested $c_1^F = 0.246$ L/kWh and $c_2^F = 0.08415$ L/kWh, which are correct for diesel generators, while for low power gasoline generators, $c_1^F = 0.5$ L/kWh and $c_2^F = 0.2$ L/kWh were suggested by Dufo-López et al. (2011).

The lifetime of a generator depends on its technology and size, from roughly 1000 h for low-power gasoline to 40,000 h for high-power diesel. In iHOGA/MHOGA, a penalization time can be added for each start (some 5 min per start). In addition, the software considers the extra O&M cost and extra aging when the generator falls out of its optimal output power (50%–80% of rated power) (Matt et al., 2005).

6.3.4.7 Hydrogen storage

Electrolysis of water is the splitting of the water molecule into hydrogen and oxygen. The electrolyzer uses DC electricity for water electrolysis, obtaining hydrogen and storing it in the hydrogen tank; the fuel cell performs the inverse electrolysis of water, using hydrogen stored and oxygen from the air to obtain DC electricity. Hydrogen production for external use can be considered (to be used in our facilities or to be sold, "green hydrogen" generation), in this case, the fuel cell will not be considered.

Different technologies for electrolyzers and fuel cells are currently used: alkaline electrolysis (AEL) and proton exchange membrane (PEM) are relatively mature technologies. Data of nominal apparent power, acquisition (CAPEX) and O&M (OPEX) costs, lifetime, minimum output power, and consumption parameters must be set for both the electrolyzer and the fuel cell. For the hydrogen tank, cost (CAPEX) and lifetime must also be set. The electrical energy needed for the compression of the hydrogen generated by the electrolyzer to the pressure needed in the hydrogen tank must also be set, as well as the power consumption of the electrolyzer in stand-by. Electrolyzer efficiency is 45%–66% of the lower heating value (LHV = 33.3 kWh/kg) of hydrogen (Abdin et al., 2020; Burton et al., 2021), while fuel cell efficiency is about 50% of LHV (Gracia et al., 2018; Huang et al., 2021). The efficiency of both devices is not constant and it must be modeled depending on the output power (fuel cell) or depending on the hydrogen produced (electrolyzer), as iHOGA/MHOGA does (unlike other software optimization tools).

6.3.4.8 Thermoelectric generator

iHOGA software can consider TEG in the system, which can be a good option in places with very low irradiation in winter, where heating is needed (Champier, 2017; Dufo-Lopez et al., 2019). Furthermore, in off-grid systems where a stove is used for heating or cooking, TEG can use the residual heat of the stove to produce electricity (with a low efficiency, around 5%). The TEG is usually composed of several thermoelectric modules (TEM) in series or parallel. The hot side of the TEM is connected to the combustion exhaust hot gasses of the stove and the cold side is connected to the ambient temperature or to a cooling circuit. For each TEG considered, the following parameters must be set: nominal power, cost (CAPEX), lifetime, number of TEMs (for each module the number of couples, electrical and thermal resistance, and also thermal resistance of the hot and cold exchangers). Data is also needed for the calculation of the output power depending on the hot and cold sources temperature: we can use the matched load output power curves of the TEM modules datasheet (Dufo-Lopez et al., 2019) or we can use the matrix equations of the simplified model (Favarel et al., 2016) to calculate the open circuit voltage and the output power calculation.

6.3.5 Control strategies

The software uses different control strategies depending on whether load must be supplied or if a grid-connected generating system without consumption load (or with low load) is considered.

6.3.5.1 Control strategies for systems with load consumption

Two main control strategies are used in systems which include renewable sources, BESS and backup generator, and/or hydrogen components (Barley & Winn, 1996):

1. Load following strategy: when power from renewable sources cannot meet the whole load, the rest is covered by the BESS. If the BESS cannot meet the whole demand, the backup generator and/or the fuel cell will run to cover the rest of the load.
2. Cycle charging strategy: when the backup generator or the fuel cell must run because the BESS cannot meet the net load, it will run at its rated power (with high efficiency), so that the extra power will be used to charge the BESS.

In both strategies, the priority to supply the net load (BESS, backup generator, or fuel cell) can be modified by specific control variables. Several variables for the control strategies can be optimized (BESS minimum SOC, backup generator minimum output power, set points for the power limits to primarily use backup generator instead of BESS, set points for SOC limits, etc.). If hydrogen storage is used, up to 12 variables can be optimized (Dufo-López et al., 2007).

6.3.5.2 Control strategies for grid-connected generating systems

In grid-connected generating systems (with or without load consumption), control strategies for the charge/discharge of the storage (BESS, PHS, and/or hydrogen) can be considered by the software: charging during the hours in which the electricity price of the AC grid is low (that is, hours when the money incomes will be low) and discharging during the hours when the price is high (hours when the incomes will be low). If the difference in prices for charging and discharging is sufficient, it can be worth charging/discharging the storage system and that can imply an economic profitability, saving on the electric bill or obtaining benefits when storage injects energy to the AC grid. Different options for the control strategy can be modeled, optimizing two variables (the minimum price for the discharge and the maximum for the charge) or three variables (following the methodology outlined in a previous work (Dufo-López, 2015)).

6.3.6 Constraints

In systems with load consumption, the following constraints must be set: maximum annual unmet load allowed, minimum number of autonomy days, minimum renewable fraction, maximum LCOE, and required ratio between the battery bank capacity and the renewable generation peak power (to ensure the battery bank will be correctly charged by the renewable sources). In grid-connected generating systems, the only constraint to be set is the maximum installation cost.

6.3.7 Other data

Financial data must also be set (nominal annual interest rate, general annual inflation, system lifetime, installation and initial variable cost, loan data, etc.). DC and AC bus voltages must be selected, and also the number of autonomy days (for off-grid systems). In the where case we want to perform multiobjective optimization considering the maximization of the HDI or the jobs created (Dufo-López et al., 2016a), the required data must be introduced.

In addition, the software can perform sensitivity analysis (it can perform automatically different projects, changing the load, irradiation, wind speed, interest rate, general inflation, or electricity or fuel price inflation rates, and components costs). There are also options for the probability analysis, considering different temporal series (following Gaussian probability distribution) of load, irradiation, wind speed, and/or water flow (and also the effect of fuel price inflation can be considered), where each combination of different time series of load and resources (and/or fuel inflation) will be performed using Monte Carlo simulation, according to the methodology presented in the previous works (Dufo-López et al., 2016b, 2016).

6.4 Presizing

After setting the load, resources, and types of components, the software can perform presizing. The software, knowing the load, resources, and available components, calculates the maximum power needed for each source (PV, wind turbines, etc.) to supply the whole annual load so that generators with higher power would not make sense (if we just want to supply the load). Furthermore, in off-grid systems, after setting the autonomy days, the software calculates the size of the storage to supply the load during that number of days. With these results, the software recommends the number of maximum components to be considered in parallel.

6.5 Optimization data

The software includes two optimization options: (i) mono-objective (minimize NPC in systems to cover a load or maximize NPV in grid-connected generating systems) or (ii) multiobjective (for systems to cover a load: minimize NPC, and/or minimize equivalent CO_2 life cycle emissions, and/or minimize unmet load, and/or maximize HID, and/or maximize job creation).

In iHOGA, there is also the possibility of optimizing the cost or the weight of the system in temporary installations (Cristóbal-Monreal & Dufo-López, 2016) (to be transported and assembled in a temporary location and, at the end of the period of use, it must be dismantled and transported back to the original place), for example the generating system of a field hospital. The software can perform the optimization by evaluating all the combinations, without GA. However, if there are many types of components to be considered (high number of possible combinations of components) or too many combinations of control strategies, the optimization times may be inadmissible; to avoid this problem, heuristic techniques can be applied, such as GA (Goldberg, 1989), particle swarm optimization technique (PSO), harmony search algorithm (HSA), simulated annealing (SA), or ant colony algorithms (ACA) (Anoune et al., 2018).

iHOGA and MHOGA make use of two GA: the main algorithm and the secondary algorithm. The variables or genes are integers. The main GA optimizes the combination of components (up to 11 genes can be considered, depending on the type of system; examples of genes are the number of PV modules/generators in parallel, the code for the PV modules/generators type, etc.), providing an optimum configuration for the components, in order to obtain the objective or objectives of the optimization. The secondary GA obtains the most appropriate control strategy (combination of control variables) for each of the components combinations provided by the main algorithm. The secondary algorithm genotype can include up to 12 genes (system control variables). For each genetic algorithm the parameters must be set: maximum number of generations, population, crossover and mutation rates, type of mutation (uniform or nonuniform), and stopping criteria. In a previous work, we found recommendations for these parameters: number of generations should be around 15, while population must be greater than 0.003% of all the possible combinations (Bernal-Agustín & Dufo-López, 2009a), mutation rate is usually within 0.5%–1%, and crossing rate is usually 70%–90% (Bernal-Agustín & Dufo-López, 2009a). The GA parameters can be selected by the user or by iHOGA/MHOGA automatically (in that case, the user must only set the maximum execution time allowed).

In multiobjective optimization (MOEA), the software considers at least two of the following objectives: minimize NPC, minimize life cycle equivalent CO_2 emissions, minimize unmet load by the stand-alone system, maximize HDI, and maximize job creation. The term "unmet load" refers to the energy that cannot be covered by the system (but it can be covered by the AC grid, if available) and it is obtained after the simulation of each combination of components and control strategy. Total equivalent CO_2 cycle life emissions include those generated by the backup generator fuel (usually diesel or gasoline), as well as those generated in the manufacture, transportation, and recycling of the components of the system, and those generated in the electricity purchased from the grid, if applicable (Dufo-López et al., 2011); the result is obtained in kg or t of CO_2 per year (total emissions during the study period from fuel and from manufacturing, transportation, and recycling of components are divided by the number of years of system life). The HDI and the jobs created are calculated following the methodology, set out in (Dufo-López et al., 2016a).

6.6 Economic calculations

In off-grid systems or in grid-connected systems where we want to cover a specific load, the NPC and the LCOE are the two main economic variables, while in grid-connected generating systems, the NPV, LCOE, and IRR are the most important economical results.

The NPC of the system includes all the costs during its lifetime (CAPEX, installation, replacement, O&M, fuel, etc., where all the cash flows are moved to the year 0 considering their annual inflation rate and the interest rate) (Dufo-Lopez et al., 2019).

NPC of each combination of components i and control strategy $k(\text{NPC}_{(i,k)})$ is obtained considering the acquisition cost (CAPEX) of all the components of the system and all the costs and incomes during the system lifetime Life_s (years). Defining $\text{Cost}_{(j)}$ as the acquisition cost (CAPEX) of component j, $\text{NPC}_{\text{rep}(j)}$ the sum of the replacement costs of component j during the system lifetime, converted to the initial moment of the system, $\text{Cost}_{O\&M(j)}$ the annual O&M cost (OPEX) of component j, Cost_F the annual cost of the fuel used by the backup generator, $\text{Cost}_{\text{Purch\_E}}$ the annual cost of the electricity purchased to the electrical grid, $\text{Inc}_{\text{Sell\_E}}$ the annual incomes of the electricity sold to the electrical grid, $\text{Inc}_{\text{Sell\_H2}}$ the annual incomes of the hydrogen sold for external use, and $\text{Cost}_{\text{INST}}$ the installation cost, NPC is calculated in Eq. (6.14) (costs positive, incomes negative):

$$\text{NPC}_{(i,k)} = \sum_j \left\{ \text{Cost}_{(j)} + \text{NPC}_{\text{rep}(j)} + \sum_{l=1}^{\text{Life}_s} \left(\text{Cost}_{O\&M(j)} \left[\frac{(1+\text{Inf}_{\text{gen}})^l}{(1+I)^l} \right]^l \right) \right\}$$
$$+ \sum_{l=1}^{\text{Life}_s} \left\{ \text{Cost}_F \left[\frac{(1+\text{Inf}_F)^l}{(1+I)^l} \right] \right\} + \sum_{l=1}^{\text{Life}_s} \left\{ \text{Cost}_{\text{Purch\_E}} \left[\frac{(1+\text{Inf}_{\text{Purch\_E}})^l}{(1+I)^l} \right] \right\}$$
$$- \sum_{l=1}^{\text{Life}_s} \left\{ \text{Inc}_{\text{Sell\_E}} \left[\frac{(1+\text{Inf}_{\text{Sell\_E}})^l}{(1+I)^l} \right] \right\} - \sum_{l=1}^{\text{Life}_s} \left\{ \text{Inc}_{\text{Sell\_H2}} \left[\frac{(1+\text{Inf}_{\text{Sell\_H2}})^l}{(1+I)^l} \right] \right\} + \text{Cost}_{\text{Inst}} \tag{6.14}$$

where Inf_{gen} is the general annual expected inflation, I the annual interest rate, Inf_F the annual expected fuel inflation, $\text{Inf}_{\text{Purch\_E}}$ the annual expected inflation for the electricity purchased to the AC grid, $\text{Inf}_{\text{Sell\_E}}$ the annual expected inflation for the electricity sold to the AC grid, and $\text{Inf}_{\text{Sell\_H2}}$ the annual expected inflation for the hydrogen sold.

The sum of replacement costs during the system lifetime (Eq. 6.15) is the total replacement cost of component j:

$$\begin{aligned} \text{PC}_{\text{rep}(j)} = & \sum_{m=1}^{N_{\text{rep}(j)}} \left\{ \text{Cost}_{(j)} \left[\frac{\left(1+\text{Inf}_{(j)}\right)^{m\times\text{Life}_{(j)}}}{(1+I)^{m\times\text{Life}_{(j)}}} \right] \right\} \\ & - \text{Cost}_{(j)} \left[\frac{\text{Life}_{(j)} - \left(\text{Life}_{HS} - N_{\text{rep}(j)} \times \text{Life}_{(j)}\right)}{\text{Life}_{(j)}} \right] \\ & \times \left[\frac{\left(1+\text{Inf}_{(j)}\right)^{\text{Life}_s}}{(1+I)^{\text{Life}_s}} \right] \end{aligned} \tag{6.15}$$

where $\text{Life}_{(j)}$ (years), $\text{Inf}_{(j)}$, and $N_{\text{rep}(j)}$ are, respectively, the lifespan of component j, the annual expected inflation of its acquisition cost, and $N_{\text{rep}(j)}$ the number of times it is replaced during the system lifetime (Eq. 6.16):

$$N_{\text{rep}(j)} = \text{Integer}\left(\frac{\text{Life}_s}{\text{Life}_{(j)}} \right) \tag{6.16}$$

The cost of the energy supplied by the system of components i and control strategy k during system lifetime, that is, the LCOE, is calculated in Eq. (6.17) (including inflation and interest rate in the equation denominator):

$$\mathrm{LCOE}_{(i,k)} = \frac{\mathrm{NPC}_{(i,k)}}{\sum_{y=1}^{\mathrm{Life}_s}\left\{E_{\mathrm{load}(y)}\left[\frac{\left(1+\mathrm{Inf}_{\mathrm{gen}}\right)^y}{(1+I)^y}\right]\right\}} \tag{6.17}$$

where $E_{\mathrm{load}(y)}$ (kWh/yr) is the annual (y) consumption load supplied by the off-grid system.

In some cases, LCOE is calculated in a simply way, not considering the factor $\left(1+\mathrm{Inf}_{\mathrm{gen}}\right)^y/(1+I)^y$, as shown in Eq. (6.18):

$$\mathrm{LCOE}_{(i,k)} = \frac{\mathrm{NPC}_{(i,k)}}{\sum_{y=1}^{\mathrm{Life}_s} E_{\mathrm{load}(y)}} \tag{6.18}$$

The calculation of NPV (for grid-connected generating systems) is the same as for NPC (Eq. 6.14) but considering incomes as positive values and costs as negative values. LCOE in this kind of system is calculated using the equation numerator NPV minus the term due to the incomes. IRR is calculated considering all the cash flow during the system lifetime.

6.7 Optimization by GA or MOEA

The two GA used by iHOGA and MHOGA are initialized by randomly generating the first population of individuals (first generation). For each individual, the simulation of the system is done (during a whole year, or during the battery lifetime if Schiffer et al. (2007) battery model is selected, or during the system lifetime for multiperiod, depending on the user selection), in steps from 1 min to 1 h (depending on the user selection). In the cases where some constraints are not met, an arbitrarily high value for NPC is assigned (0 for NPV).

In mono-objective optimization, the NPC or NPV for each valid individual is calculated, and all the valid individuals are sorted from best (lowest NPC or highest NPV) to worst (highest NPC or lowest NPV). Next the fitness of each individual q is specified by applying Eq. (6.19):

$$F_{(q)} = \frac{(Q+1)-q}{\sum_{q=1}^{Q}\{(Q+1)-q\}};\, q = 1,\ldots,Q \tag{6.19}$$

In the following step, reproduction, crossing, and mutation are applied, evaluating the new individuals generated. Elitism is applied (the best individual always passes to the next generation). The procedure is iterated until the stopping criteria are met or a maximum number of generations have appeared. The individual with the lowest NPC or highest NPV of the last generation will be the optimal one.

In multiobjective optimization, the software uses an MOEA based on the SPEA2 algorithm. A Pareto front is obtained for each generation, and at the end of the optimization process (when the last generation of the MOEA is evaluated), the nondominated solutions of the Pareto front are the best, considering, simultaneously, all the objectives. This will be considered the optimal Pareto front (Dufo-López & Bernal-Agustín, 2008).

6.8 Examples of application

iHOGA and MHOGA software has many possibilities of types of projects: off-grid low- or high-power systems with different load consumption (AC, DC, water pumped, hydrogen), different power generators (PV, wind, Diesel, etc., and combinations) and different storage technologies (BESS, PHS, hydrogen, and combinations), and the same for grid-connected systems. Furthermore, different control strategies can be applied (for off-grid systems or for grid-connected systems). In the iHOGA/MHOGA software website (Dufo-López, 2022) download area, a getting started guide is available for iHOGA and another one for MHOGA, where many different example projects are shown.

In this section, we briefly show three different projects: mono-objective optimization of an off-grid PV-wind-diesel-BESS system, multiobjective optimization of a grid-connected PV-wind-diesel-BESS system, and mono-objective optimization of a grid-connected hybrid PV-wind generator with BESS.

6.8.1 iHOGA project, mono-objective, low-power off-grid

The optimization of a low-power off-grid PV-wind-diesel-BESS system is shown in this section. The system is located in Mali, latitude 19.467 degree (North), longitude −1.512 degree (West), altitude 328 m. Average daily horizontal irradiation is 6.42 kWh/m^2/day, average wind speed at 10 m height is 4.16 m/s, average temperature is 28.05 °C (NASA monthly average data, year 2019). The same load profile is considered for the entire year, with an average of 3.63 kWh/day. The optimal tilt angle obtained by iHOGA to maximize the minimum ratio PV generation/load during the year is 15 degree for azimuth 0 degree, which will be the angle used. A total annual irradiation of 2503 kWh/m^2 is obtained for that angle (using NASA hourly data of 2019 as input).

We have selected different component types for PV modules, wind turbines, BESS, etc., with different maximum numbers of allowed components in parallel. In addition, the two main control strategies (load following and cycle charging) will be tested for each combination of components, and the only control strategy variable to be optimized will be the minimum battery SOC (more variables would increase the computation time). Number of possible combinations is 1296 for the components (product of the types of components and the maximum allowed in parallel) and 12 for the control strategy (2 main strategies x 6 possibilities for min SOC), yielding a total of 15,552 possible combinations. 5 min is selected as time step and Schiffer et al. (2007) battery model is selected; therefore, a high simulation time for each combination is needed, roughly 7 s per case (in a computer Intel(R) Core(TM) i5-6500 CPU@3.20 GHz 3.19 GHz, 16 GB RAM); with other battery models, the simulation time can be reduced by a factor of about 15; with 1 h time step it could also be reduced by a factor of about 20).

In our case, evaluating all the combinations would need an execution time of 28 h. Using GA, we have chosen for the main GA 15 generations, population 20, crossover rate 90%, and nonuniform mutation rate 1%, while for the control strategies we have selected to evaluate the 12 combinations (no GA). After 3 h 23′ of execution time, the optimal system is composed of 2 serial × 2 parallel PV modules of 325 Wp, 2 wind turbines of 550 W, 1 diesel generator of 1.9 kVA, 24 serial × 1 parallel OPzS lead-acid batteries of 180 Ah, 2 V, inverter of 1.6 kVA, and PV battery charge controller of 40 A. Optimal control strategy is load following, optimal minimum SOC is 20%. NPC of the optimal system is 21,639.7€, with LCOE 0.83€/kWh. Fig. 6.15 shows the evolution of the main GA during the generations while Figs. 6.16–6.18 show, respectively, the report, costs report (with detailed cash flow during the years for each component), and simulation screen of iHOGA for the optimal solution. In Fig. 6.18 we can see that in that day, the AC net load (AC load, gray curve, minus wind generation, green curve) is very low; the net AC load must be supplied by the BESS (blue curve, DC) by means of the inverter; as net AC load is very low, the inverter efficiency (Fig. 6.14) is very low, needing high power from the BESS to supply the net AC load, showing the importance of the correct modeling of the inverter efficiency vs the output power.

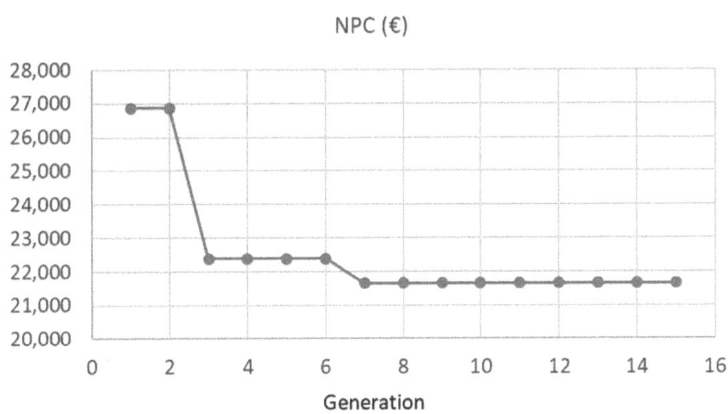

Fig. 6.15 iHOGA project, mono-objective, low-power off-grid. Evolution of the main GA.

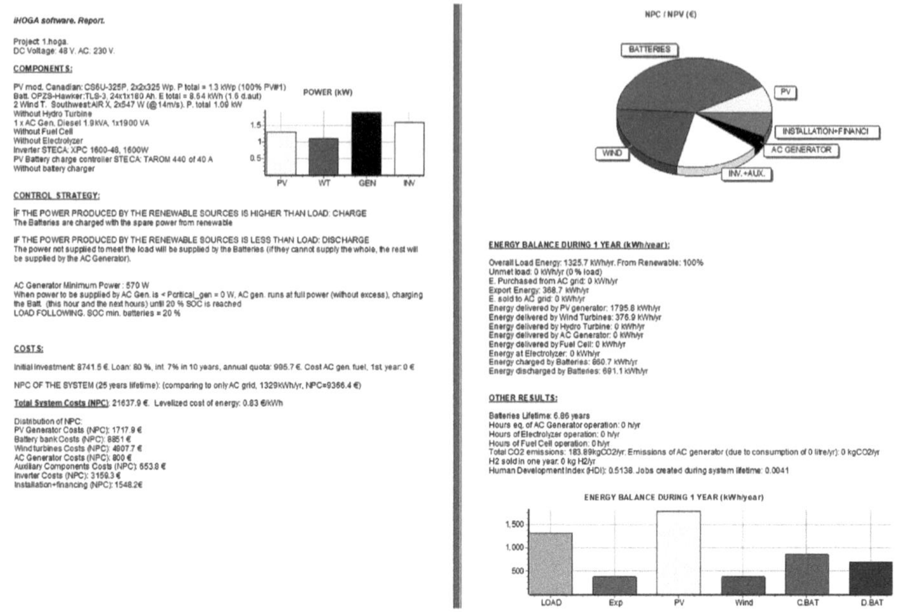

Fig. 6.16 iHOGA project, mono-objective, low-power off-grid. Report of the optimal solution.

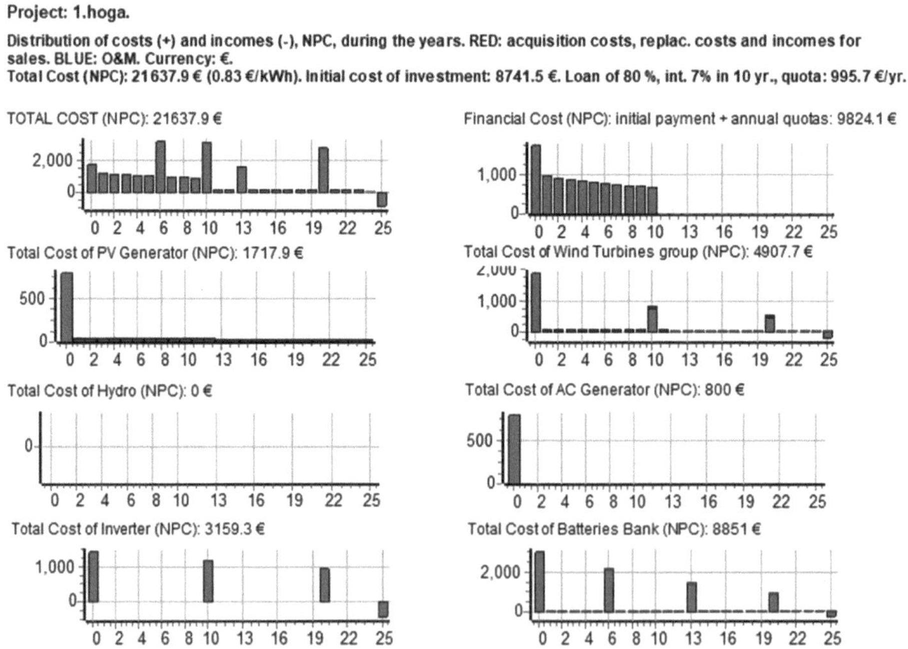

Fig. 6.17 iHOGA project, mono-objective, low-power off-grid. Costs report of the optimal solution.

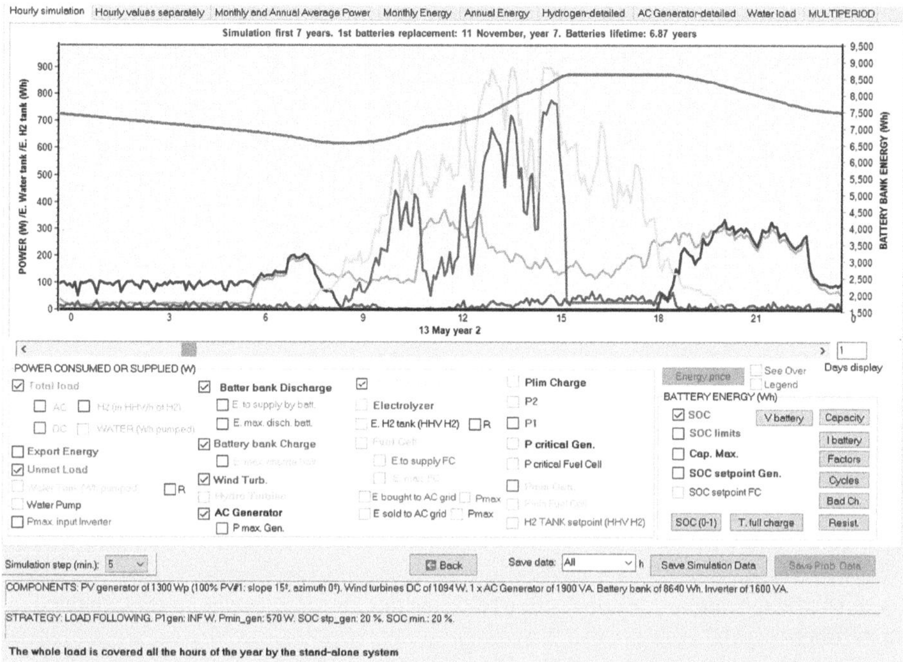

Fig. 6.18 iHOGA project, mono-objective, low-power off-grid. Simulation screen showing a specific day.

6.8.2 iHOGA project, multiobjective, low-power grid-connected system

The same project as the previous one has been optimized in multiobjective (in this case, minimize NPC and unmet load by the standalone system). In this case, we suppose there is AC grid connection; therefore, the unmet load (not covered by the standalone system) will be covered by the AC grid. A maximum 100% of unmet load by the standalone systems is allowed. The results of the multiobjective optimization are shown in Fig. 6.19 (optimal Pareto front with 24 different solutions). We can see the simulation, report, etc., for each solution of the optimal Pareto front.

6.8.3 MHOGA project, mono-objective, high-power grid-connected

The optimization of a high-power grid-connected wind-BESS generating system is shown in this section. It is a generating system without load consumption (therefore, the maximization of the NPV will be the objective), and the control strategies of the BESS will be optimized (to determine the electricity price limits to charge/discharge the BESS). The electricity sold to the AC grid varies during each hour (real-time tariff). The system is located in Gran Canaria Island, Spain, latitude 28.06 degree

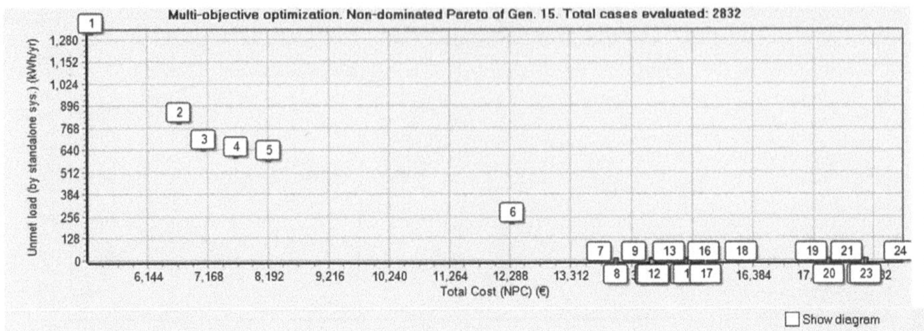

#	Dom. by	Total Cost (NPC)(€)	Emission (kgCO2/yr)	Unmet(kWh/yr)	Unmet(%)	D.aut	Cn(Wh)/(Ppv+Pw)(W)	Ren(%)	LCOE(€/kWh)	Simulate	Re
1	0	5171.4	529.98	**1324.9**	100	INF	0	0	0.2	SIMULATE..	RI
2	0	6672.2	335.8	**807.5**	60.95	INF	0	39.1	0.26	SIMULATE..	RI
3	0	7090.6	286.23	**651.6**	49.18	INF	0	50.8	0.27	SIMULATE..	RI
4	0	7635	282.29	**609.7**	46.02	INF	0	54	0.29	SIMULATE..	RI
5	0	8200.7	286.06	**587.2**	44.31	INF	0	55.7	0.32	SIMULATE..	RI
6	0	12322.3	144.32	**230.8**	17.42	INF	13.3	82.6	0.47	SIMULATE..	RI
7	0	13785.3	78.6	**0**	0	INF	8.8	100	0.53	SIMULATE..	RI
8	0	14071.9	78.6	**0**	0	INF	8.8	100	0.54	SIMULATE..	RI
9	0	14376	91.4	**0**	0	INF	6.6	100	0.55	SIMULATE..	RI

Fig. 6.19 iHOGA project, multiobjective NPC—unmet load, low-power system. Optimal Pareto front.

(North), longitude −15.5 degree (West). The average wind speed at 100 m height is 8.51 m/s with a Weibull form factor of 3.7 (using Renewables Ninja hourly data of 2019 as input). Wind turbines of 2 MW are considered, from 6 to 15 in parallel. BESS Li-ion batteries of 0, 2.5 or 5 MWh are considered, with 0, 2.5 or 5 MW inverter-charger. The maximum power that can be injected to the AC grid is 23 MW and the unique constraint is that the investment cannot be higher than 40 M€. Two variables are considered for the optimization of the BESS control (the minimum price of electricity for the discharge and the maximum for the charge), considering 8 values (linearly from 0 to maximum electricity cost) for each one.

The number of possible combinations of components is 90, while the possible control strategies are 64; therefore, the total number of possible combinations is 5760. Multiperiod optimization is selected, considering the reduction in 1%/year of the power curve of the wind turbines, and different annual variations in the wind speed and in the inflation of the price of electricity sold to the grid. The evaluation of all the possible combinations of components and control strategies took more than 4 h (enumerative method), while using GA (main and secondary), in less than 0.5 h we obtained the same optimal combination: a system composed of 15 wind turbines of 2 MW, BESS of 5 MWh, and inverter/charger of 5 MW, with the following optimal control strategy: charge when electricity price is lower than 0.0729 €/kWh; discharge when price is higher than 0.1451 €/kWh. Fig. 6.20 shows the simulation of 3 days (1st to 3rd January of the 3rd year) of the optimal solution: in the top graph the electricity price is shown, with the limits to charge and discharge, while in the bottom graph, the wind turbines generation, the charge/discharge of the BESS, and the electricity sold to the grid are shown.

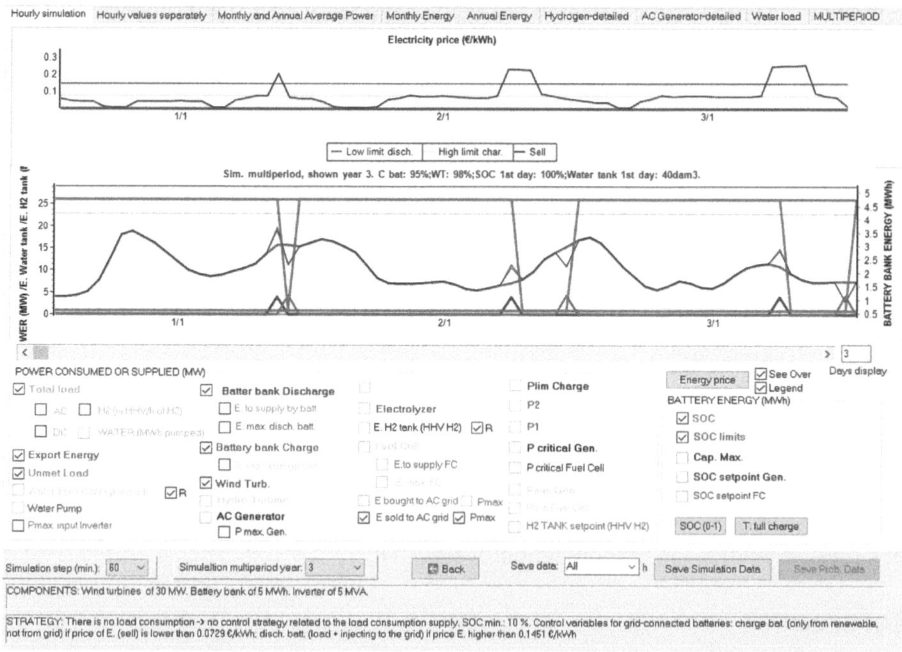

Fig. 6.20 MHOGA project, mono-objective (max. NPV). Simulation of the optimal system during 3 days of January.

6.9 Conclusions

In this chapter, the software tools iHOGA and MHOGA are shown. They are two software tools based in the same architecture: iHOGA is for low- and medium-power systems and MHOGA is used for high-power systems (MW scale). The software uses a GA to efficiently optimize the combination of components and control strategies of renewable-based systems. The systems to be optimized can include PV generator, wind turbine, hydro turbine (including PHS), inverter/charger, BESS, backup generator (diesel, gasoline, etc.), TEG, and/or hydrogen components (electrolyzer, hydrogen tank, and fuel cell). The software tries to obtain the optimal combination of the different components and control strategies to supply a specific load (AC and/or DC electrical load, and/or external hydrogen load, and/or water load consumption pumped form river or reservoir), satisfying all the constraints (unmet load lower than a maximum percentage, renewable fraction higher than a minimum value, energy price lower than a maximum value, and others).

The software can consider off-grid systems (systems where all the load must be covered by the stand-alone system) or grid-connected systems (where electricity can be imported/exported to the AC electrical distribution grid), considering a load consumption system (where a specific load must be supplied) or considering a generating system (where all or almost all of the electricity produced is injected to the AC grid, that is, there is no load or a very low load consumption). Green hydrogen generating systems (grid-connected renewable generators that export electricity to the AC grid and also generate hydrogen in electrolyzers) can also be considered. In systems with load consumption, the optimization objective is to minimize the NPC of the system, while in generating systems (without load) the objective is to maximize the NPV.

In off-grid systems, storage and/or a backup generator is needed and the control strategy can be optimized to minimize the NPC. The control strategy decides how the net load (the load not covered by the renewable power) must be primarily covered (by the battery, by the fuel cell, or by the backup generator), using set points for different control variables. In grid-connected systems, different prices can be considered for the electricity imported/exported to the distribution electrical grid, including the TOU tariffs and the RTP, where control strategies can be used to optimize the storage (charging batteries from the grid when electricity price is low and discharging them when the price is high, supplying the load and/or injecting electricity to the grid). In off-grid systems, MOEA can be applied to obtain an optimal Pareto front, not only considering the minimization of NPC as an objective but also the minimization of life cycle equivalent CO_2 emissions, the minimization of unmet load, the maximization of jobs created, and/or the maximization of the HDI.

References

Abdin, Z., Zafaranloo, A., Rafiee, A., Merida, W., Lipinski, W., & Khalilpour, K. R. (2020). Hydrogen as an energy vector. *Renewable and Sustainable Energy Reviews, 120.* https://doi.org/10.1016/j.rser.2019.109620, 109620.

Aguiar, R., & Collares-Pereira, M. (1992). TAG: A time-dependent, autoregressive, Gaussian model for generating synthetic hourly radiation. *Solar Energy*, *49*(3), 167–174. https://doi.org/10.1016/0038-092X(92)90068-L.

Al-falahi, M. D. A., Jayasinghe, S. D. G., & Enshaei, H. (2017). A review on recent size optimization methodologies for standalone solar and wind hybrid renewable energy system. *Energy Conversion and Management*, *143*, 252–274. https://doi.org/10.1016/j.enconman.2017.04.019.

Anoune, K., Bouya, M., Astito, A., & Ben Abdellah, A. (2018). Sizing methods and optimization techniques for PV-wind based hybrid renewable energy system: A review. *Renewable and Sustainable Energy Reviews*, *93*, 652–673. https://doi.org/10.1016/j.rser.2018.05.032.

Armand, M., Axmann, P., Bresser, D., Copley, M., Edstrom, K., Ekberg, C., Guyomard, D., Lestriez, B., Novak, P., Petranikova, M., Porcher, W., Trabesinger, S., Wohlfahrt-Mehrens, M., & Zhang, H. (2020). Lithium-ion batteries—Current state of the art and anticipated developments. *Journal of Power Sources*, *479*. https://doi.org/10.1016/j.jpowsour.2020.228708, 228708.

Barley, C. D., & Winn, C. B. (1996). Optimal dispatch strategy in remote hybrid power systems. *Solar Energy*, *58*(4–6), 165–179. https://doi.org/10.1016/S0038-092X(96)00087-4.

Bernal-Agustín, J. L., & Dufo-López, R. (2009a). Efficient design of hybrid renewable energy systems using evolutionary algorithms. *Energy Conversion and Management*, *50*(3), 479–489. https://doi.org/10.1016/j.enconman.2008.11.007.

Bernal-Agustín, J. L., & Dufo-López, R. (2009b). Multi-objective design and control of hybrid systems minimizing costs and unmet load. *Electric Power Systems Research*, *79*(1), 170–180. https://doi.org/10.1016/j.epsr.2008.05.011.

Bernal-Agustín, J. L., & Dufo-López, R. (2009c). Simulation and optimization of stand-alone hybrid renewable energy systems. *Renewable and Sustainable Energy Reviews*, *13*(8), 2111–2118. https://doi.org/10.1016/j.rser.2009.01.010.

Bernal-Agustín, J. L., Dufo-López, R., & Rivas-Ascaso, D. M. (2006). Design of isolated hybrid systems minimizing costs and pollutant emissions. *Renewable Energy*, *31*(14), 2227–2244. https://doi.org/10.1016/j.renene.2005.11.002.

Burton, N. A., Padilla, R. V., Rose, A., & Habibullah, H. (2021). Increasing the efficiency of hydrogen production from solar powered water electrolysis. *Renewable and Sustainable Energy Reviews*, *135*. https://doi.org/10.1016/j.rser.2020.110255, 110255.

Champier, D. (2017). Thermoelectric generators: A review of applications. *Energy Conversion and Management*, *140*, 167–181. https://doi.org/10.1016/j.enconman.2017.02.070.

Coello, C. A., Veldhuizen, D. A. V., & Lamont, G. B. (2002). *Evolutionary Algorithms for Solving Multi-Objective Problems*. Springer. https://doi.org/10.1007/978-0-387-36797-2.

Collares Pereira, M., & Rabl, A. (1979). The average distribution of solar radiation-correlations between diffuse and hemispherical and between daily and hourly insolation values. *Solar Energy*, *22*(2), 155–164. https://doi.org/10.1016/0038-092X(79)90100-2.

Cristóbal-Monreal, I. R., & Dufo-López, R. (2016). Optimisation of photovoltaic–diesel–battery stand-alone systems minimising system weight. *Energy Conversion and Management*, *119*, 279–288. https://doi.org/10.1016/j.enconman.2016.04.050.

Downing, S., & Socie, D. (1982). Simple rainflow counting algorithms. *International Journal of Fatigue*, *4*(1), 31–40. https://doi.org/10.1016/0142-1123(82)90018-4.

Dufo-López, R. (2015). Optimisation of size and control of grid-connected storage under real time electricity pricing conditions. *Applied Energy*, *140*, 395–408. https://doi.org/10.1016/j.apenergy.2014.12.012.

Dufo-López, R. (2022). *Software iHOGA/MHOGA*. https://ihoga.unizar.es/en/.

Dufo-López, R., & Bernal-Agustín, J. L. (2005). Design and control strategies of PV-diesel systems using genetic algorithms. *Solar Energy*, *79*(1), 33–46. https://doi.org/10.1016/j.solener.2004.10.004.

Dufo-López, R., & Bernal-Agustín, J. L. (2008). Multi-objective design of PV-wind-diesel-hydrogen-battery systems. *Renewable Energy*, *33*(12), 2559–2572. https://doi.org/10.1016/j.renene.2008.02.027.

Dufo-López, R., & Bernal-Agustín, J. L. (2012). New methodology for the generation of hourly wind speed data applied to the optimization of stand-alone systems. *Energy Procedia*, *14*, 1973–1978. https://doi.org/10.1016/j.egypro.2011.12.1196.

Dufo-López, R., Bernal-Agustín, J. L., & Contreras, J. (2007). Optimization of control strategies for stand-alone renewable energy systems with hydrogen storage. *Renewable Energy*, *32*(7), 1102–1126. https://doi.org/10.1016/j.renene.2006.04.013.

Dufo-López, R., Bernal-Agustín, J. L., Yusta-Loyo, J. M., Domínguez-Navarro, J. A., Ramírez-Rosado, I. J., Lujano, J., & Aso, I. (2011). Multi-objective optimization minimizing cost and life cycle emissions of stand-alone PV–wind–diesel systems with batteries storage. *Applied Energy*, *88*(11), 4033–4041. https://doi.org/10.1016/j.apenergy.2011.04.019.

Dufo-Lopez, R., Champier, D., Gibout, S., Lujano-Rojas, J. M., & Dominguez-Navarro, J. A. (2019). Optimisation of off-grid hybrid renewable systems with thermoelectric generator. *Energy Conversion and Management*, *196*, 1051–1067. https://doi.org/10.1016/j.enconman.2019.06.057.

Dufo-López, R., Cristóbal-Monreal, I. R., & Yusta, J. M. (2016a). Optimisation of PV-wind-diesel-battery stand-alone systems to minimise cost and maximise human development index and job creation. *Renewable Energy*, *94*, 280–293. https://doi.org/10.1016/j.renene.2016.03.065.

Dufo-López, R., Cristóbal-Monreal, I. R., & Yusta, J. M. (2016b). Stochastic-heuristic methodology for the optimisation of components and control variables of PV-wind-diesel-battery stand-alone systems. *Renewable Energy*, *99*, 919–935. https://doi.org/10.1016/j.renene.2016.07.069.

Dufo-López, R., Lujano-Rojas, J. M., & Bernal-Agustín, J. L. (2014). Comparison of different lead–acid battery lifetime prediction models for use in simulation of stand-alone photovoltaic systems. *Applied Energy*, *115*, 242–253. https://doi.org/10.1016/j.apenergy.2013.11.021.

Dufo-López, R., Pérez-Cebollada, E., Bernal-Agustín, J. L., & Martínez-Ruiz, I. (2016). Optimisation of energy supply at off-grid healthcare facilities using Monte Carlo simulation. *Energy Conversion and Management*, *113*, 321–330. https://doi.org/10.1016/j.enconman.2016.01.057.

Durusoy, B., Ozden, T., & Akinoglu, B. G. (2020). Solar irradiation on the rear surface of bifacial solar modules: A modeling approach. *Scientific Reports*, *10*(13300). https://doi.org/10.1038/s41598-020-70235-3.

Erbs, D., Klein, S., & Beckman, W. (1983). Estimation of degree-days and ambient temperature bin data from monthly-average temperatures. *ASHRAE Journal*, *25*(6), 60–65.

Erbs, D., Klein, S., & Duffie, J. (1982). Estimation of the diffuse radiation fraction for hourly, daily and monthly-average global radiation. *Solar Energy*, *28*(4), 293–302. https://doi.org/10.1016/0038-092X(82)90302-4.

PVGIS Photovoltaic Geographical Information System. (2020). European Commission. https://joint-research-centre.ec.europa.eu/pvgis-photovoltaic-geographical-information-system_en.

Favarel, C., Bedecarrats, J. P., Kousksou, T., & Champier, D. (2016). Experimental analysis with numerical comparison for different thermoelectric generators configurations. *Energy Conversion and Management*, *107*, 114–122. https://doi.org/10.1016/j.enconman.2015.06.040.

Goldberg, D. (1989). *Genetic Algorithms in Search, Optimization, and Machine Learning* (1st ed.). Addison-Wesley.

Gracia, L., Casero, P., Bourasseau, C., & Chabert, A. (2018). Use of hydrogen in off-grid locations, a techno-economic assessment. *Energies*, *11*(11), 3141. https://doi.org/10.3390/en11113141.

Graham, V. A., & Hollands, K. G. T. (1990). A method to generate synthetic hourly solar radiation globally. *Solar Energy*, *44*(6), 333–341. https://doi.org/10.1016/0038-092X(90)90137-2.

Groot, J., Swierczynski, M., Stan, A. I., & Kær, S. K. (2015). On the complex ageing characteristics of high-power LiFePO4/graphite battery cells cycled with high charge and discharge currents. *Journal of Power Sources*, *286*, 475–487. https://doi.org/10.1016/j.jpowsour.2015.04.001.

Hay, J. E., & Davies, J. A. (1978). In J. E. Hay, & T. K. Won (Eds.), *Calculations of the Solar Radiation Incident on an Inclined Surface* (1st ed.). Ministry of Supply and Services Canada.

Huang, P. H., Kuo, J. K., Jiang, W. Z., & Wu, C. B. (2021). Simulation analysis of hydrogen recirculation rates of fuel cells and the efficiency of combined heat and power. *International Journal of Hydrogen Energy*, *46*(31), 16823–16835. https://doi.org/10.1016/j.ijhydene.2020.08.010.

Krieger, E. M., Cannarella, J., & Arnold, C. B. (2013). A comparison of lead-acid and lithium-based battery behavior and capacity fade in off-grid renewable charging applications. *Energy*, *60*, 492–500. https://doi.org/10.1016/j.energy.2013.08.029.

Leloux, J., Lorenzo, E., García-Domingo, B., Aguilera, J., & Gueymard, C. A. (2014). A bankable method of assessing the performance of a CPV plant. *Applied Energy*, *118*, 1–11. https://doi.org/10.1016/j.apenergy.2013.12.014.

Liu, B. Y. H., & Jordan, R. C. (1960). The interrelationship and characteristic distribution of direct, diffuse and total solar radiation. *Solar Energy*, *4*(3), 1–19. https://doi.org/10.1016/0038-092X(60)90062-1.

Manwell, J., McGowan, J., & Rogers, A. L. (2009). *Wind Energy Explained. Theory, Design and Application* (2nd ed.). John Wiley & Sons Ltd.

Matt, C. F., Vieira, L. S. R., Soares, G. F. W., & De Faria, L. P. T. (2005). Optimization of the operation of isolated industrial diesel stations. In *6th World Congress on Structural and Multidisciplinary Optimization Rio de Janeiro*. 30 May – 03 June 2005, Brazil.

NASA Prediction of Worldwide Energy Resources. (2022). *The POWER Project*. https://power.larc.nasa.gov/.

Petit, M., Prada, E., & Sauvant-Moynot, V. (2016). Development of an empirical aging model for Li-ion batteries and application to assess the impact of vehicle-to-grid strategies on battery lifetime. *Applied Energy*, *172*, 398–407. https://doi.org/10.1016/j.apenergy.2016.03.119.

Pfenninger, S., & Staffell, I. (2022). *Renewable Ninja*. https://www.renewables.ninja.

Rietveld, M. (1978). A new method for estimating the regression coefficients in the formula relating solar radiation to sunshine. *Agricultural Meteorology*, *19*(2–3), 243–252. https://doi.org/10.1016/0002-1571(78)90014-6.

Saxena, S., Hendricks, C., & Pecht, M. (2016). Cycle life testing and modeling of graphite/LiCoO2 cells under different state of charge ranges. *Journal of Power Sources*, *327*, 394–400. https://doi.org/10.1016/j.jpowsour.2016.07.057.

Schiffer, J., Sauer, D. U., Bindner, H., Cronin, T., Lundsager, P., & Kaiser, R. (2007). Model prediction for ranking lead-acid batteries according to expected lifetime in renewable energy systems and autonomous power-supply systems. *Journal of Power Sources*, *168*(1), 66–78. https://doi.org/10.1016/j.jpowsour.2006.11.092.

Sinha, S., & Chandel, S. S. (2014). Review of software tools for hybrid renewable energy systems. *Renewable and Sustainable Energy Reviews*, *32*, 192–205. https://doi.org/10.1016/j.rser.2014.01.035.

Skarstein, O., & Ulhen, K. (1989). Design considerations with respect to long-term diesel saving in wind/diesel plants. *Wind Engineering*, *13*(2), 72–87. https://www.jstor.org/stable/43749369.

Swierczynski, M., Stroe, D. I., Stan, A. I., Teodorescu, R., & Kær, S. K. (2015). Lifetime estimation of the nanophosphate LiFePO4/C battery chemistry used in fully electric vehicles. *IEEE Transactions on Industry Applications*, *51*(4), 3453–3461. https://doi.org/10.1109/TIA.2015.2405500.

Wang, J., Liu, P., Hicks-Garner, J., Sherman, E., Soukiazian, S., Verbrugge, M., Tataria, H., Musser, J., & Finamore, P. (2011). Cycle-life model for graphite-LiFePO4 cells. *Journal of Power Sources*, *196*(8), 3942–3948. https://doi.org/10.1016/j.jpowsour.2010.11.134.

Zitzler, E., & Thiele, L. (1999). Multiobjective evolutionary algorithms: A comparative case study and the strength Pareto approach. *IEEE Transactions on Evolutionary Computation*, *3*(4), 257–271. https://doi.org/10.1109/4235.797969.

Zubi, G., Dufo-López, R., Carvalho, M., & Pasaoglu, G. (2018). The lithium-ion battery: State of the art and future perspectives. *Renewable and Sustainable Energy Reviews*, *89*, 292–308. https://doi.org/10.1016/j.rser.2018.03.002.

Operation and management of modern electrical systems

7

7.1 Energy policy and environmental problems

The world is undergoing substantial changes due to many environmental and sociopolitical events. Among the environmental problems, climate change is the most relevant issue. Climate change produces dramatic variations in meteorological patterns, impacting food production, raising sea levels, and consequently, the probability of flooding. Thus, drastic changes in human behavior toward a sustainable society are urgently required.

Greenhouse gas (GHG) emissions play a vital role in the ecosystem by making the planet an appropriate living place. They preserve part of the warmth coming from the sun, making the earth a suitable place for the development of life. However, industrialization, deforestation, and extensive agriculture has increased GHG emissions to overwhelming values, never previously observed in 3 million years (United Nations, 2022a).

Economic growth, the increase in the human population, and rising living standards drive the increase in GHG emissions, which is directly related to the average global temperature. GHG emissions and global temperature have been growing since the Industrial Revolution and carbon dioxide (CO_2), which accounts for most GHGs, results from the usage of fossil fuels (United Nations, 2022a).

Based on this reasoning, Fig. 7.1 presents the industrialization intensity index per country for 2018, and Fig. 7.2 shows the evolution of CO_2 emissions per country for 1960, 1979, 1999, and 2018 (The World Bank, 2022a, 2022b). Several countries such as China (CHN), the United States (USA), India (IND), and Russia (RUS) have been emphasized to improve information understanding.

According to Fig. 7.1, the United States and China were countries with a high industrialization intensity in 2018. In Fig. 7.2, we can observe how CO_2 emissions increased in the United States, China, and India. In the case of Russia, the data shows a sudden reduction in 1999, 8 years after the collapse of the Soviet Union. Another decline in CO_2 emissions occurred during the COVID-19 pandemic. Between 2019 and 2020, China reduced CO_2 emissions by 1.4%, the United States by 12.9%, the European Union and the United Kingdom by 7.7%, India by 8%, and Russia by 2.8%. In total, the global reduction was 6.4%. In general, economic and political events are closely related to industrial performance and GHG emissions (Tollefson, 2021).

Scientists worldwide are working to understand the effects of climate change on environmental variables such as ambient temperature, wind speed, and precipitation to develop projections and strategies to reduce GHG emissions effectively.

Genetic Optimization Techniques for Sizing and Management of Modern Power Systems
https://doi.org/10.1016/B978-0-12-823889-9.00006-0

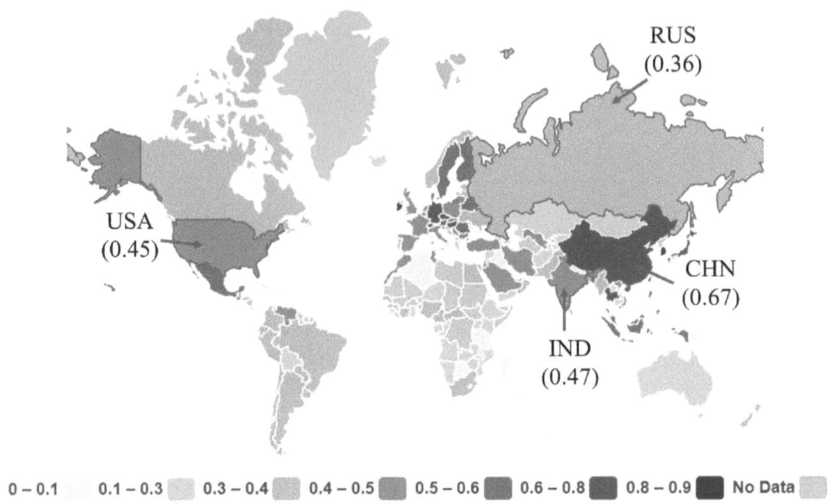

Fig. 7.1 Industrialization intensity index.

The Intergovernmental Panel on Climate Change (IPCC) performs studies and predictions on GHG emissions and their impact on the weather pattern. IPCC disseminates its findings periodically to provide policymakers with unbiased scientific data. According to the IPCC Sixth Assessment Report (AR6) (IPCC Working Group I, 2021), industrial activities have caused the atmosphere, ocean, and land warming. Consequently, the atmosphere, ocean, cryosphere, and biosphere are changing rapidly and extensively.

The magnitude and scale of the changes currently observed in the climate system have not been seen in many centuries or thousands of years. The changes in the weather due to human actions are exhibited in extreme meteorological events such as heatwaves, heavy precipitation, droughts, and tropical cyclones (IPCC Working Group I, 2021).

Adopting updated information, including paleoclimate evidence and radiative forcing, results in a more reliable estimation of the equilibrium climate sensitivity, which is 3 °C. Experts estimate GHG emissions have increased since 1750 due to human activities. The growing trend has continued, reaching 410 ppm of CO_2, 1866 parts per billion (ppb) of methane (CH_4), and 332 ppb of nitrous oxide (N_2O) in 2019. Over the last 60 years, the land and ocean have absorbed around 56% per year of the CO_2 emissions (IPCC Working Group I, 2021).

Industrial activities have transformed the climate system profoundly to a state never seen before. The highest amounts of CO_2, CH_4, and N_2O were registered in 2019. Thus, CO_2, CH_4, and N_2O concentrations increased by 47%, 156%, and 23%, respectively. The increments of CO_2 and CH_4 are higher, and the N_2O is similar to that observed during the transition from the glacial to the interglacial period (IPCC Working Group I, 2021).

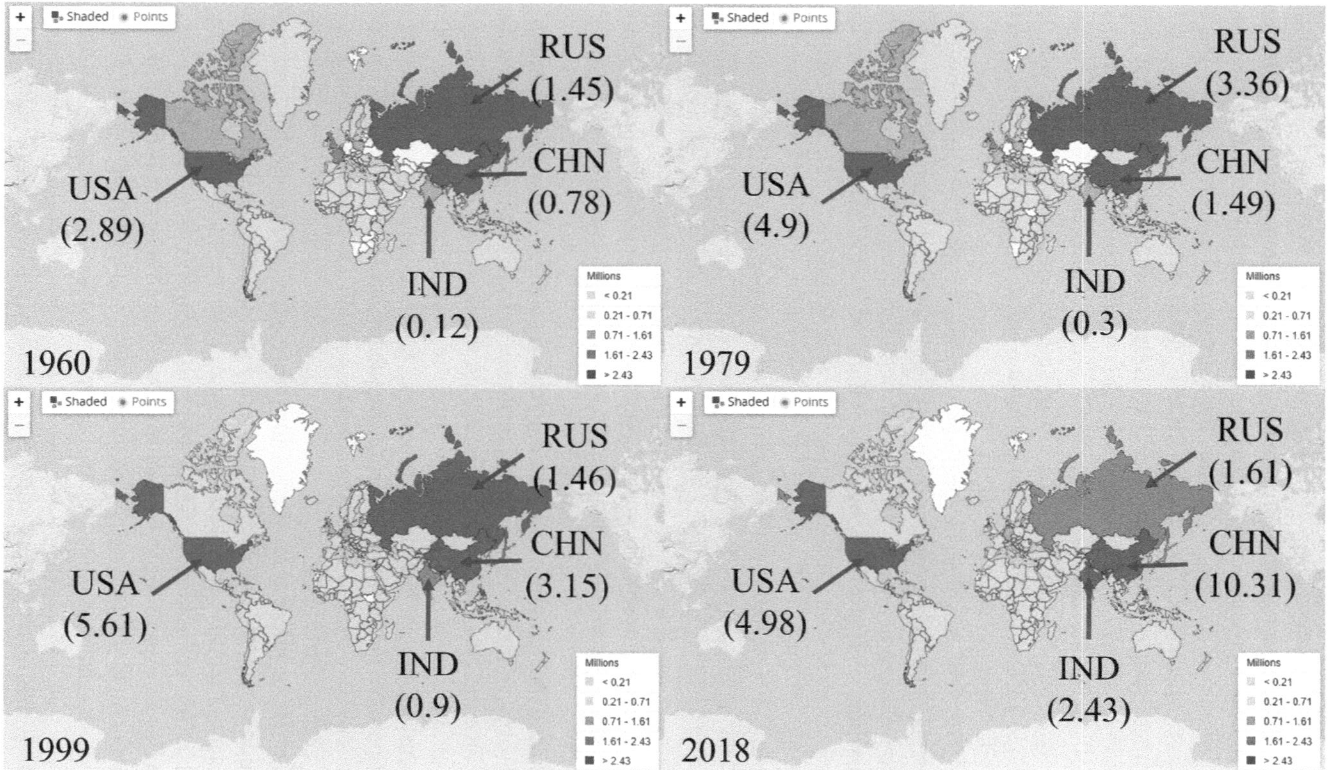

Fig. 7.2 Carbon dioxide emissions.

The radiative forcing induced by human activities has warmed the climate system by incrementing GHG emissions and is reduced in part by the cooling phenomena related to aerosol concentrations. The radiative forcing reported in 2019 was $2.72\,W/m^2$. The net radiative forcing (including the cooling associated with the aerosol concentration) produces energy accumulation, which is transformed into heat, warming the climate system. However, the warming effect is reduced by increasing energy loss to space (IPCC Working Group I, 2021).

The climate system heating from 1971 to 2006 was $0.5\,W/m^2$, while that observed in 2006–2018 was $0.79\,W/m^2$, representing a considerable increment. Consequently, the ocean, land, ice loss, and atmospheric warming represented 91%, 5%, 3%, and 1%, respectively. From 2011 to 2020, the global surface temperature (GST) was 1.09 °C higher than in 1850–1900. These changes have a higher impact on the land than on the ocean (IPCC Working Group I, 2021).

Since 1970, GST has been increasing faster than in any other period over the last 2000 years. Thus, the temperature from 2011 to 2020 was higher than that registered during the multicentury warm period that occurred approximately 6500 years ago. GHG emissions changed GST by a magnitude of 1–2 °C, and aerosols related to human activities reduced the temperature between 0 and 0.8 °C. Natural factors modified GST by −0.1 to 0.1 °C, and the internal variability changed the temperature by −0.2 to 0.2 °C. Furthermore, GHG emissions have caused tropospheric warming since 1979, and the stratospheric ozone depletion was the main factor in lower stratosphere cooling from 1979 to 1990 (IPCC Working Group I, 2021).

Figs. 7.3 and 7.4 (IPCC Working Group I, 2022) present the mean temperature and wind speed changes for several warming scenarios. We can observe drastic changes in the environmental conditions that would make living even more difficult than it is now.

Concerning precipitation, it has been increasing since 1950, with a faster rate since 1980. Probably, human behavior has contributed to the modifications of precipitation patterns since the mid-20th century, including the near-surface ocean salinity. Mid-latitude storm tracks have shifted poleward since 1980. Additionally, the extratropical jet in the austral summer has moved poleward. Environmental events involving heavy precipitation have increased in intensity, frequency, and area since 1950. In addition, human actions have increased land evapotranspiration and associated droughts (IPCC Working Group I, 2021).

Northern Hemisphere aerosol emissions have decreased the global land monsoon precipitation from 1950 to 1980. However, the increment of GHG concentrations and the internal variability observed in decadal and multidecadal time scales have increased such precipitation. In South and East Asia, and West Africa, monsoon precipitation has increased due to GHG emissions. Nevertheless, it is neutralized by the aerosol concentrations emitted during the 20th century. Tropical cyclones of significant categories have occurred more frequently in the last 40 years. Furthermore, the location where the tropical cyclones have reached their maximum intensity has moved northward. Due to human influence, heavy precipitations related to tropical cyclones have increased (IPCC Working Group I, 2021).

Industrial activities have also motivated the global retreat of glaciers since 1990 and the decrease in Arctic sea ice since 1979. In this regard, relevant changes were

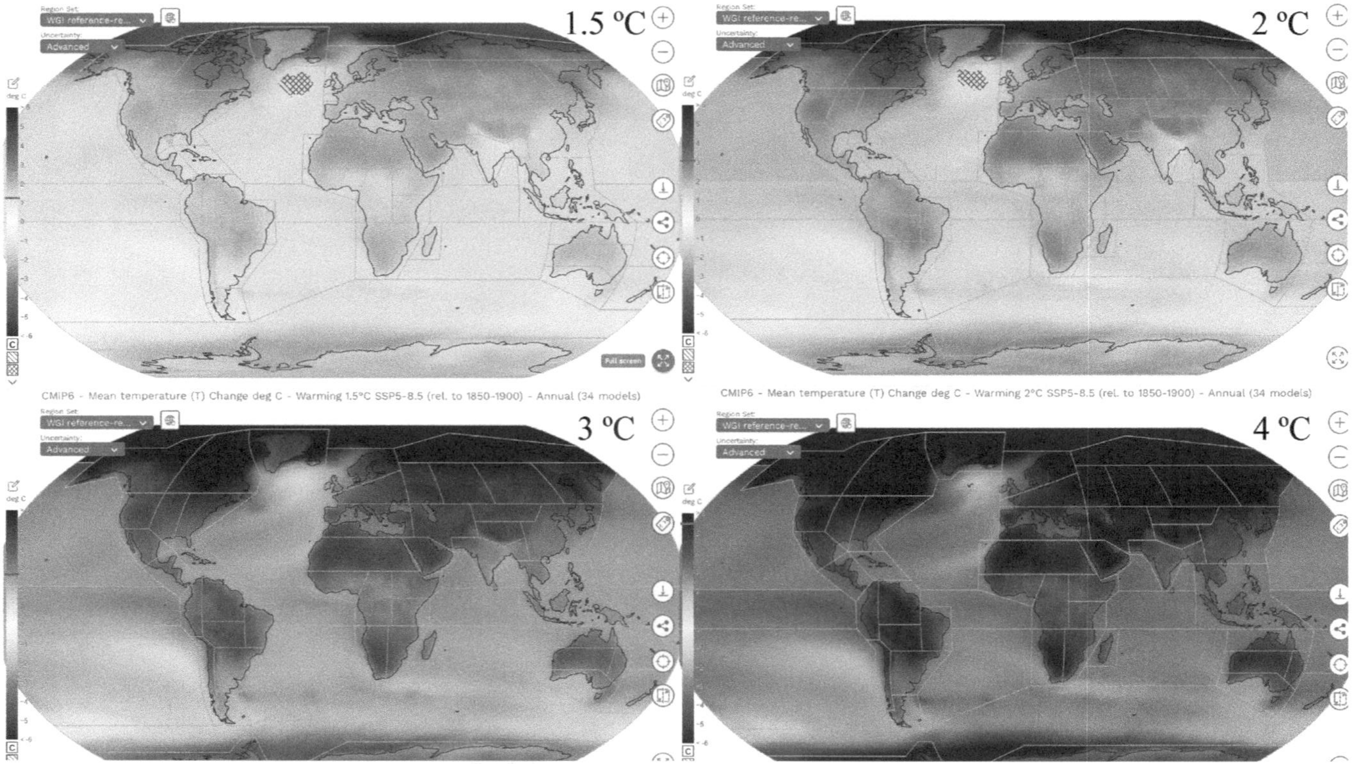

Fig. 7.3 Mean temperature change for different warming conditions.

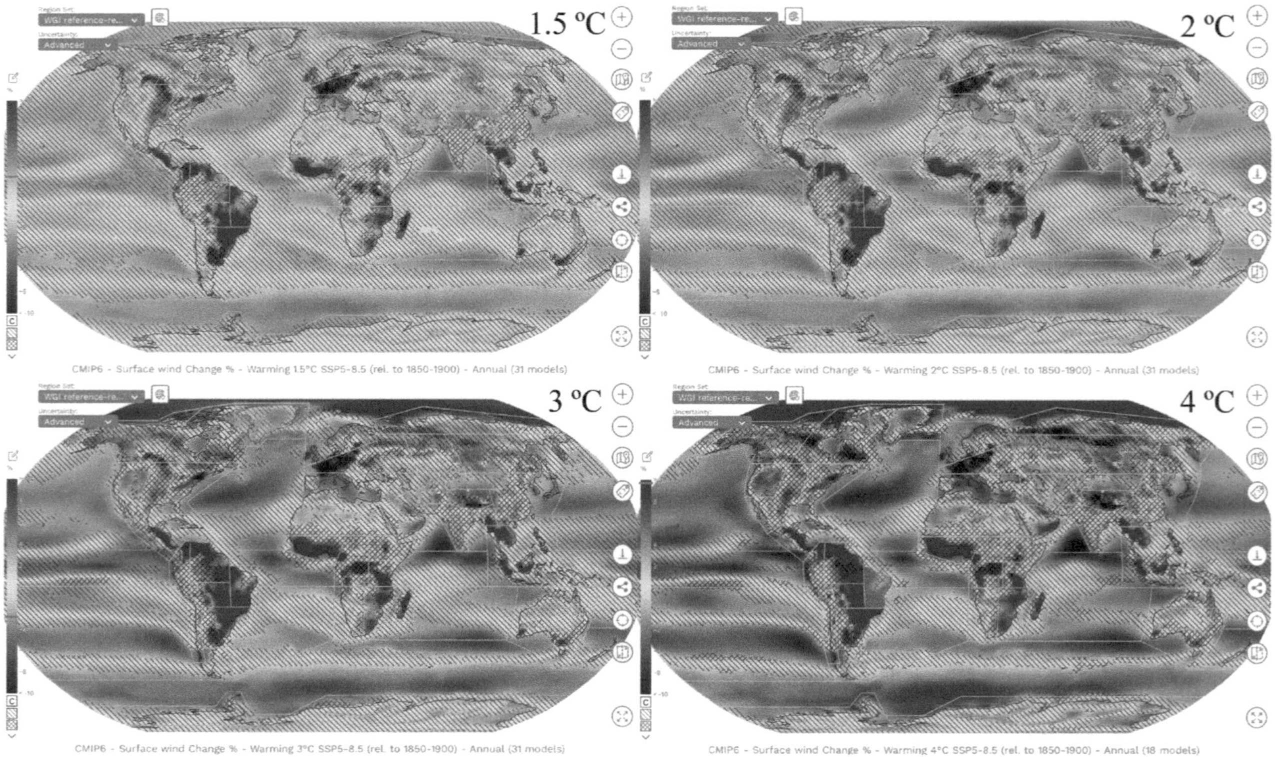

Fig. 7.4 Surface wind change for different warming conditions.

registered from 1979 to 2020 due to opposing trends and considerable internal variability regarding the Antarctic sea ice. Moreover, human behavior has also contributed to reducing Northern Hemisphere spring snow cover, the melting of the Greenland Ice Sheet, and the Antarctic Ice Sheet mass loss. From 2011 to 2020, the Arctic sea ice area reached its minimum value in 1000 years. The glaciers have retreated since 1950 worldwide, almost simultaneously, in a manner never seen in 2000 years (IPCC Working Group I, 2021).

CO_2 emissions have also warmed the global upper ocean, contributing to the global acidification of the surface open ocean and the fall of oxygen levels in the upper ocean regions. From 1901 to 2018, the global mean sea (GMS) level increased between 0.15 and 0.25 m; specifically, between 0.6 and 2.1 mm/yr during the period 1901–1971, between 0.8 and 2.9 mm/yr during the period 1970–2006, and between 3.2 and 4.2 mm/yr during the period 2006–2018 (IPCC Working Group I, 2021).

Since 1900, the GMS level has increased at the highest rate in the last 3000 years. Globally, the ocean has warmed in the past century at a higher rate than observed since the last deglacial transition. Additionally, the surface open ocean pH has been showing lower values than in the previous 2 million years, which is uncommon. The increment in the sea level that occurred from 1971 to 2018 is related to thermal expansion, ice loss in glaciers and ice sheets, and variations in land and water structure by 50%, 22% and 20%, and 8%, respectively (IPCC Working Group I, 2021).

In both hemispheres, climate change has alarmingly modified the climate zones. Concretely, climate zones have moved poleward, while in the Northern Hemisphere extratropics, the growing season has been extended by up to 2 days per decade. Since 1950, sweltering temperatures—heatwaves—have occurred with higher frequency, while freezing temperatures have reduced their frequency and intensity. Marine heatwaves have also increased since 1980 due to human influence. Consequently, the frequency of and correlation between heatwaves and droughts on a global scale has increased, resulting in fire weather and compound flooding, depending on the specific location (IPCC Working Group I, 2021).

Scientists employ equilibrium climate sensitivity to understand the relationship between climate behavior and radiative forcing. Equilibrium climate sensitivity is probably between 2 and 5 °C, while the best estimation is 3 °C (IPCC Working Group I, 2021).

AR6 also presents pertinent information on the expected climate transformation in the near term, mid-term, and long term, reaching the year 2100. Five scenarios with different GHG emissions and pollutant conditions are described, considering the shared socioeconomic path (SSP) and the radiative forcing. The authors of the AR6 document describe each scenario as SSPx-y, where x indicates SSP and y refers to the corresponding radiative forcing. Researchers expect that GST will continuously increase in all of the scenarios considered. They warn humanity that if practical actions to reduce GHG emissions are not taken shortly, the temperature threshold of 1.5 and 2 °C of global warming will be surpassed. Tables 7.1–7.3 (IPCC Working Group I, 2021) present the variations of GST for the 2021–2040, 2041–2060, and 2081–2100 periods. Besides, investigators report the GST temperature variations

Table 7.1 Variations of the global surface temperature for 2021–2040 (near term).

Scenario	Estimation (°C)	Minimum (°C)	Maximum (°C)
SSP1–1.9	1.5	1.2	1.7
SSP1–2.6	1.5	1.2	1.8
SSP2–4.5	1.5	1.2	1.8
SSP3–7.0	1.5	1.2	1.8
SSP5–8.5	1.6	1.3	1.9

Table 7.2 Variations of the global surface temperature for 2041–2060 (mid-term).

Scenario	Estimation (°C)	Minimum (°C)	Maximum (°C)
SSP1–1.9	1.6	1.2	2.0
SSP1–2.6	1.7	1.3	2.2
SSP2–4.5	2.0	1.6	2.5
SSP3–7.0	2.1	1.7	2.6
SSP5–8.5	2.4	1.9	3.0

Table 7.3 Variations of the global surface temperature for 2081–2100 (long-term).

Scenario	Estimation (°C)	Minimum (°C)	Maximum (°C)
SSP1–1.9	1.4	1.0	1.8
SSP1–2.6	1.8	1.3	2.4
SSP2–4.5	2.7	2.1	3.5
SSP3–7.0	3.6	2.8	4.6
SSP5–8.5	4.4	3.3	5.7

using the magnitudes registered from 1850 to 1900 as a reference for comparison (IPCC Working Group I, 2021).

To summarize, the increment of global warming directly influences modifications of the climate system. The alterations in the climate system appear as an increment in the frequency and intensity of extreme heat, marine heatwaves, heavy precipitation, droughts, tropical cyclones, and a decline in Arctic sea ice, snow cover, and permafrost. In addition, global warming will affect the variability of the water cycle and monsoon precipitation globally and the heaviness of wet and dry climatological incidents. GHG accumulation has negative consequences for the ocean, ice sheets, and the global sea level that could be irrevocable for centuries or even millennia. Ocean and land carbon sinks are less efficient than reducing the CO_2 accumulation in the atmosphere (IPCC Working Group I, 2021).

Under this panorama, governments and environmental agencies worldwide are working on designing a set of practical actions to mitigate global warming and its effects. The United Nations has adopted 17 sustainable development goals (SDGs) (United Nations, 2022b) to transform society, providing peace and prosperity to the population of developed and developing countries. The topics covered by the SDGs include water, energy, climate, oceans, urbanization, transport, and science, among many others. As can be observed, it is a very ambitious plan to transform how humanity interacts with the planet and other living beings.

Table 7.4 summarizes the SDGs. Goals 7, 9, and 12 are related to energy production, industrial and technological innovation, societal development, and rational usage and consumption of the available resources. These goals aim to develop a human society in which electricity and other products are produced cost-effectively and in an environmentally friendly manner (United Nations, 2022b).

The 7th goal refers to the necessity of guaranteeing access to clean energy at a reasonable price. Energy access is a widely discussed issue because, in many countries, a portion of the population does not have electricity. In the case of developing countries, some communities located in rural areas only have electricity using hybrid energy systems of small capacity using batteries, a photovoltaic (PV) panel or a wind turbine, and a diesel generator. In the case of developed countries, the electricity price could be so high that a portion of the population cannot pay it, so they do not have access to electricity despite living in a developed society. Additionally, the electricity price is constrained by geopolitical factors and events such as armed conflicts and political policies, which influence the relationship between offers and demand, altering the energy price.

The 9th goal refers to maintaining an updated technological infrastructure, which is vital to improving energy conversion efficiency. A clear example is PV generation, where the efficiency of PV cells is continuously increasing. Another crucial topic is infrastructure resilience. As stated before, the meteorological conditions are going to change due to the global warming produced by the GHG emissions of industrial activities and, therefore, extreme weather conditions will increase in frequency and intensity, so an electrical system capable of dealing with these conditions is an urgent requirement.

The 12th goal refers to the fact that some critical natural resources, such as petroleum, are running out. Therefore, humanity needs to spend natural resources more rationally and efficiently. Concerning modern power systems, adopting a demand response program is an option to improve how consumers use electricity. Consequently, by 2030, the United States aims to reduce GHG emissions by between 50% and 52% from the emission levels of 2005. Later, by 2035, the United States will be free of pollutants in the power generation system, reaching total environmental sustainability by 2050 (The White House, 2022).

China also has ambitious environmental goals. China aims to reduce CO_2 emissions by 65% by 2030 with respect to the emission levels of 2005. By 2030, China plans to increase power generation from nonfossil resources by 25%. Concerning the forest stock, China aims to increase it by 6 billion cubic meters by 2030 with respect to the magnitude of 2005. Finally, China expects to increase the

Table 7.4 United Nations sustainable development goals.

Goal	Title	Description
1	No poverty	End poverty in all its forms everywhere
2	Zero hunger	End hunger, achieve food security, improve nutrition, and promote sustainable agriculture
3	Good health and wellbeing	Ensure healthy lives and promote wellbeing for all at all ages
4	Quality education	Ensure inclusive and equitable quality education and promote lifelong learning opportunities for all
5	Gender equality	Achieve gender equality and empower all women and girls
6	Clean water and sanitation	Ensure availability and sustainable management of water and sanitation for all
7	Affordable and clean energy	Ensure access to affordable, reliable, sustainable, and modern energy for all
8	Decent work and economic growth	Promote sustained, inclusive, and sustainable economic growth, full and productive employment, and decent work for all
9	Industry, innovation, and infrastructure	Build resilient infrastructure, promote inclusive and sustainable industrialization, and foster innovation
10	Reduced inequalities	Reduce inequality within and among countries
11	Sustainable cities and communities	Make cities and human settlements inclusive, safe, resilient, and sustainable
12	Responsible consumption and production	Ensure sustainable consumption and production patterns
13	Climate action	Take urgent action to combat climate change and its impacts
14	Life below water	Conserve and achieve sustainable use of the oceans, seas, and marine resources for sustainable development
15	Life on land	Protect, restore, and promote sustainable use of terrestrial ecosystems, sustainably manage forests, combat desertification, halt and reverse land degradation, and halt biodiversity loss
16	Peace, justice, and strong institutions	Promote peaceful and inclusive societies for sustainable development, provide access to justice for all, and build effective, accountable, and inclusive institutions at all levels
17	Partnerships for the goals	Strengthen the means of implementation and revitalize the global partnership for sustainable development

wind and solar generation capacity by 2030 to 1.2 billion kW (The State Council Information Office of the People's Republic of China, 2022).

The European Union aims to reduce GHG by 55% by 2030 with respect to the emission levels of 1990 (European Commission, 2022). Other developed and developing countries not mentioned here also have their own goals for reducing GHG emissions and the negative influence of the industrial and manufacturing process on the environment.

7.2 Energy sources and commodities

There is a vast family of technologies and devices for energy conversion, which allow us to exploit the energy sources locally available. This section briefly comments on some characteristics of some of the most widely implemented technologies (Black & Veatch & National Renewable Energy Laboratory, 2012).

The energy conversion process of a commercial nuclear plant is based on fission. Specifically, the heat obtained from the conversion process is transformed into steam, which is later used to drive a turbine, producing electricity. The nuclear plant uses uranium as fuel, which is located at the heart of the reactor (US Energy Information Administration, 2021b).

Nuclear power plants are widely employed. According to information from 2019, 31 countries use this technology, and 14 have sufficient capacity to satisfy at least 20% of their annual energy requirements. The United States has a power capacity of 98.12 million kW and produces 809.41 billion kWh, representing 19% of the yearly electricity demand. Thus, the United States has the highest installed capacity, and France is the country with the second-highest power capacity: France has 63.13 million kW and generates 382.4 billion kWh, satisfying 70% of its needs. The generation capacity of China is 45.52 million kW, producing 330.12 billion kWh, supplying 5% of its demand. Russia has 28.37 million kW and generates 195.54 billion kWh, representing 18%. Finally, South Korea can generate 23.09 million kW, generating 138.31 billion kWh to satisfy 25% of its electricity demand (US Energy Information Administration, 2021b).

Nuclear technology has been undergoing constant development to adapt to the actual situation. According to the Generation IV International Forum (GIF) (Generation IV International Forum (GIF), 2020) there are six rising technologies: gas-cooled fast reactor (GFR), lead-cooled fast reactor (LFR), molten salt reactor (MSR), supercritical water-cooled reactor, sodium-cooled fast reactor (SFR), and very high-temperature reactor (VHTR). These technologies satisfy demanding criteria in reliability, physical protection, and security. They have been designed not only for power generation but also for those industrial applications that require high-temperature conditions. VHTR, GFR, LFR, and MSR can be adopted for processes requiring temperatures between 700 and 950 °C. Moreover, SFR can supply industrial operations of 550 °C. The heat obtained from the nuclear reactor can be employed in hydrogen production or oil refining, among other applications. In addition to this progress, GIF has identified the necessity of improving nuclear generation to increase its operability with other renewable sources and fuels, enhance its scalability, and simplify its construction process (National Renewable Energy Laboratory & Nuclear Innovation: Clean Energy Future (NICE), 2020).

The idea of redesigning and updating nuclear technologies to contribute to the sustainability of industrial activities has motivated the concept of a nuclear-renewable hybrid energy system (NRHES), in which, heat and power could be obtained from one or two different energy sources. Integrating various energy sources on a single device for power and heat generation is not only an operational concept but also an organizational idea. In this regard, the Nuclear Innovation: Clean Energy Future (NICE future) was created to rethink the role of nuclear energy in techno-economic sustainability. One of the objectives of NICE Future is to involve decision-makers (DMs) in a multilateral discussion about the contributions of nuclear energy to the society of the future. Another objective is to comment on the opportunities and barriers to achieving the aforementioned clean energy production goals, including techno-economic and corporative perspectives (Ruth et al., 2019).

The working approach previously described intends to analyze the transition of nuclear energy from independent operation to interacting successfully with other power sources such as wind and PV generation. An advantage of adopting nuclear generation is that it does not require changes in the fuel with high frequency. Currently, nuclear technology requires refueling with annual or lower frequency. This feature, combined with the fact that renewable power sources do not require any fuel type, makes the combination of nuclear with solar PV or wind an attractive option. Another reason to develop the NRHES is the intrinsic variability of energy prices due to the structure and behavior of the oil and natural gas markets. In this context, the NRHES concept would be beneficial due to its limited fuel dependence (Ruth et al., 2019).

The behavior of NRHES can be enhanced by incorporating the water electrolysis process for hydrogen generation. This could be understood as a responsive load or an energy storage capability, increasing NRHES operational flexibility (Ruth et al., 2019). It is essential to highlight that even when nuclear generation has difficulties increasing or decreasing its power production or maintaining its output power at a low value, nuclear technology has unique features that contribute to power system flexibility (National Renewable Energy Laboratory & Nuclear Innovation: Clean Energy Future (NICE), 2020).

Scientists and technicians can build the previously mentioned nuclear reactor technologies in a reduced size to be installed in a distributed manner. Furthermore, they can change their output power relatively fast. These properties make nuclear technology susceptible to being combined with other generation technologies, such as wind and solar PV power, and energy storage systems, such as hydrogen production, to satisfy electric and thermal loads (National Renewable Energy Laboratory & Nuclear Innovation: Clean Energy Future (NICE), 2020).

In summarizing, the existing experience and knowledge related to reactor physics, thermal hydraulics, and material science indicates that nuclear generators can vary their power generation. When combined with other technologies such as hydrogen production and thermal energy storage, the flexibility of nuclear generation can be increased to supply electric and other types of loads. The advanced reactors previously described can be built to provide power from 1 to 100 MW, which allows them to be installed in isolated zones with low electricity demand (National Renewable Energy Laboratory & Nuclear Innovation: Clean Energy Future (NICE), 2020).

Another important topic related to nuclear technology is fusion. Currently, it is a research topic in which its applicability for power production needs to be demonstrated. A crucial topic is the use of tritium as fuel, which is unavoidable in the middle term. Using tritium as fuel for a fusion reactor is not an option developed enough to be applied in the industrial sector. Many countries have been working on constructing experimental devices to analyze fusion reactions. Some examples are the Joint European Torus in the United Kingdom, the Tokamak Fusion Test Reactor in the United States, the Japan Tokamak in Japan, and the Tokamak T3 in Russia (Perevezentsev & Rozenkevich, 2021).

Oil, coal, and natural gas are critical resources for power generation and other key industries. Natural gas prices are sensitive to production and storage capacity, the imported or exported amount, the season of the year, economic performance, and the prices of other fuels and commodities (US Energy Information Administration, 2021a).

The production and storage capacity directly impact natural gas prices due to the interaction between supply and demand. Thus, high natural gas demands in a region or country with low production capacity results in high prices (US Energy Information Administration, 2021a). The amount of natural gas imported or exported influences domestic prices. In the United States, natural gas production increased between 2005 and 2019, reducing the prices and their volatility (US Energy Information Administration, 2021a).

Meteorological events such as hurricanes can interrupt natural gas production, increasing its price. A representative example was observed in 2005 during the occurrence of storms Katrina and Rita in the Gulf of Mexico. Additionally, during the winter, the demand for natural gas for heating typically increases (US Energy Information Administration, 2021a). During periods of economic growth, the demand for goods and services increases, and consequently, so does the industrial activity to manufacture or provide such goods and services. Thus, the demand for natural gas in power generation and chemical industries increases (US Energy Information Administration, 2021a).

Finally, if other resources such as coal and petroleum have low prices and industrial consumers can use different fuels, the demand for and consequently the price of natural gas reduces (US Energy Information Administration, 2021a).

Russia is a crucial producer of energy commodities. In 2021, Russia was the country with the largest natural gas exports, and it was among the oil and condensates exporters with the highest amounts, after only Saudi Arabia. Most of the Russian energy commodities are consumed by European countries (US Energy Information Administration, 2022b). In the year mentioned above, Russia exported 4.7 million barrels per day of oil and condensate, or approximately 45% of its total production, to some European countries. Additionally, countries in Asia and Oceania received 38% of the oil and condensate exports. Specifically, China received 1.4 million barrels per day, while the Netherlands and Germany obtained 1.1 million barrels per day. Finally, the United States only imported 199,000 barrels per day from Russia (US Energy Information Administration, 2022b).

Regarding natural gas, Russia exported 8.9 trillion cubic feet in 2021, and most of the production, about 75%, was sold to European countries. Specifically, Germany, Turkey, Italy, Belarus, and France were the most critical consumers. In addition, China and Japan consume 882 billion cubic feet, representing around 10% of Russian exports (US Energy Information Administration, 2022b).

Regarding coal exports, Russia exported 262 million short tons in 2021, of which China consumed 63 million short tons. Other Asian countries such as South Korea, Japan, and Taiwan jointly consumed 22%, and European countries such as Germany, the Netherlands, Turkey, and Poland consumed 24%. It is essential to highlight that 90% of Russian exported coal is used for power generation (US Energy Information Administration, 2022b).

European countries are susceptible to changes in the energy commodity markets due to their massive imports from Russia. For this reason, the armed conflict in Ukraine has resulted in a sudden increase in electricity prices and other services.

The massive adoption of renewable energies and carbon capture and sequestration technology, combined with many other strategies such as demand management, supports the reduction of GHG emissions. In addition, governmental incentives and federal tax credits are crucial to meet the ambitious goal of building a sustainable and modern society in the middle term (The White House, 2021). The levelized cost of energy (LCOE) is a relevant indicator of power production cost. Concerning this topic, Table 7.5 (National Renewable Energy Laboratory, 2022) presents the estimation of LCOE for 2050, considering the market conditions and governmental policies.

7.3 Energy system models

Environmental problems related to GHG emissions have driven the adoption of renewable power sources on a large scale. DMs have been using simulation models to analyze likely scenarios in the middle and long term to optimize the available resources. Such models are based on the relationship between GHG emissions and

Table 7.5 Levelized cost of energy for some generation technologies.

Technology	LCOE ($/MWh)
Land-based wind	17.39
Offshore wind	40.84
Utility solar photovoltaic	16.37
Concentrating solar power	51.28
Geothermal	49.37
Hydropower	67.10

their associated costs, which are not easy to estimate. Several factors, such as technological evolution and the preferences of consumers and businesses, influence the GHG-emission reduction costs. These factors are highly uncertain and complex to predict in the middle and long terms (Jaccard et al., 2003; Murphy et al., 2007; Rivers & Jaccard, 2006).

Implementing a specific policy for GHG emission reduction consists of selecting determined environmentally-friendly equipment, incentivizing the population to change their usual behavior, or modifying a concrete stage in an industrial process to reduce GHG emissions. In its more general definition, the costs related to implementing a specific policy are the calculation between maintaining everything as usual and the associated costs of implementing the new policy. However, other approaches could also be applied (Jaccard et al., 2003; Murphy et al., 2007; Rivers & Jaccard, 2006).

Energy system models can be classified into bottom-up, top-down, and hybrid models. Bottom-up models are used to analyze the influence of specific factors such as energy efficiency, fuel type, and the adoption of emission control systems on GHG emission rates. Scientists usually perform the economic assessment by converting the capital, operating, and maintenance costs to the present, assuming a determined discount rate. This approach frequently results in high profitability or low cost for those technologies employed in GHG emission reduction (Jaccard et al., 2003; Murphy et al., 2007; Rivers & Jaccard, 2006.

One drawback of bottom-up models is that they do not consider that the population does not entirely accept all the products required for GHG emission reduction. Another disadvantage is that the financial model applied is not accurate enough because it cannot consider the viewpoint of consumers and businesses when selecting a determined technology (Jaccard et al., 2003; Murphy et al., 2007; Rivers & Jaccard, 2006).

The simplified financial model cannot consider the option value related to the financial risk of adopting new technology and the costs associated with installing such technology in a specific location. Furthermore, the bottom-up models cannot realize the extra weight of a particular product beyond the financial one. Thus, the results of bottom-up implementation can overestimate the willingness of the society to adopt the required policy to minimize the GHG emission rate, which is not realistic nor appropriate for DMs (Jaccard et al., 2003; Murphy et al., 2007; Rivers & Jaccard, 2006).

The top-down models employ historical information to represent the relationship between the energy sector and the economic structure. Relevant concepts such as the elasticity of substitution (ESUB) and the autonomous energy efficiency index (AEEI) are incorporated into top-down models (Jaccard et al., 2003; Murphy et al., 2007; Rivers & Jaccard, 2006).

On the one hand, ESUB designates the substitution capability of capital by energy. On the other hand, AEEI represents how technological progress enhances energy productivity (Jaccard et al., 2003; Murphy et al., 2007; Rivers & Jaccard, 2006). A high ESUB indicates a high interchange rate of capital and energy, and a high AEEI shows a high interchange rate of devices with high and low GHG emissions. Consequently, this scenario results in a low cost related to the GHG emission reduction. As explained before, ESUB and AEEI are coefficients obtained from the analysis of historical information, which mathematically represents the preferences of consumers and businesses (Jaccard et al., 2003; Murphy et al., 2007; Rivers & Jaccard, 2006).

Incorporating the interaction between consumers and businesses allows DMs to estimate specific policy implementation effectiveness. For example, DMs could analyze the effect of implementing a subsidy to a determined technology to reduce GHG emissions. The top-down models consider the interaction between energy supply and demand and the relationship between these factors and the economic structure, suggesting high costs of GHG emission reduction (Jaccard et al., 2003; Murphy et al., 2007; Rivers & Jaccard, 2006).

The main disadvantage of the top-down models is that they cannot incorporate long-term technological changes and their effects on GHG emissions. DMs need to estimate how a specific policy can improve technological development and reduce capital, operation, and maintenance costs, and GHG emissions. The uncertainty related to ESUB and AEEI estimation is also a drawback. Even when estimated from chronological data, there is no guarantee about their reliability in the future (Jaccard et al., 2003; Murphy et al., 2007; Rivers & Jaccard, 2006).

The bottom-up and top-down models have significant aspects that can be improved. The bottom-up approach does not consider the option values and the preferences of the consumer, whereas the top-down method includes these factors. However, it does not incorporate the technological changes in the long term and their relationship with the economic structure and GHG emissions, which is highly appreciated by DMs worldwide (Jaccard et al., 2003; Murphy et al., 2007; Rivers & Jaccard, 2006).

The hybrid or CIMS model was proposed to overcome the weaknesses of the bottom-up and top-down approaches. Hybrid models can simulate economic behavior, including the effect of DM policies on technological development, the preferences of consumers and businesses, and the interaction between energy consumers and suppliers. Technological explicitness, preference incorporation, and equilibrium feedback (relationship between the economy and energy generation and consumption) are the essential features of hybrid models (Jaccard et al., 2003; Murphy et al., 2007; Rivers & Jaccard, 2006).

The bottom-up models have been widely analyzed in the literature from the simulation and optimization perspectives. A representative example of a simulation analysis was presented by Elliston et al. (2012), considering scenarios with high penetration of renewable energies in the Australian power system for 2010. The authors considered concentrating solar thermal generation with thermal storage, wind and solar PV, hydropower technology, and biofuel-based generation. Solar irradiance and wind speed time series were obtained using satellite data and information provided by meteorological stations (Elliston et al., 2012).

It is essential to highlight that the simulation model does not consider some operating constraints of conventional generators, such as the minimum output power, minimum up and down operating times, and ramping restrictions. These assumptions could lead to good results if the system under analysis is not provided with coal, nuclear, and combined-cycle gas generators. The simulation process informs the

DM about the total excess of energy, the energy not supplied, and its frequency. Additionally, the authors performed a sensitivity analysis to enhance reliability. Specifically, the investigators analyzed the scenario of minimizing the number of unmet hours and peak demand by applying demand management and examined the effects of increasing solar thermal and PV rated capacity (Elliston et al., 2012).

From their analysis, the authors observed that the massive incorporation of variable generation results in excess energy production during low load periods and energy shortage during periods of high demand. In other words, the time series of renewable generation and electricity demand are not correlated, which results in the loss of available resources. This situation can be mitigated by implementing a demand management program to incentivize consumers to increase their power consumption during times of high wind speed and solar irradiation and reduce their consumption at peak-load hours (Elliston et al., 2012).

Seasonal behavior of electricity demand is also relevant. Concerning this topic, the authors discussed the option of incrementing the solar thermal generation and storage capacity to meet the load demand during winter, compensating for the limited solar resources during this season. In other words, the strategy consists of producing excess energy in summer and storing it so that it is available during the winter when the solar resource is minimal. Under this scenario, thermal energy storage capacity and location become crucial. Finally, the authors reiterated the importance of the peaking plant required to supply high demand levels and the marginal cost increment related to this technology (Elliston et al., 2012).

Later, the same authors (Elliston et al., 2013) integrated all of this knowledge and experience into an optimization model to determine the type and capacity of each generating unit, minimizing the annual cost. The generating units were dispatched following a predefined merit order or priority list for each hour of the year during the simulation process. Then, the excess energy was stored in a pump storage hydropower device or a thermal energy storage system. If no storage technology is under consideration, the extra power is spilled. The merit order was determined considering the generating technology and the region in which the unit was located. The target parameter during the optimization process was the total annual cost, considering the annualized capital cost and the fixed and variable operating and maintenance costs. In addition, a fixed discount rate was considered (Elliston et al., 2013).

The optimization was carried out using a genetic algorithm (GA). A vector represented each GA-individual, and each element of such vector was the capacity of a specific unit. As stated before, the authors analyzed the Australian power system. Concretely, they dimensioned 25 generators, or in other words, they estimated the capacity of 25 generators to minimize the total annualized cost (Elliston et al., 2013).

The bottom-up models have been implemented for simulation and optimization in a large family of commercial software. Researchers have identified the main conceptual barriers to bottom-up model evolution. Most of these barriers are related to the resolution from different viewpoints (temporal, special, techno-economical, and sectorial resolution). Most available models do not simultaneously offer high resolution in all of the perspectives mentioned above (Prina et al., 2020).

Bottom-up models are classified as short and long-term models, and then the long-term models are classified depending on their prediction capability as provided with perfect forecasting and myopic models. Additionally, essential characteristics, such as the energy sectors covered, the geographical coverage, the temporal resolution, the adopted methodology, and the programming technique, are recognized to enhance the classification process (Prina et al., 2020).

The concept of energy sectors covered aims to identify whether the model has been designed to work with a specific sector. It allows recognition of the energy commodity (electricity, heat, hydrogen, and fuels) and the demand sector (building, industry, and transport) for which the model has been designed (Prina et al., 2020).

The geographical coverage definition allows us to recognize if the model works with a single node or if it can incorporate the limitations of the transport infrastructure. A model with only one node cannot represent bottlenecks or losses in electricity or natural gas transport (Prina et al., 2020).

Time resolution refers to the temporal horizon. The horizon is the period for which the model was designed, for example, estimating GHG emissions for 2050. Additionally, it is crucial to recognize the number of periods considered until the horizon year is reached. Frequently, the system behavior is not simulated yearly until the horizon year because the computational resources are limited. On the contrary, the system performance is analyzed in steps of 2, 3, or 5 years (Prina et al., 2020). The methodology consists of identifying if the model is for simulation or optimizing dispatch and investments. On the one hand, a simulation model allows us to estimate the total annualized costs, annual CO_2 emissions, energy savings, and excess. On the other hand, the dispatch optimization models usually employ a simplified technique based on a merit order or priority list for power dispatch, reducing the computational burden. The investment optimization models are designed to determine the capacity expansion appropriately (Prina et al., 2020).

The programming technique lets us know which optimization technique is being employed by the model. Most of the available tools are based on linear programming (LP), mixed-integer linear programming, dynamic programming, and heuristic approaches such as GA (Prina et al., 2020). According to recently published studies, about 37% of the bottom-up models available are designed to analyze insular systems, 47% are prepared to study nationwide systems, and 16% are designed to investigate micro-grids. EnergyPLAN (2020) and HOMER Energy (UL, 2020) are among the most widely used bottom-up commercial tools (Prina et al., 2021).

Participatory methods are another framework for the analysis of energy systems, in which the perspective of academic and nonacademic actors and stakeholders on local and regional scales are considered. Traditionally, creating energy policy incorporated a limited set of concepts related to economic viability, the security of supply, and environmental sustainability. However, public acceptance has been recognized as a vital factor in energy transition due to the democratization process and the open discussion and implementation of energy solutions. Thus, social acceptance and engagement have relevant weight on energy system analysis (McGookin et al., 2021).

A qualitative manner of registering the stakeholder viewpoint is through interviews and surveys. This option has been adopted in 46% of the available studies. Another approach consists of creating some likely scenarios based on the available information about the stakeholder. This method is observed in 53% of the available studies (McGookin et al., 2021). Cognitive mapping and serious games are techniques usually employed in the qualitative analysis of the perception of stakeholders. Cognitive maps allow us to understand the causal relationships related to the system operation, and serious games consist of proposing alternative realities to improve our understanding of the system (McGookin et al., 2021). Moreover, quantitative analyses are carried out using simulation and optimization tools, multicriteria decision analysis, system dynamics modeling, and resource assessment (McGookin et al., 2021).

7.4 Optimal power flow

In its simplest form, solving optimal power flow (OPF) consists of finding the power production of each unit to minimize the total operating costs, satisfying the power demand and transmission network constraints. Researchers have proposed a broad family of techniques to solve this crucial problem. Several authors have used the Newton approach.

Dommel and Tinney (1968) proposed a method that incorporates the active and reactive power and transformer ratios as optimization variables, resulting in a feasible solution. The methodology uses the Newton approach for power flow solution and the gradient adjustment method to estimate the optimal solution, including the penalty functions frequently added to consider the operational constraints. The approach employs the Lagrangian multipliers method incorporating the objective function and the auxiliary variables related to the restrictions. Then, the authors identify a beneficial relationship between the Jacobian matrix to solve power flow and one of the necessary conditions for optimality (Dommel & Tinney, 1968).

The optimization process requires applying an iterative algorithm. First, the control variables are initialized. Second, the Newton method solves the power flow, paying attention to the Jacobian matrix. Third, the value of the auxiliary variables is estimated. Fourth, the gradient is calculated. Fifth, if the gradient is neglectable, the minimum has been reached. Otherwise, the values of the control variables previously assumed are updated. The authors recognized that this aspect is critical because, depending on how the control variables are updated, the algorithm efficiency is or is not enhanced (Dommel & Tinney, 1968).

Sun et al. (1984)) developed another technique based on the Newton method, in which a quadratic approximation of the Lagrangian function is minimized in each iteration. Thus, the algorithm converges toward the Kuhn-Tucker conditions during the iterative process. The algorithm implementation consists of a problem definition, identifying the objective function, the control variables, and constraints. Then, a decoupling process is performed. After that, specific methods are applied to deal with the inequality constraints and binding inequalities. Finally, sparsity techniques are employed. The authors assessed their technique using a part of the Northeastern United States power system with 912 buses and 1637 branches. OPF was implemented to minimize the active power losses, controlling 255 reactive power sources and 41 controllable transformers. According to the conclusions reported, the proposed method offers a linear convergence to the Kuhn-Tucker conditions without requiring any particular tuning factor (Sun et al., 1984).

Weber (1997) took advantage of all the experience previously described and implemented Newton-based OPF in a commercial optimization program.

Alsac and Stott (1974) improved the model proposed by Dommel and Tinney previously described. Concretely, Alsac and Stott extended the Newton-based OPF by incorporating the security constraints. Thus, the controllable variables are optimized for the operation under normal and contingency conditions, guaranteeing the feasibility of the solution. The implementation requires several fulfilling several steps. First, OPF is solved using the method of Dommel and Tinney, considering the normal operating conditions. Second, the system security is analyzed by solving the power flow problem. Third, the outage effects are included on the optimization process by incorporating the information obtained in the second step into the gradient calculation procedure. Then, the algorithm searches for security and optimal system operation (Alsac & Stott, 1974).

This technique offers high accuracy in representing the operational outage restrictions, which allows us to use this approach as a benchmark to assess the performance of other methodologies or to determine the security limits of the system. Regarding the computational burden, it increases by less than 10% compared to the original method of Dommel and Tinney (Alsac & Stott, 1974).

Granville (1994) developed a technique using an interior point method based on the primal-dual logarithmic barrier algorithm, a methodology with high computational efficiency applied to linear and quadratic programming problems. The authors focused on the optimal reactive dispatch (ORD) problem, a specific version of OPF. ORD is a challenging problem because it is nonlinear and nonconvex. Thus, the authors aimed to overcome these characteristics by exploiting the high computational capabilities of the primal-dual method. Furthermore, the researchers studied relevant topics related to selecting the barrier parameter and its relationship with the duality gap used in linear programming (Granville, 1994). The investigators considered two different ORD problem versions. The first formulation has its primary objective to minimize the costs related to the reactive power provision, which is required to guarantee the correct system operation. The second formulation has an objective function built by the combination of reactive power provision and active power loss. Such a combination is introduced to assess the algorithm performance in the presence of two conflicting objectives (Granville, 1994).

The method performance was evaluated by analyzing two large-scale systems in Brazil and the United States. The authors observed that the number of iterations is not highly dependent on the system size and number of control variables. Additionally, the method effectively solves large-scale problems and ill-conditioned systems, optimizing the reactive power distribution and power loss (Granville, 1994).

Bakirtzis et al. (2002) presented a methodology based on a GA to solve OPF. The authors employed a binary GA to consider continuous variables related to the output power of each unit and discrete variables associated with the transformer-tap settings and the reactive power injection. The authors incorporated advanced and problem-specific

GA operators in addition to the basic ones (selection, crossover, and mutation) to increase computational efficiency. The advanced operators are the fitness scaling, the elitism, and the hill climbing. Fitness scaling is applied to promote fair competition among the individuals, avoiding the overwhelming influence of those individuals with high capabilities. Elitism is incorporated to prevent the loss of the individual with the highest fitness from one generation to the next. Sequential implementation of GA-operators may result in the loss of the best solution, however, the elitism corrects this problem. Hill climbing enhances the search capabilities, specifically in the zones of the searching space with a smooth landscape (Bakirtzis et al., 2002).

The problem-specific operators are the gene swap operator (GSO), gene cross-swap operator (GCSO), gene copy operator (GCO), gene inverse operator (GIO), and gene max-min operator (GMMO) (Bakirtzis et al., 2002). Concretely, all operators work over randomly chosen individuals and their elements. Thus, GSO works over two components of an individual and exchanges their respective values. GCSO operates over two individuals and two elements and exchanges their respective values. GCO is applied over an element from an individual; it copies its information to the predecessor or the successor element within the same control variable. GIO is an improved version of the classical mutation operator. Finally, GMMO operates over a component of an individual and completes its area with ones or zeros with 50% probability (Bakirtzis et al., 2002).

According to the reported experience, the authors concluded that GA offers flexibility to incorporate challenging features such as the nonconvex cost functions, prohibited operating zones, and discrete control variables related to transformer operation. However, it is not possible to guaranty the optimality of the obtained solution from a GA. Additionally, the mathematical complexity increases with the system size, which demands more computational resources (Bakirtzis et al., 2002).

Particle swarm optimization (PSO) is another heuristic algorithm widely used. Abido (2002) applied PSO to the OPF problem to minimize the fuel consumption cost, paying particular attention to the voltage profile quality and stability. The author considered the restrictions in the power generation, the constraints related to the transformer-tap settings, the reactive power provision, and system security. According to the results obtained from the analysis of a benchmark system with 30 buses, the author concluded that PSO is a highly flexible technique, not only from the modeling perspective but also from the viewpoint of the exploration capabilities of the searching landscape. In other words, PSO can perform a global and local analysis of the searching space, offering an excellent solution to an OPF problem of a relatively small size (Abido, 2002).

An essential disadvantage of PSO implementation is that discrete control variables such as transformer-tap settings were modeled as continuous variables and rounded off to obtain a discrete solution. The author suggests the modification of this procedure to avoid the associated errors (Abido, 2002).

Based on the concept of an energy hub, Geidl and Andersson (2007) studied the operation and management of an energy system provided with the infrastructure to integrate different energy sources, such as electricity and natural gas (combined heat and power), and heat (heat exchanger) in a consolidated manner. The energy hub is a versatile concept that can be adapted to a wide family of facilities such as power and industrial plants, large-scale buildings such as airports and hospitals, or limited areas such as rural locations and districts (Geidl & Andersson, 2007).

Concerning its working principle, an energy hub is provided with input and output ports, allowing their interconnection and energy exchange. Energy conversion is carried out within each hub, and the energy transmission is performed through the input or output ports. The simulation and optimization analyses are carried out assuming the steady-state condition. In addition, the converter devices are modeled by their efficiency, which is the principal mean of power loss. Finally, the power flow has only one direction, from input to output ports (Geidl & Andersson, 2007). The energy transport is incorporated considering the relationship between the voltage and current (electricity network) and the pressure and flow (pipeline network). As aforementioned, the electrical and thermal relationships are built according to the physical laws under steady-state conditions (Geidl & Andersson, 2007). Concerning the optimization model, the authors derived the linear model of the energy hub to solve the OPF problem, considering the Kuhn-Tucker conditions (Geidl & Andersson, 2007).

Riffonneau et al. (2011) developed a supervisory control for an energy system based on solar PV generation and a battery system. The management system is composed of three stages, the forecasting stage, the predictive optimization stage, and finally, the local command stage. During the forecasting stage, the prediction of the variables of interest such as ambient temperature, solar irradiation, electricity demand, and prices is carried out. Then, during the predictive optimization, the solution of the OPF is carried out. Finally, the local control implements the solution previously obtained, assigning the corresponding signals to the power electronic devices (Riffonneau et al., 2011).

The authors formulated the OPF to minimize the electricity bill (peak shaving) of the owner constrained to the operational limits of each device. Then, they applied the dynamic programming (DP) approach, considering the decision process as a multistage phenomenon (Riffonneau et al., 2011). The investigators formulated the DP formulation regarding the battery operation. Concretely, each DP algorithm state was considered a discrete variation of the battery state of charge over the forecasting horizon. The effectiveness of the supervisory management system was verified using a hardware-in-the-loop architecture, estimating a reduction in the electricity bill of 13% under the specific conditions of the experiment. Additionally, the authors commented on the importance of extending the management algorithm to control reactive power to enhance interaction with the power system (Riffonneau et al., 2011).

The semidefinite programming (SDP) approach is another technique employed to solve the OPF problem. Bai et al. (2008) proposed a model based on expressing the OPF in the SDP form to find the SDP problem solution using an interior point method. The benefit of this approach is that the original nonlinear and nonconvex optimization problem is transformed into a convex SDP problem solved by primal-dual interior-point algorithms, finding the optimal global solution.

Following this direction of research, Lavaei and Low (2012) analyzed the mathematical properties required to ensure the existence of the no-duality-gap condition to solve the OPF. The authors examined the power system structure and its optimization models to minimize power loss. They considered DC (resistive networks without reactive loads) and AC systems, successfully dealing with the problem of convexification due to the physical properties of the transmission system. However, the computational efficiency of the SDP is sensitive to the system size.

Dall'Anese et al. (2013) adapted SDP to the working conditions of the modern smart grid by solving the OPF in a distributed manner. Implementing the SDP consists of decomposing the optimization problem into several subcases solved in parallel, increasing the computational efficiency.

Farivar and Low (2013a, 2013b) developed a theoretical framework to conveniently model the system branch flow so that the SDP always reaches the global optimum. Concretely, the authors proposed using a specially designed model and relaxation techniques (angle and conic relaxations) to obtain the global optimum or a high-quality solution for radial and meshed systems.

The OPF solution has been incorporated into many computing tools. Some examples are PowerWorld (PowerWorld Corporation, 2022), PSAT (Milano, 2005), and MATPOWER (Zimmerman et al., 2011), adapted to the specific conditions and rules of liberalized electricity markets. Solving the OPF problem is a research topic in continuous evolution. Skolfield and Escobedo (2022) have just published information about the recent techniques suggested in the technical literature concerning the solution of the OPF problem.

7.4.1 Optimal power flow for vertically integrated systems

Vertically integrated systems refers to those utilities that possess and manage power generation, transmission, and distribution. Under these circumstances, the optimization goal is to minimize the operating cost.

Assuming a power system with generators, the vector of Eq. (7.1) represents the power production, and Eqs. (7.2)–(7.4) represent the minimum and maximum power generation constraints:

$$\boldsymbol{p} = \begin{bmatrix} p_{(1)} & \cdots & p_{(i)} & \cdots & p_{(I)} \end{bmatrix} \tag{7.1}$$

$$\boldsymbol{p}_{\mathbf{min}} = \begin{bmatrix} p_{(1)}^{\min} & \cdots & p_{(i)}^{\min} & \cdots & p_{(I)}^{\min} \end{bmatrix} \tag{7.2}$$

$$\boldsymbol{p}_{\mathbf{max}} = \begin{bmatrix} p_{(1)}^{\max} & \cdots & p_{(i)}^{\max} & \cdots & p_{(I)}^{\max} \end{bmatrix} \tag{7.3}$$

$$p_{(i)}^{\min} \leq p_{(i)} \leq p_{(i)}^{\max} \forall i = 1, \dots, I \tag{7.4}$$

The OPF problem considers the influence of the transmission system, which is estimated by solving the power flow with the Newton–Raphson method (Zhu, 2009).

Assuming a power system with R nodes and letting r be the index for each system bus; during the power flow implementation, the nodes are numbered in a specific manner, including first the load nodes (PQ nodes) from 1 to y, the generation nodes (PV nodes) without the slack bus from $y+1$ to $R-1$, and finally, the slack bus with the number R. Thus, the index $r=1, \ldots, y, y+1, \ldots, R-1, R$ considers different node types.

Power flow is a nonlinear problem consisting of finding out the unknown variables (voltages and angles, depending on the node type) to obtain a neglectable value of active and reactive power mismatch. Power flow formulation is summarized in Eqs. (7.5)–(7.7), where $\Delta p_{(r)}$ and $\Delta q_{(r)}$ are the active and reactive power mismatch, and $p_{(r)}^{spc}$ and $q_{(r)}^{spc}$ are the active and reactive power specified by the generators and consumers. The consumers specify the active and reactive power, while the generators only specify their active power injection:

$$\Delta p_{(r)} = p_{(r)}^{spc} - V_{(r)} \sum_{m=1}^{R} V_{(r)} \left(G_{(r,m)} \cos\left(\theta_{(r,m)}\right) + B_{(r,m)} \sin\left(\theta_{(r,m)}\right) \right)$$

$$\forall r = 1, \ldots, R-1 \tag{7.5}$$

The variable $V_{(r)}$ is the voltage magnitude of node r, and the parameters $G_{(r,m)}$ and $B_{(r,m)}$ are the conductance and susceptance of the branch between nodes r and m, respectively. The variable $\theta_{(r)}$ is voltage angle of bus r.

$$\Delta q_{(r)} = q_{(r)}^{spc} - V_{(r)} \sum_{m=1}^{R} V_{(r)} \left(G_{(r,m)} \sin\left(\theta_{(r,m)}\right) - B_{(r,m)} \cos\left(\theta_{(r,m)}\right) \right)$$

$$\forall r = 1, \ldots, y \tag{7.6}$$

$$\theta_{(r,m)} = \theta_{(r)} - \theta_{(m)} \tag{7.7}$$

Once power flow is solved, the main results are the voltage magnitude and angles of the system (depending on the node type). Then, the power system engineers can calculate the power or current through transformers and transmission lines to verify whether they are overloaded. The overloading conditions for the branch between nodes r and m are represented in Eq. (7.8), where $A_{(r,m)}$ is the current or power obtained from the power flow solution, and $A_{(r,m)}^{\max}$ is the maximum current or power limit:

$$A_{(r,m)} \leq A_{(r,m)}^{\max} \forall r \neq m \,|\, r = 1, \ldots, R;\ m = 1, \ldots, R \tag{7.8}$$

Voltage is a crucial factor in the successful operation of a power system. For this reason, it is necessary to maintain the voltage magnitudes within a specific interval. To this end, the constraint of Eq. (7.9) is enforced:

$$V_{(r)}^{\min} \leq V_{(r)} \leq V_{(r)}^{\max} \forall r = 1, \ldots, y \tag{7.9}$$

As aforementioned, in this work, the OPF is formulated to minimize the fuel consumption cost, which is shown in Eq. (7.10). As can be observed, the effects of the prohibited operating zones and valve-point loading have not been considered.

$$F(p) = \sum_{i=1}^{I}\left\{a_{(i)} + b_{(i)}\left(p_{(i)}\right) + c_{(i)}\left(p_{(i)}\right)^2\right\} \quad (7.10)$$

7.4.2 Optimal power flow under fair-trade conditions

Under liberalized conditions, the electricity market coordinator determines the energy prices, analyzing the offers of generators and consumers concerning their quantities and prices and the power system structure to maximize social welfare. In general, the coordinator of the electricity market uses locational marginal prices (LMPs) to determine the price at which the generators are paid and the price to be paid by the loads, considering their corresponding location (bus) in the power system. LMP is calculated by adding three components: the system marginal price (SMP), the congestion-related price, and the marginal loss-related fee (PJM State & Mem Training Department, 2017).

SMP represents the price of increasing the system load in a specific quantity (1 MWh) in the reference bus, considering the actual dispatch conditions. Consequently, SMP is intrinsically related to social welfare maximization. SMP is equal for all system nodes, so it does not consider the locational characteristic of LMP. Additionally, SMP does not consider the transmission system losses or congestions (PJM State & Mem Training Department, 2017).

SMP results from the equilibrium point between generation and demand curve prices once the social welfare has been maximized and the energy balance is guaranteed. It is assumed that generation and demand offers are expressed in a discrete form using energy blocks. Regarding generation offers, they are represented in a matrix with a total number of rows equal to the number of blocks (K) and a total number of columns equal to the number of generators (I). Modeling the generation prices follows a similar approach. Thus, generation offers (MWh) and prices (\$/MWh) are symbolized by the matrices $E_{(k,i)}^{\text{gen}}$ and $\lambda_{(k,i)}^{\text{gen}}$, respectively, where $k = 1, \ldots, K$ is the index for each block, and $i = 1, \ldots, I$ is the index for each generator.

Concerning the load offers, they are expressed using a matrix with a total number of rows equal to the number of blocks (K) and a total number of columns equal to the number of buses (R). Modeling the load prices follows a similar approach. Thus, load offers (MWh) and prices (\$/MWh) are expressed through the matrices $E_{(k,r)}^{\text{dem}}$ and $\lambda_{(k,r)}^{\text{dem}}$, respectively, where $k = 1, \ldots, K$ is the index for each block, and $r = 1, \ldots, R$ is the index for each bus.

The generation (load) offers and prices are the available energy magnitudes and prices from each side. From these, the electricity market organizer has to determine the energy to take from the supply and demand to maximize social welfare, maintaining the energy balance. The energy accepted by the market organizer from the supply is represented by the matrix $e_{(k,i)}^{\text{gen}}$ and the energy accepted from the demand side is expressed through the matrix $e_{(k,r)}^{\text{dem}}$.

The methodology followed by the electricity market coordinator is mathematically presented in the optimization problem of Eqs. (7.11)–(7.15) (Gómez-Expósito et al., 2018) and graphically illustrated in Fig. 7.5.

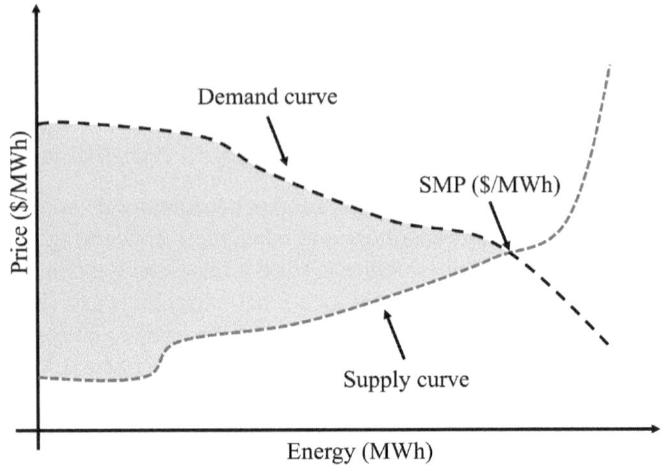

Fig. 7.5 Supply–demand curve.

Eq. (7.11) is the maximization of social welfare, considering the offered prices of the demand and supply energies. In Fig. 7.5, the shadow area shows the subtraction of demand and supply curves, which is to be maximized. Eq. (7.12) is an alternative form of Eq. (7.11), written as a minimization problem, which is convenient for computational implementation (Gómez-Expósito et al., 2018; Zavala et al., 2017). The equality constraint of Eq. (7.13) represents the energy balance between generation and demand. The restrictions of Eqs. (7.14) and (7.15) express that for each accepted offer ($e^{\text{gen}}_{(k,i)}$ and $e^{\text{dem}}_{(k,r)}$), which has to be lower than or equal to that presented by the generator or load ($E^{\text{gen}}_{(k,i)}$ and $E^{\text{dem}}_{(k,r)}$).

$$\max\left\{\sum_{r=1}^{R}\sum_{k=1}^{K} e^{\text{dem}}_{(k,r)} \times \lambda^{\text{dem}}_{(k,r)} - \sum_{i=1}^{I}\sum_{k=1}^{K} e^{\text{gen}}_{(k,i)} \times \lambda^{\text{gen}}_{(k,i)}\right\} \tag{7.11}$$

$$\min\left\{-\left[\sum_{r=1}^{R}\sum_{k=1}^{K} e^{\text{dem}}_{(k,r)} \times \lambda^{\text{dem}}_{(k,r)} - \sum_{i=1}^{I}\sum_{k=1}^{K} e^{\text{gen}}_{(k,i)} \times \lambda^{\text{gen}}_{(k,i)}\right]\right\} \tag{7.12}$$

$$\sum_{i=1}^{I}\sum_{k=1}^{K} e^{\text{gen}}_{(k,i)} = \sum_{r=1}^{R}\sum_{k=1}^{K} e^{\text{dem}}_{(k,r)} \tag{7.13}$$

$$0 \le e^{\text{gen}}_{(k,i)} \le E^{\text{gen}}_{(k,i)} \tag{7.14}$$

$$0 \le e^{\text{dem}}_{(k,r)} \le E^{\text{dem}}_{(k,r)} \tag{7.15}$$

In Fig. 7.5, the supply curve is built by sorting all energy-price pairs ($E^{\text{gen}}_{(k,i)}$ $E^{\text{gen}}_{(k,i)}$, $\lambda^{\text{gen}}_{(k,i)}$) in ascending order by price and then calculating the cumulative energy values. Then, a constantly growing curve is obtained. A similar procedure is applied over the accepted offers using the load results in descending order. Finally, the interception of both curves is the SMP.

The next step consists of performing a power flow analysis by the Newton–Raphson method to estimate the system power loss. Power flow consists of solving the equations system of Eqs. (7.5)–(7.7). Then, the restrictions relating to the transmission line ampacity (Eq. 7.8) and voltage profile (Eq. 7.9) are verified.

During the power flow analysis, the marginal generator (the generator in charge of providing the energy in case the system load increases) is adopted as the slack unit. The remainder generating units are PV type, and the load buses are PQ type. It is crucial to highlight that once the power flow study is carried out, the output power of the marginal unit increases. Thus, we have to update the accepted energy supply matrix $e^{gen}_{(k,i)}$, where i is the slack unit, so that the addition of its elements equals the power production considering the system power loss. To deal with this issue, we formulate a minimization LP problem, taking as objective the product of the prices of the slack bus (i) by the power in each block ($k = 1, \ldots, K$) and forcing the addition of the power generation of each block to equal the value previously obtained from the Newton–Raphson study.

If an operational constraint is violated, the electricity market coordinator could perform a system reconfiguration or a re-dispatch, among other actions. We focus on the generation re-dispatch using a GA, which will be explained later in the following subsection. The constraint has been avoided once the generation system has been re-dispatched by GA; however, the total cost increases due to the necessity of operating expensive generators conveniently located to overcome the restriction. Thus, we need to calculate the savings associated with the relaxation of the constrained under analysis. The price associated with such savings is known as the shadow price.

Considering the case in which a determined transmission line is overloaded, we specify the loading level in terms of the apparent power (MVA). To estimate the shadow price, we calculate the total cost (TC) using Eq. (7.16) on the re-dispatched condition:

$$TC = \sum_{i=1}^{I} \sum_{k=1}^{K} e^{gen}_{(k,i)} \times \lambda^{gen}_{(k,i)} \tag{7.16}$$

Then, we relax the thermal limit of the overloaded line in 1 MVA ($A^{max}_{(r,m)} \leftarrow A^{max}_{(r,m)} + 1$) and apply the re-dispatching process by GA. After that, we calculate the associated cost of this condition using Eq. (7.16) again. Finally, the shadow price is defined as subtracting the total cost after the relaxation process from the total cost before.

Once the shadow price has been calculated, we need to estimate the shift factors. Shift factors analyze the variation of the power flowing through the line of interest when the power injected at each node of the system increases by 1 MW. Essentially, the shift factor allows us to test the sensitivity of the constraint concerning the power injection of a generation unit or load control.

We use the initial power flow study as the base case. Then, we select one node (r) at a time (except the slack bus for it is zero) and inject 1 MW on it, perform a power flow analysis, and register the power flowing through the line of interest (the line with thermal limit problems). Finally, the shift factor would be the subtraction between the just registered value and that observed in the base case for the corresponding line, divided by the power injection (Zhu, 2009). Finally, the congestion-related price for each node would be the shadow price multiplied by the shift factor of each bus.

A similar approach is followed to estimate the marginal loss-related price. Considering the initial power flow study as the base case, we select one node (r) at a time (except the slack bus for it is zero) and inject 1 MW on it, perform a power flow analysis, and register the power loss. Then, the loss sensitivity of the corresponding node ($LS_{(r)}$) is the subtraction between the just registered loss value and that observed in the base case. We calculate the penalty factor ($PF_{(r)}$) according to Eq. (7.17), using the loss sensitivity ($LS_{(r)}$). Finally, we calculate the marginal loss-related price as shown in Eq. (7.18) (PJM State & Mem Training Department, 2007).

$$PF_{(r)} = \frac{1}{\left(1 - LS_{(r)}\right)} \tag{7.17}$$

$$SMP \times \left(\frac{1}{PF_{(r)}} - 1\right) \tag{7.18}$$

We can calculate the LMP as the addition of the SMP, which is the same for all nodes, the congestion-related and the marginal loss-related prices. It is essential to highlight that the congestion-related and the marginal loss-related prices are different for each system node, representing the locational characteristic of energy costs.

7.5 Solving the optimal power flow by genetic algorithms

As aforementioned, the OPF consists of finding the output power of each unit ($p_{(i)}$ or $e^{gen}_{(k,i)}$) to minimize the total operational costs, considering the operating limits of the generation units and transmission system.

This concept has slightly different implementation depending on the system type. In the case of vertically integrated systems, the system coordinator knows the cost parameters so that the total cost is directly minimized considering the function F of Eq. (7.19):

$$\begin{aligned} F = &\sum_{i=1}^{I}\left\{a_{(i)} + b_{(i)}\left(p_{(i)}\right) + c_{(i)}\left(p_{(i)}\right)^2\right\} \\ &+ \sum_{\Omega}\left\{b_A\left(\left|A^{\max}_{(r,m)} - A_{(r,m)}\right|\right) + c_A\left(\left|A^{\max}_{(r,m)} - A_{(r,m)}\right|\right)^2\right\} \\ &+ \sum_{\Gamma}\left\{b_V\left(\left|V^{\max}_{(r)} - V_{(r)}\right|\right) + c_V\left(\left|V^{\max}_{(r)} - V_{(r)}\right|\right)^2\right\} \\ &+ \sum_{\Theta}\left\{b_W\left(\left|V_{(r)} - V^{\min}_{(r)}\right|\right) + c_W\left(\left|V_{(r)} - V^{\min}_{(r)}\right|\right)^2\right\} \end{aligned} \tag{7.19}$$

The function F is extended to penalize those solutions with transmission system overloading and voltages out of the required limits (Ω is the set of overloaded lines, Γ is the set of nodes with overvoltage, and Θ is the set of nodes with undervoltage). Other constraints can also be incorporated.

In the case of systems operating in competitive markets, the function F to be minimized during system re-dispatch is shown in Eq. (7.20):

$$\begin{aligned} F = & \sum_{i=1}^{I}\sum_{k=1}^{K} e_{(k,i)}^{\text{gen}} \times \lambda_{(k,i)}^{\text{gen}} \\ & + \sum_{\Omega}\left\{ b_A\left(\left|A_{(r,m)}^{\max} - A_{(r,m)}\right|\right) + c_A\left(\left|A_{(r,m)}^{\max} - A_{(r,m)}\right|\right)^2 \right\} \\ & + \sum_{\Gamma}\left\{ b_V\left(\left|V_{(r)}^{\max} - V_{(r)}\right|\right) + c_V\left(\left|V_{(r)}^{\max} - V_{(r)}\right|\right)^2 \right\} \\ & + \sum_{\Theta}\left\{ b_W\left(\left|V_{(r)} - V_{(r)}^{\min}\right|\right) + c_W\left(\left|V_{(r)} - V_{(r)}^{\min}\right|\right)^2 \right\} \end{aligned} \tag{7.20}$$

If a transmission line between the buses r and m is overloaded, the parameters b_A and c_A are adjusted to a high value ($b_A \rightarrow \infty$ and $c_A \rightarrow \infty$), whereas, the parameters b_A and c_A are neglected ($b_A \leftarrow 0$ and $c_A \leftarrow 0$). A similar approach is applied to the voltage constraints and the parameters b_V, b_W, c_V, and c_W. If the restriction is fulfilled, the corresponding parameters are set to zero; otherwise, they are set to an artificially high value.

The voltage and ampacity constraints are checked using a power flow analysis by the Newton–Raphson approach. In the case of vertically integrated systems, we have direct access to the output power of each generator ($p_{(i)}$), so that we can evaluate the power production directly on the Newton–Raphson method. In the case of systems operating in competitive markets, we specify the power generation of each unit using Eq. (7.21):

$$p_{(i)} = \sum_{k=1}^{K} e_{(k,i)}^{\text{gen}} \tag{7.21}$$

Once the Newton–Raphson method is applied, the column of the matrix $e_{(k,i)}^{\text{gen}}$ of the slack generator is found by solving an LP problem in which the addition of the product of $e_{(k,i)}^{\text{gen}}$ by $\lambda_{(k,i)}^{\text{gen}}$ for $k=1, \ldots, K$ is minimized, subject to the condition that the addition of $e_{(k,i)}^{\text{gen}}$ for $k=1, \ldots, K$ is equal to the power production resulting from the Newton–Raphson approach for the slack unit. Remember that for PV type units, the elements of the matrix $e_{(k,i)}^{\text{gen}}$ are already known, and they do not change during the Newton–Raphson study.

Concerning the energy balance constraint, it is directly guaranteed by the Newton–Raphson method, and it is not necessary to incorporate a weighting parameter in the objective function.

The real-coded GA developed by Chuang et al. (2015, 2016) will be adopted to solve the ED problem. The algorithm has three operators, the ranking selection

(RS), the direction-based crossover (DBX), and the dynamic random mutation (DRM). These operators are applied considering the crossover rate (p_c), the mutation rate (p_m), and the proportional parameter (p_r) (y_{obj} is the objective function F, x_{pop} is the population matrix with I rows and N columns).

RS is in charge of extracting the best individuals from the population. This operator is implemented using the procedure explained as follows:

Step 1: Individuals are sorted according to fitness. In this regard, the order in which the individuals should be grouped is saved in a list ($\mathcal{L}_{\mathbf{A}}$). The index for each individual is $n = 1, \ldots, N$.

Step 2: Calculate the number of individuals that have relatively worse fitness. It is made by rounding down the product $N \times p_r$(floor($N \times p_r$)).

Step 3: Select the first $N -$ floor($N \times p_r$) elements of the list $\mathcal{L}_{\mathbf{A}}$ and save the result in another list ($\mathcal{L}_{\mathbf{B}}$). These are the indices of the relatively best individuals to be kept. The remaining are the index of the individuals to be discarded.

Step 4: Select the first floor($N \times p_r$) elements of the list $\mathcal{L}_{\mathbf{A}}$ and save them in another list ($\mathcal{L}_{\mathbf{C}}$). These are the indices of the individuals used to replace relatively worse individuals.

Step 5: Create a list ($\mathcal{L}_{\mathbf{D}}$) by joining or merging the lists $\mathcal{L}_{\mathbf{B}}$ and $\mathcal{L}_{\mathbf{C}}$ previously built.

Step 6: Sort the individuals in the list $\mathcal{L}_{\mathbf{D}}$ according to the fitness and save the results in another list ($\mathcal{L}_{\mathbf{E}}$), this is the result of the RS process.

DBX divides the family of individuals obtained from the RS into two sets: the individuals with a high ranking and those with a lower position. These two groups are crossed sequentially, improving the capabilities of the traditional crossover operator. The DBX is implemented by following the procedure:

Step 1: Set $u \leftarrow$ floor($N/2$).

Step 2: Start the sequential crossing of pairs of individuals. Go to Step 2.1.

Step 2.1: Set $m \leftarrow 1$.
Step 2.2: If $m \leq$ floor($N/2$), go to Step 2.3, else end.
Step 2.3: Set $u \leftarrow u + 1$.
Step 2.4: Select the individual m of $\mathcal{L}_{\mathbf{E}}\left(\mathcal{L}_{E(m)}\right)$, it will be called $x^a_{\text{pop}}\left(x^a_{\text{pop}} \leftarrow x_{\text{pop}}\left(\mathcal{L}_{E(m)}\right)\right)$.
Step 2.5: Select the individual u of $\mathcal{L}_{\mathbf{E}}\left(\mathcal{L}_{E(u)}\right)$, it will be called $x^a_{\text{pop}}\left(x^a_{\text{pop}} \leftarrow x_{\text{pop}}\left(\mathcal{L}_{E(u)}\right)\right)$.
Step 2.6: If $RAND > 1 - p_c$ and $(\max\{y_{\text{obj}(n)} \forall n = 1, \ldots, N\} - \min\{y_{\text{obj}(n)} \forall n = 1, \ldots, N\}) > \varepsilon$, calculate the variable S_c using Eq. (7.22). $RAND$ is a uniform random number in the interval [0,1], and ε is a small number (tolerance).

$$S_c = \frac{\left|y_{\text{obj}}\left(x^a_{\text{pop}}\right) - y_{\text{obj}}\left(x^b_{\text{pop}}\right)\right|}{\max\left\{y_{\text{obj}(n)} \forall n = 1, \ldots, N\right\} - \min\left\{y_{\text{obj}(n)} \forall n = 1, \ldots, N\right\}} \tag{7.22}$$

Step 2.7: Calculate the element of the corrective vector (Δx_{pop}) which has I elements using Eq. (7.23). $RAND$ is a random number from a uniform distribution within the interval [0,1].

$$\Delta x_{\text{pop}(i)} = \begin{cases} 0; & RAND < 0.5 \\ x^a_{\text{pop}(i)} - x^b_{\text{pop}(i)}; & RAND \geq 0.5 \end{cases} \tag{7.23}$$

Step 2.8: Calculate the offspring of individuals a and b using Eq. (7.24).

$$\begin{aligned} x^a_{\text{pop}} &\leftarrow x^a_{\text{pop}} + S_c \Delta x_{\text{pop}} \\ x^b_{\text{pop}} &\leftarrow x^b_{\text{pop}} + S_c \Delta x_{\text{pop}} \end{aligned} \tag{7.24}$$

Step 2.9: Perform DBX in the next pair of individuals by setting $m \leftarrow m+1$ and going to Step 2.2.

DRM is incorporated to increase the diversity among the individuals in the population. This operator is implemented by following the steps as follows:

Step 1: Considering the generation under study (j), calculate the variable S_m using Eq. (7.25), where J is the total number of generations ($j=1, \ldots, J$).

$$S_m = \left(\frac{1-j}{J}\right)^2 \tag{7.25}$$

Step 2: Apply DRM over the population. Go to Step 2.1.

Step 2.1: Set $n \leftarrow 1$.
Step 2.2: If $n \leq N$, go to Step 2.3, else end.
Step 2.3: If $RAND < p_m$, go to Step 2.4.
Step 2.4: Set $x_{\text{pop}(n)} \leftarrow x_{\text{pop}(n)} + S_m \boldsymbol{RAND}(p^{\max}_{(i)} - p^{\min}_{(i)} \forall i=1,\ldots,I)$, where $\boldsymbol{RAND}$ is a random vector with I elements taken from a uniform distribution within the interval [0,1].
Step 2.5: Set $n \leftarrow n+1$ and go to Step 2.2.

Some individuals may violate the constraints of Eq. (7.4); these individuals are adjusted to the closest minimum or maximum values in such a case.

Elitism is incorporated into this implementation by inserting the best individual in a position randomly selected from the interval [1,N].

7.6 Case studies

This section illustrates the solution of the OPF by a GA with two different cases. The first case considers the management of a vertically integrated electrical system, and the second case analyzes the operation of an electrical system in a market-based context. We have implemented the optimization algorithms previously described in a computer with i7-3630QM CPU at 2.40GHz, 8GB of RAM, and a 64-bit operating system. The computing language used was MATLAB®.

7.6.1 Analyzing a vertically integrated electrical system

In this subsection, we study a system previously analyzed by Weber (1997) using a Newton-based technique and also analyzed by Osman et al. (2004) using a real-coded GA but with different characteristics than the algorithm previously described in Section 7.5. The system has six buses, the one-line diagram is publically available (Weber, 1997), and will not be repeated here.

The system has four generation nodes. Node one is assumed as the slack bus, nodes 2, 3, and 4 are PV-type nodes, and nodes 5 and 6 are PQ-type nodes. We have adopted the minimum and maximum voltages of 0.95 and 1.05 per unit (pu), respectively. Additionally, we have considered a base of 100 MVA and 13.8 kV.

The power flow has been solved using the Newton–Raphson approach, considering and tolerance of 0.0001 (power mismatch) with a maximum number of iterations equal to 150. We directly used the power flow solver offered by MATPOWER (Zimmerman et al., 2011) with the settings mentioned above. Tables 7.6–7.8 show information concerning the cost curves of each generator, transmission line characteristics, and electricity demand (Weber, 1997).

The OPF problem has been solved using the real-coded GA of Section 7.5. We have used the MATPOWER interior point solver (MIPS) for comparative purposes. The real-coded GA explained in Section 7.5 was implemented considering a proportional parameter of 0.3 (p_r), crossover rate of 0.95 (p_c), and a mutation rate of 0.1 (p_m). The population size was set to 100 individuals ($N = 100$), and the generations considered were 300 ($J = 300$).

Fig. 7.6 shows the GA evolution converging to a high-quality solution in a reduced number of generations. Table 7.9 compares the active power generation obtained from GA and MIPS, observing similar results. Table 7.10 presents a similar comparison but with reactive power generation. Important differences can be observed because the

Table 7.6 Data of generation units.

Bus	Maximum (MW)	Minimum (MW)	*c* ($/MW2h)	*b* ($/MWh)	*a* ($/h)
1	250	50	0.012	12	105
2	250	50	0.0096	9.6	96
3	250	50	0.013	13	105
4	250	50	0.0094	9.4	94

Table 7.7 Data of transmission lines.

From	To	Resistance (pu)	Reactance (pu)	Susceptance (pu)	Capacity (MVA)
1	2	0.04	0.08	0.02	100
1	5	0.04	0.08	0.02	100
2	4	0.04	0.08	0.02	100
3	5	0.04	0.08	0.02	100
3	6	0.04	0.08	0.02	100
4	5	0.04	0.08	0.02	50
4	6	0.04	0.08	0.02	100

Table 7.8 Power system electricity demand.

Bus	Active power (MW)	Reactive power (MVAr)
1	100	20
2	100	20
3	100	20
4	100	20
5	100	50
6	100	10

MIPS also incorporates the voltage profile in the optimization problem, which has not been considered here.

Table 7.11 shows the apparent power through the transmission system, where similar values are observed from both methods.

Tables 7.12 and 7.13 present the voltage profile comparison. We have fixed the voltage settings to one in all generators by default. However, MIPS optimizes such settings, offering a solution of higher quality. The differences between the optimization-problem formulation described in Section 7.5 and that developed in MATPOWER software explain the slight contrast observed in the results reported in Tables 7.12 and 7.13.

Finally, Table 7.14 presents the cost obtained from each method. The solutions are very similar; however, the computational resources employed by the implemented GA are considerably higher. The computational time could be reduced by using stopping

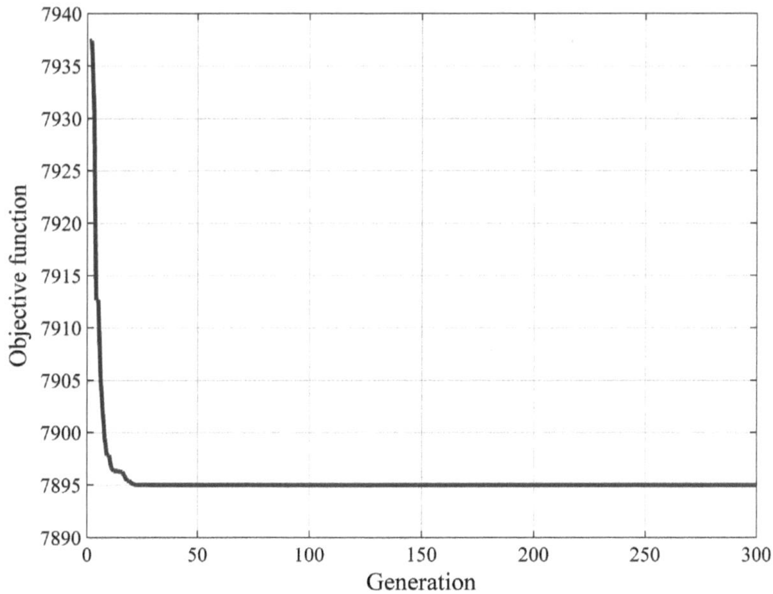

Fig. 7.6 GA evolution for solving OPF.

Table 7.9 Active power generation.

Bus	GA (MW)	MIPS (MW)
1	122.20	123.34
2	190.24	195.77
3	116.41	109.52
4	179.18	178.91

Table 7.10 Reactive power generation.

Bus	GA (MVAr)	MIPS (MVAr)
1	43.36	62.45
2	−24.59	3.51
3	75.54	54.48
4	48.02	19.97

Table 7.11 Transmission system ampacity.

From	To	GA (MVA)	MIPS (MVA)
1	2	42.95	44.60
1	5	60.78	68.70
2	4	56.82	54.24
3	5	37.34	23.79
3	6	29.77	24.60
4	5	50.00	50.00
4	6	80.14	82.22

Table 7.12 Voltage magnitudes.

Bus	GA (pu)	MIPS (pu)
1	1	1.05
2	1	1.05
3	1	1.02
4	1	1.03
5	0.97	1.00
6	0.97	1.00

criteria based on the objective-function behavior. In other words, the optimization process could be detained when no improvements are observed. In this way, the significant periods of stagnation shown in Fig. 7.6, for example, between generation 50 and 300, are avoided. Another option is adopting parallel computing.

Table 7.13 Voltage angle.

Bus	GA (degree)	MIPS (degree)
1	0	0
2	2.22	2.10
3	−3.94	−3.23
4	−0.67	−0.14
5	−2.74	−2.24
6	−4.56	−3.83

Table 7.14 Computational time and objective function.

Performance	GA	MIPS
Cost ($)	7895.000	7872.240
Time (min)	5.55	0.45

7.6.2 Analyzing an electrical system with a market-based operation

In this subsection, we illustrate the implementation of the OPF solution by GA in the competitive electricity market. Fig. 7.7 shows the scheme of the energy system of interest, where the power production of 10 units needs to be optimized to maximize the social welfare.

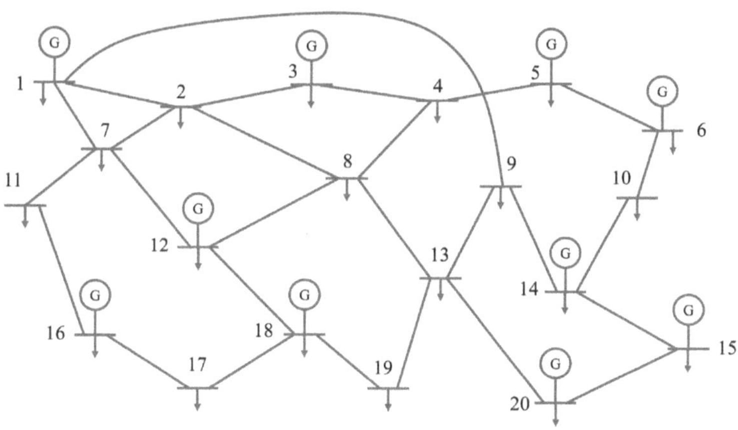

Fig. 7.7 One-line diagram of the system under study.

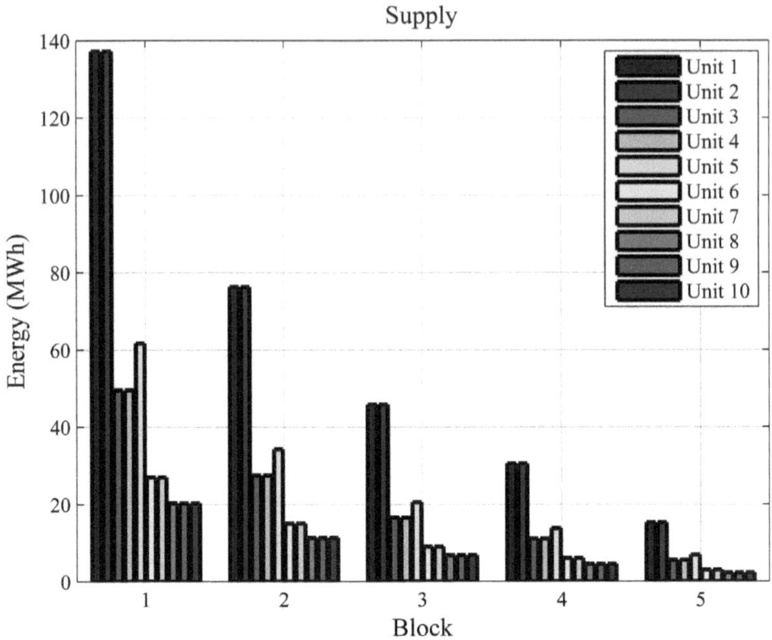

Fig. 7.8 Energy per block (generation).

As previously mentioned, supply and demand offers (energy and prices) are represented using the matrices $E^{\mathrm{gen}}_{(k,i)}$ and $\lambda^{\mathrm{gen}}_{(k,i)}$, for generation, and the matrices $E^{\mathrm{dem}}_{(k,r)}$ and $\lambda^{\mathrm{dem}}_{(k,i)}$, for demand. The power system has 20 nodes ($R = 20$), and 10 units ($I = 10$). We have considered five offer blocks ($K = 5$).

Concerning the supply, the information on the energy matrix ($E^{\mathrm{gen}}_{(k,i)}$) is presented in Fig. 7.8, and the corresponding prices ($\lambda^{\mathrm{gen}}_{(k,i)}$) are shown in Fig. 7.9.

Concerning the electricity demand, the values of the energy matrix ($E^{\mathrm{dem}}_{(k,r)}$) are shown in Fig. 7.10, and the associated prices ($\lambda^{\mathrm{dem}}_{(k,r)}$) are shown in Fig. 7.11.

The optimization problem of Eqs. (7.12)–(7.15) was formulated using the information on matrices $E^{\mathrm{gen}}_{(k,i)}$ and $\lambda^{\mathrm{gen}}_{(k,i)}$, for generation, and the matrices $E^{\mathrm{dem}}_{(k,r)}$ and $\lambda^{\mathrm{dem}}_{(k,r)}$, for demand. The LP approach solved the problem, obtaining a total demand of 1206.25 MWh.

The energy price curve of the system is built by considering the available information about the energy and prices from the supply perspective. Concretely, the pairs of data composed of the energy offer and its price are sorted in ascending order, from the cheapest to the most expensive. Then, we calculate the cumulative values of energy. Fig. 7.12 shows the resulting curve for our illustrative case study.

Finally, we can estimate the SMP and the marginal unit by evaluating the accepted electricity demand (1206.25 MWh) in the curve of Fig. 7.12. For this case, the SMP equals 55.338 $/MWh, and the marginal unit is connected to node 18 of Fig. 7.7.

The results obtained from the LP problem of Eqs. (7.12)–(7.15) were incorporated into the power flow problem solution by Newton–Raphson. The impedances of each

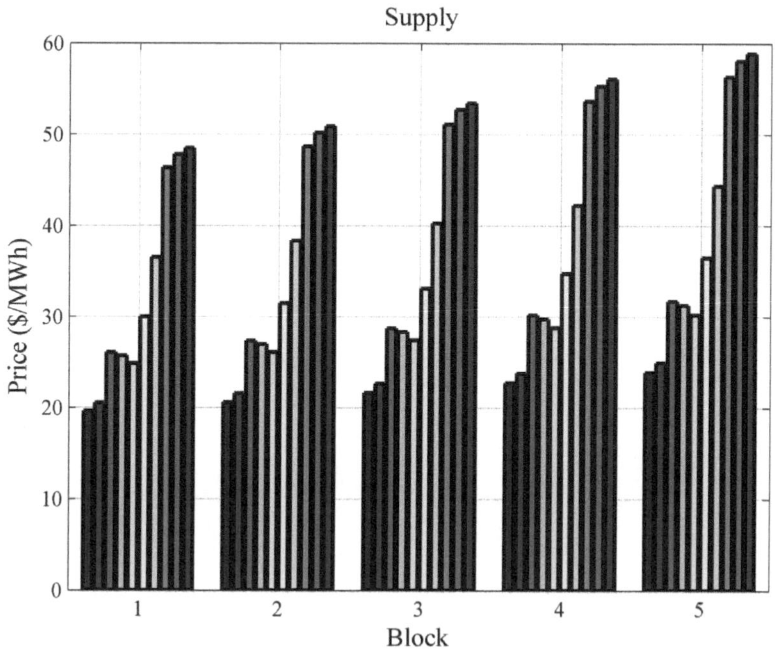

Fig. 7.9 Energy price per block (generation).

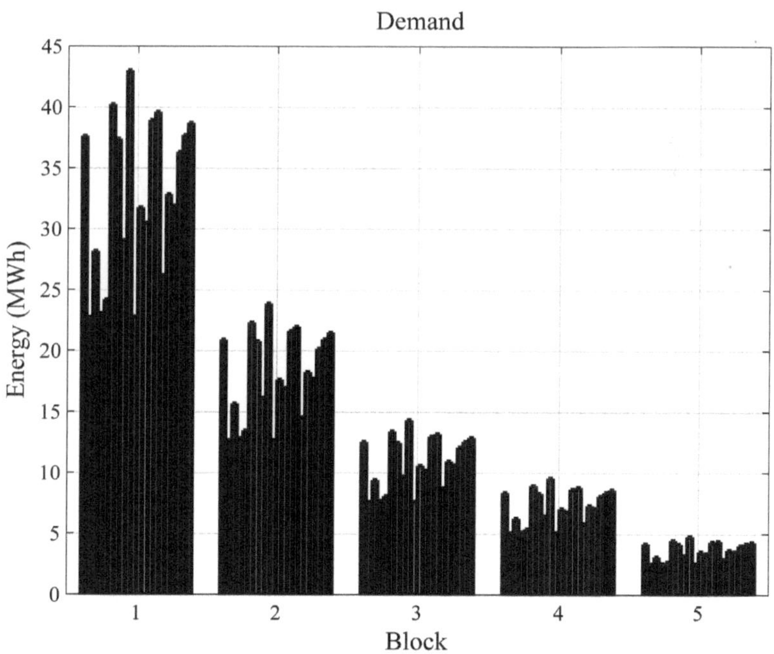

Fig. 7.10 Energy per block (demand).

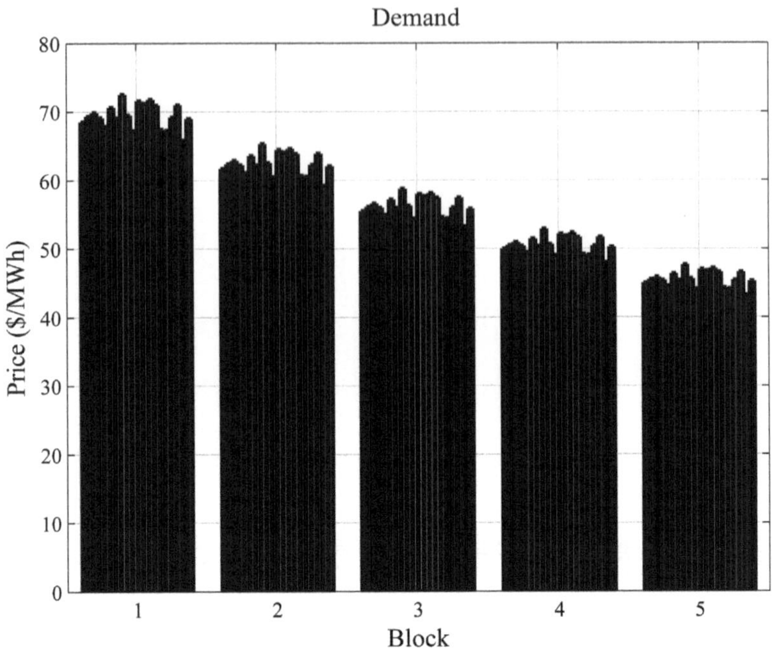

Fig. 7.11 Energy price per block (demand).

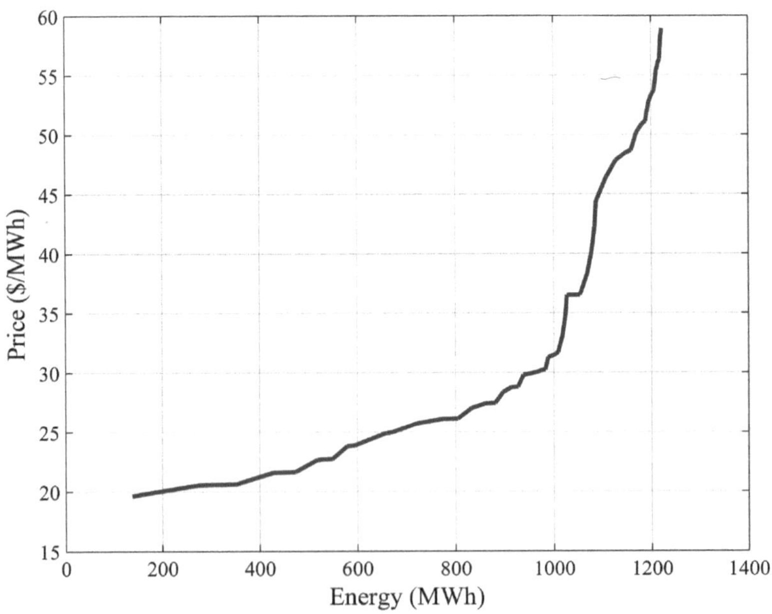

Fig. 7.12 Energy price curve of the system.

Table 7.15 Transmission system impedances.

From	To	Resistance (pu)	Reactance (pu)	Susceptance (pu)
1	2	0.0038	0.0418	0.2863
1	7	0.0041	0.0447	0.2677
2	7	0.0018	0.0199	0.6003
2	3	0.0041	0.0449	0.2663
2	8	0.0033	0.0360	0.3324
3	4	0.0017	0.0190	0.6299
4	5	0.0023	0.0247	0.4835
4	8	0.0030	0.0333	0.3595
1	9	0.0042	0.0463	0.2582
5	6	0.0019	0.0209	0.5723
6	10	0.0043	0.0467	0.2559
7	11	0.0029	0.0313	0.3820
7	12	0.0038	0.0413	0.2895
8	12	0.0019	0.0204	0.5864
8	13	0.0027	0.0293	0.4083
9	13	0.0041	0.0450	0.2659
9	14	0.0042	0.0464	0.2579
10	14	0.0034	0.0367	0.3257
11	16	0.0016	0.0170	0.7026
12	18	0.0039	0.0429	0.2790
13	19	0.0042	0.0456	0.2625
13	20	0.0034	0.0374	0.3193
14	15	0.0036	0.0400	0.2993
16	17	0.0036	0.0395	0.3028
17	18	0.0026	0.0283	0.4218
18	19	0.0034	0.0367	0.3257
15	20	0.0019	0.0213	0.5608

branch are shown in Table 7.15. The reactive power of electricity demand at each node was calculated assuming a power factor of 95%.

The power flow results were used to estimate the power loss and verify the fulfillment of power system constraints (voltage and ampacity). The analysis revealed that the ampacity of the branch between nodes 4 and 8 was violated. A re-dispatch process was carried out to correct this problem. The re-dispatch process consists of solving the OPF by GA, as explained in Section 7.5.

Fig. 7.13 and Table 7.16 show the energy production after the re-dispatch process. The system generates 1210.008 MWh, including the power loss.

Table 7.17 presents the flow of apparent power through the system. We can observe the apparent power between buses 4 and 8 is at its limit (120 MVA). Table 7.18 shows the voltage magnitudes and angles and the LMP.

Concerning the LMP estimation, we have already calculated the SMP. Thus, we only need to calculate the transmission system congestion and power loss components.

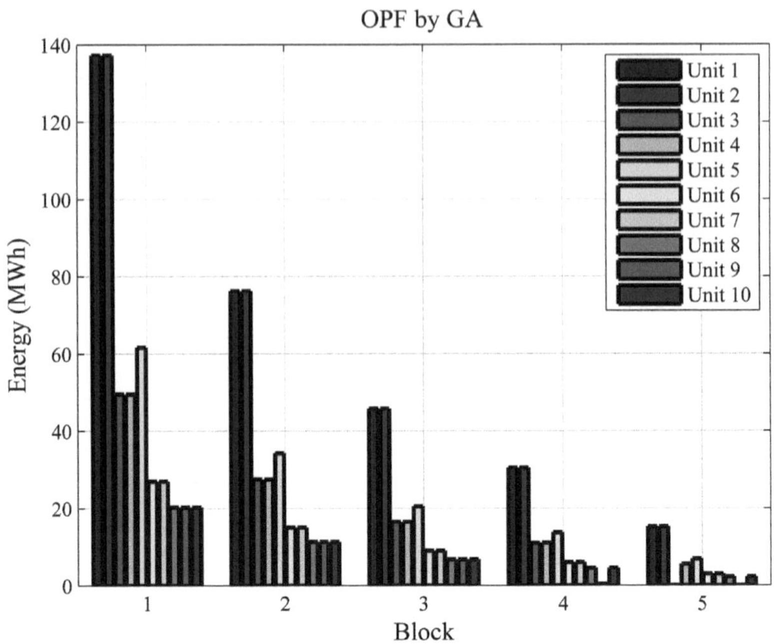

Fig. 7.13 Energy production per unit and block.

Table 7.16 Energy production per generation unit.

Unit	Energy (MWh)
1	305
2	305
3	104.44
4	110
5	137
6	60
7	60
8	45
9	38.57
10	45

The shadow price is a crucial concept behind estimating the price related to the system congestion (branch between buses 4 and 8). According to Section 7.4.2, we need to relax the ampacity constraint in one unit (MVA) and evaluate the cost reduction. Fig. 7.14 shows the evolution of the GA solved in constrained and relaxed conditions:

Table 7.17 Transmission system ampacity.

From	To	Ampacity (MVA)	Capacity (MVA)
1	2	64.45	164.18
1	7	90.69	190.88
2	7	87.72	188.84
2	3	115.19	216.53
2	8	41.55	141.42
3	4	157.88	256.62
4	5	20.85	124.66
4	8	120	120.00
1	9	106.52	206.59
5	6	49.81	151.26
6	10	76.05	177.73
7	11	97.16	198.71
7	12	7.86	107.91
8	12	17.32	117.63
8	13	91.56	192.87
9	13	36.23	136.52
9	14	7.41	107.57
10	14	38.54	140.24
11	16	35.31	136.87
12	18	86.43	187.23
13	19	42.05	142.26
13	20	3.96	104.00
14	15	24.25	125.38
16	17	32.56	132.00
17	18	37.75	138.39
18	19	34.42	134.29
15	20	41.64	142.70

The cost of relaxing the ampacity limit (branch of nodes 4–8) is 32,926.975$, and the cost related to the constrained situation (ampacity of Table 7.17) is 32,970.3143$. Therefore, the difference (shadow price) is −43.339$.

Once we have determined the shadow price, we calculate the shift and sensitivity factors as previously mentioned (Section 7.4.2), obtaining the LMP of Table 7.18. Details about shift and sensitivity factors have been carefully explained by Zhu (2009). To improve the understanding of the results, we have combined the information in Fig. 7.7 and Table 7.18, obtaining Fig. 7.15.

We can observe that the branch between buses 4 and 8 is critical. The lowest LMP is in node 4, and the highest is in node 8. Thus, the benefit of a determined generation company is affected by its location in the system. The units upstream of the constrained branch have their benefits reduced, while those generators downstream of the constrained section have higher revenues (PJM State & Mem Training Department, 2017).

Table 7.18 Voltage profile and energy prices.

Bus	Voltage (pu)	Angle (degree)	LMP ($/MWh)
1	1	4.09	51.35
2	1.01	2.88	50.73
3	1	5.84	36.31
4	1.01	4.30	32.58
5	1	4.63	33.83
6	1	4.17	36.72
7	1.01	1.96	52.60
8	1.01	2.12	57.06
9	1.00	1.33	51.69
10	1.00	2.21	43.92
11	1.01	0.28	53.56
12	1	2.08	56.09
13	1.01	0.65	54.38
14	1	1.45	48.37
15	1	1.04	51.02
16	1	−0.04	53.93
17	1.00	−0.60	54.87
18	1	0	55.34
19	1.00	−0.40	55.01
20	1	0.69	55.34

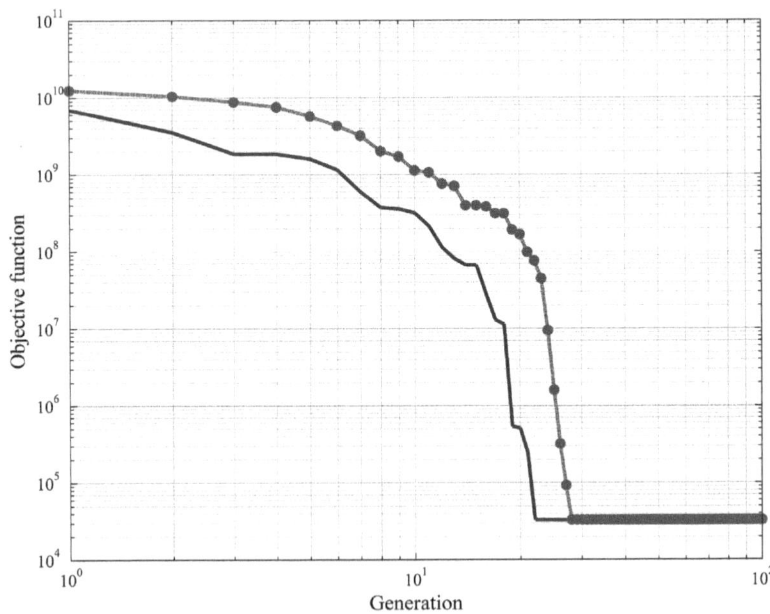

Fig. 7.14 GA evolution for constrained and relaxed OPF problems.

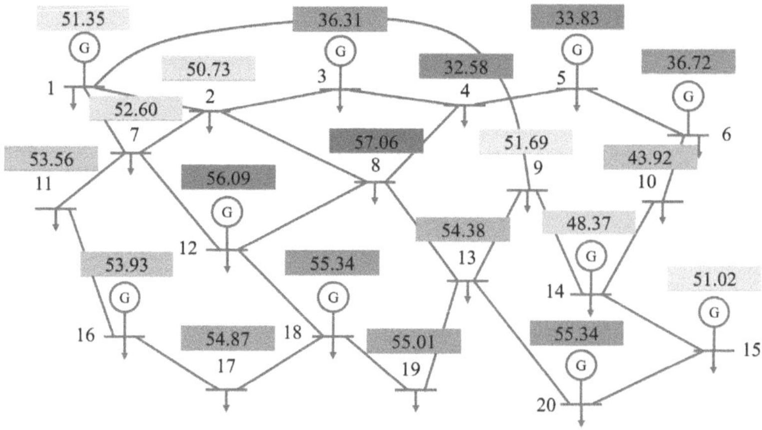

Fig. 7.15 LMP of the energy system.

In Sections 7.1 and 7.2, we emphasized the relevance of the actual environmental problems, the necessity of creating an effective energy policy, market commodities' behavior, and sociopolitical circumstances. Reducing CO_2 emissions has significant effects on the cost of energy. To illustrate this idea, Table 7.19 (US Energy Information Administration, 2022a) presents the capital, variable, and fixed operating and maintenance (O&M) costs and the heat rate for some selected technologies.

As previously discussed, natural gas and coal are crucial energy commodities. Thus, the selected technologies are ultra-supercritical coal (USC), ultra-supercritical coal with 30% carbon capture and sequestration (USC-30CCS), ultra-supercritical coal with 90% carbon capture and sequestration (USC-90CCS), combined cycle gas turbine (single-shaft) (CCGT-SS), combined cycle gas turbine (multishaft) (CCGT-MS), combined cycle with 90% carbon capture and sequestration (CCGT-90CCS), and nuclear—small modular reactor (N-SMR).

According to Table 7.19, reducing GHG emissions requires the adoption of costly equipment. In the case of coal, Table 7.20 shows the cost increment regarding USC, and Table 7.21 presents the cost increment relating to CCGT, which is based on natural gas.

Table 7.19 Cost of some power generation technologies.

Technology	Capital cost ($/kW)	Variable O&M cost ($/MWh)	Fixed O&M cost ($/kW-yr)	Heat rate (Btu/kWh)
USC	4074	4.71	42.49	8638
USC-30CCS	5045	7.41	56.84	9751
USC-90CCS	6495	11.49	62.34	12,507
CCGT	1062	1.96	12.77	6370
CCGT-90CCS	2736	6.11	28.89	7124
N-SMR	6861	3.14	99.46	10,443

Table 7.20 Cost increment regarding USC.

Technology	Capital cost (%)	Variable O&M cost (%)	Fixed O&M cost (%)	Heat rate (%)
USC-30CCS	23.83	57.32	33.77	12.88
USC-90CCS	59.43	143.95	46.72	44.79

Table 7.21 Cost increment regarding CCGT.

Technology	Capital cost (%)	Variable O&M cost (%)	Fixed O&M cost (%)	Heat rate (%)
CCGT-90CCS	157.63	211.73	126.23	11.84

In the case of coal, reducing CO_2 emissions by 30% requires investing almost 24% more capital, while a 90% reduction implies nearly 60%. In the case of natural gas, reducing the emissions by 90% requires increasing the investment by almost 160%. Adopting state-of-the-art nuclear technology also requires a considerable investment. Additionally, the transmission system needs relevant investments to avoid congestion because it creates shallow and high values of LMP, affecting the benefits of the units installed upstream of the congested zone.

The option of massively adopting wind and solar generation has significant limitations because the undesirable effects of forecasting error increase with the renewables penetration, requiring more reserve generation. However, energy storage devices could provide some controllability to the renewable generators, increasing the generating costs.

7.7 Conclusions

This chapter commented on several crucial aspects of power production. We started by introducing the reader to the actual environmental problems related to human activities and industrial development. Then, we described some of the political actions promoted on a global scale, such as the SDGs of the United Nations. After that, we described the actual circumstances of the energy market commodities, including the evolution of coal, natural gas, and nuclear energy. Energy system models frequently employed in developing energy policy were briefly described.

We carefully explained the context in which the OPF is formulated. Thus, the reader has a wide perspective on OPF applicability. The LMP and its components were analyzed, considering the conditions of a typical liberalized electricity market.

The effects of transmission system congestion on the LMP were explained by analyzing an illustrative case study.

Finally, the economic barriers to reaching a climate-neutral energy system were briefly discussed, observing that adopting carbon capture and sequestration equipment combined with a transmission system adaptation is required to avoid high emissions rates and extreme values of LMP. However, it requires an increased investment due to the high costs of zero-emission technologies and devices.

References

Abido, M. A. (2002). Optimal power flow using particle swarm optimization. *International Journal of Electrical Power and Energy Systems*, *24*(7), 563–571. https://doi.org/10.1016/S0142-0615(01)00067-9.

Alsac, O., & Stott, B. (1974). Optimal load flow with steady-state security. *IEEE Transactions on Power Apparatus and Systems*, *PA93*(3), 745–751. https://doi.org/10.1109/TPAS.1974.293972.

Bai, X., Wei, H., Fujisawa, K., & Wang, Y. (2008). Semidefinite programming for optimal power flow problems. *International Journal of Electrical Power and Energy Systems*, *30*(6–7), 383–392. https://doi.org/10.1016/j.ijepes.2007.12.003.

Bakirtzis, A. G., Biskas, P. N., Zoumas, C. E., & Petridis, V. (2002). Optimal power flow by enhanced genetic algorithm. *IEEE Transactions on Power Systems*, *17*(2), 229–236. https://doi.org/10.1109/TPWRS.2002.1007886.

Black & Veatch, & National Renewable Energy Laboratory. (2012). *Cost and performance data for power generation technologies*. https://refman.energytransitionmodel.com/publications/1921.

Chuang, Y. C., Chen, C. T., & Hwang, C. (2015). A real-coded genetic algorithm with a direction-based crossover operator. *Information Sciences*, *305*, 320–348. https://doi.org/10.1016/j.ins.2015.01.026.

Chuang, Y. C., Chen, C. T., & Hwang, C. (2016). A simple and efficient real-coded genetic algorithm for constrained optimization. *Applied Soft Computing*, *38*, 87–105. https://doi.org/10.1016/j.asoc.2015.09.036.

Dall'Anese, E., Zhu, H., & Giannakis, G. B. (2013). Distributed optimal power flow for smart microgrids. *IEEE Transactions on Smart Grid*, *4*(3), 1464–1475. https://doi.org/10.1109/TSG.2013.2248175.

Dommel, H. W., & Tinney, W. F. (1968). Optimal power flow solutions. *IEEE Transactions on Power Apparatus and Systems*, *PAS-87*(10), 1866–1876. https://doi.org/10.1109/TPAS.1968.292150.

Elliston, B., Diesendorf, M., & MacGill, I. (2012). Simulations of scenarios with 100% renewable electricity in the Australian National Electricity Market. *Energy Policy*, *45*, 606–613. https://doi.org/10.1016/j.enpol.2012.03.011.

Elliston, B., MacGill, L., & Diesendorf, M. (2013). Least cost 100% renewable electricity scenarios in the Australian National Electricity Market. *Energy Policy*, *59*, 270–282. https://doi.org/10.1016/j.enpol.2013.03.038.

EnergyPLAN. (2020). *Department of development and planning*. Aalborg University. https://www.energyplan.eu/.

European Commission. (2022). *2030 climate target plan*. https://ec.europa.eu/clima/eu-action/european-green-deal/2030-climate-target-plan_en.

Farivar, M., & Low, S. H. (2013a). Branch flow model: Relaxations and convexification—Part I. *IEEE Transactions on Power Systems*, *28*(3), 2554–2564. https://doi.org/10.1109/TPWRS.2013.2255317.

Farivar, M., & Low, S. H. (2013b). Branch flow model: Relaxations and convexification—Part II. *IEEE Transactions on Power Systems*, *28*(3), 2565–2572. https://doi.org/10.1109/TPWRS.2013.2255318.

Geidl, M., & Andersson, G. (2007). Optimal power flow of multiple energy carriers. *IEEE Transactions on Power Systems*, *22*(1), 145–155. https://doi.org/10.1109/TPWRS.2006.888988.

Generation IV International Forum (GIF). (2020). *GIF 2020 annual report*. https://www.gen-4.org/gif/.

Gómez-Expósito, A., Conejo, A. J., & Cañizares, C. A. (2018). *Electric energy systems analysis and operation* (2nd ed.). Taylor & Francis Group.

Granville, S. (1994). Optimal reactive dispatch through interior-point methods. *IEEE Transactions on Power Systems*, *9*(1), 136–142. https://doi.org/10.1109/59.317548.

IPCC Working Group I. (2021). *Climate change 2021. The physical science basis. Summary for policymakers*. https://www.ipcc.ch/report/sixth-assessment-report-working-group-i/.

IPCC Working Group I. (2022). *IPCC working group I atlas*. https://interactive-atlas.ipcc.ch/.

Jaccard, M., Nyboer, J., Bataille, C., & Sadownik, B. (2003). Modeling the cost of climate policy: Distinguishing between alternative cost definitions and long-run cost dynamics. *The Energy Journal*, *24*(1), 49–73. https://www.jstor.org/stable/41323420.

Lavaei, J., & Low, S. H. (2012). Zero duality gap in optimal power flow problem. *IEEE Transactions on Power Systems*, *27*(1), 92–107. https://doi.org/10.1109/TPWRS.2011.2160974.

McGookin, C., Gallachoir, B. O., & Byrne, E. (2021). Participatory methods in energy system modelling and planning—A review. *Renewable and Sustainable Energy Reviews*, *151*. https://doi.org/10.1016/j.rser.2021.111504, 111504.

Milano, F. (2005). An open source power system analysis toolbox. *IEEE Transactions on Power Systems*, *20*(3), 1199–1206. https://doi.org/10.1109/TPWRS.2005.851911.

Murphy, R., Rivers, N., & Jaccard, M. (2007). Hybrid modeling of industrial energy consumption and greenhouse gas emissions with an application to Canada. *Energy Economics*, *29*, 826–846. https://doi.org/10.1016/j.eneco.2007.01.006.

National Renewable Energy Laboratory. (2022). *Annual technology baseline*. https://atb.nrel.gov/electricity/2021/changes_in_2021.

National Renewable Energy Laboratory, & Nuclear Innovation: Clean Energy Future (NICE). (2020). *Flexible nuclear energy for clean energy systems*. https://www.nrel.gov/docs/fy20osti/77088.pdf.

Osman, M. S., Abo-Sinna, M. A., & Mousa, A. A. (2004). A solution to the optimal power flow using genetic algorithm. *Applied Mathematics and Computation*, *155*, 391–405. https://doi.org/10.1016/S0096-3003(03)00785-9.

Perevezentsev, A., & Rozenkevich, M. (2021). *Tritium technologies for thermonuclear fusion reactors*. Academic Press. https://doi.org/10.1016/B978-0-12-824322-0.00030-9.

PJM State & Mem Training Department. (2007). *Marginal losses implementation training*. https://www.pjm.com/~/media/training/new-initiatives/ip-ml/marginal-losses-implementation-training.ashx.

PJM State & Mem Training Department. (2017). *Locational marginal pricing components*. https://www.pjm.com/-/media/training/nerc-certifications/markets-exam-materials/mkt-optimization-wkshp/locational-marginal-pricing-components.ashx.

PowerWorld Corporation. (2022). *PowerWorld*. https://www.powerworld.com/.

Prina, M. G., Groppi, D., Nastasi, B., & Garcia, D. A. (2021). Bottom-up energy system models applied to sustainable islands. *Renewable and Sustainable Energy Reviews*, *152*. https://doi.org/10.1016/j.rser.2021.111625, 111625.

Prina, M. G., Manzolini, G., Moser, D., Nastasi, B., & Sparber, W. (2020). Classification and challenges of bottom-up energy system models—A review. *Renewable and Sustainable Energy Reviews*, *129*. https://doi.org/10.1016/j.rser.2020.109917, 109917.

Riffonneau, Y., Bacha, S., Barruel, F., & Ploix, S. (2011). Optimal power flow management for grid connected PV systems with batteries. *IEEE Transactions on Sustainable Energy*, *2*(3), 309–320. https://doi.org/10.1109/TSTE.2011.2114901.

Rivers, N., & Jaccard, M. (2006). Useful models for simulating policies to induce technological change. *Energy Policy*, *34*, 2038–2047. https://doi.org/10.1016/j.enpol.2005.02.003.

Ruth, M., Spitsen, P., & Boardman, R. (2019). *Opportunities and challenges for nuclear-renewable hybrid energy systems*. https://www.nrel.gov/docs/fy19osti/72004.pdf.

Skolfield, J. K., & Escobedo, A. R. (2022). Operations research in optimal power flow: A guide to recent and emerging methodologies and applications. *European Journal of Operational Research*, *300*, 387–404. https://doi.org/10.1016/j.ejor.2021.10.003.

Sun, D. I., Ashley, B., Brewer, B., Hughes, A., & Tinney, W. F. (1984). Optimal power flow by Newton approach. *IEEE Transactions on Power Apparatus and Systems*, *103*(10), 2864–2880. https://doi.org/10.1109/TPAS.1984.318284.

The State Council Information Office of the People's Republic of China. (2022). *Full text: Responding to climate change: China's policies and actions*. http://www.scio.gov.cn/.

The White House. (2021). *Council on environmental quality report to congress on carbon capture, utilization, and sequestration*. https://www.whitehouse.gov/wp-content/uploads/2021/06/CEQ-CCUS-Permitting-Report.pdf.

The White House. (2022). *FACT SHEET: President Biden renews US leadership on world stage at UN climate conference (COP26)*. https://www.whitehouse.gov/.

The World Bank. (2022a). *CO2 emissions (kt)*. https://data.worldbank.org/indicator/EN.ATM.CO2E.KT?view=map.

The World Bank. (2022b). *Industrialization intensity index*. https://tcdata360.worldbank.org/indicators/mva.ind.int?indicator=3793&viz=choropleth&years=2018&indicators=944.

Tollefson, J. (2021). COVID curbed carbon emissions in 2020—But not by much. *Nature*. https://doi.org/10.1038/d41586-021-00090-3.

UL. (2020). *HOMER energy*. https://www.homerenergy.com/.

United Nations. (2022a). *Climate change*. https://www.un.org/en/global-issues/climate-change.

United Nations. (2022b). *The 17 goals*. https://sdgs.un.org/goals.

US Energy Information Administration. (2021a). *Natural gas explained—Factors affecting natural gas prices*. https://www.eia.gov/energyexplained/natural-gas/.

US Energy Information Administration. (2021b). *Nuclear explained*. https://www.eia.gov/energyexplained/nuclear/nuclear-power-plants.php.

US Energy Information Administration. (2022a). *Cost and performance characteristics of new generating technologies*. https://www.eia.gov/outlooks/aeo/assumptions/pdf/table_8.2.pdf.

US Energy Information Administration. (2022b). *Europe is a key destination for Russia's energy exports*. https://www.eia.gov/todayinenergy/detail.php?id=51618.

Weber, J. D. (1997). *Implementation of a Newton-based optimal power flow into a power system simulation environment*. University of Illinois at Urbana-Champaign. https://www.powerworld.com/files/Newton-based-Optimal-Power-Flow.pdf.

Zavala, V. M., Kim, K., Anitescu, M., & Birge, J. (2017). A stochastic electricity market clearing formulation with consistent pricing properties. *Operations Research*, *65*(3), 557–576. https://doi.org/10.1287/opre.2016.1576.

Zhu, J. (2009). *Optimization of power system operation* (1st ed.). John Wiley & Sons, Inc.

Zimmerman, R. D., Murillo-Sánchez, C. E., & Thomas, R. J. (2011). MATPOWER: Steady-state operations, planning, and analysis tools for power systems research and education. *IEEE Transactions on Power Systems*, *26*(1), 12–19. https://doi.org/10.1109/TPWRS.2010.2051168.

Index

Note: Page numbers followed by *f* indicate figures and *t* indicate tables.

H

S

T

U

V

W

MIX
Papier | Fördert gute Waldnutzung
FSC® C083411

Kolibri 360 GmbH
Ferdinand-Jühlke-Straße 7
99095 Erfurt, Deutschland
produktsicherheit@kolibri360.de